Fabrication of Wear and Corrosion Resistant Materials

Fabrication of Wear and Corrosion Resistant Materials presents cutting-edge surface processing methods, incorporating concepts from the fields of metallurgy, chemistry, physics, and nanotechnology. It covers developments in manufacturing processes that aim to improve materials' and components' resistance to wear and corrosion.

Investigating techniques such as plasma treatment, laser cladding, chemical vapor deposition, and nanostructuring, the book evaluates each technique by its capability of ensuring durability and resistance to corrosion to prolong the life of materials. In addition, the book explores smart surfaces, digital surface design, machine learning tools, and biomedical applications. It discusses new materials, coatings, surface treatments, and additive manufacturing techniques using an interdisciplinary approach.

This book is intended for researchers and graduate students studying surface engineering, coatings and claddings, and corrosion and wear resistance of materials.

Advances in Surface Engineering and Tribology

Series Editors: Hitesh Vasudev, Jashanpreet Singh, and Jeyaprakash Natarajan

Advances in Surface Engineering and Tribology explores the interdisciplinary aspects of surface science and engineering, as well as tribology discussions pertaining to the examination of friction, wear, and lubrication. It covers topics ranging from surface modification techniques, nanomaterials and their applications in nanotechnology, catalytic and photocatalytic processes, synthesis of materials, colloids and surfaces, thin film deposition processes, methods for characterizing surfaces, and surface engineering for diverse industrial purposes. The volumes comprehensively explore the complex mechanisms that regulate surface phenomena, presenting both established principles and emerging developments in the field of surface science and engineering. The series offers a forum for professionals, engineers, and researchers in the energy, automotive, aerospace, and biomedical sectors to delve into innovative developments in these interrelated disciplines.

Should you have an idea for this series, please get in contact with Kyra Lindholm (kyra.lindholm@taylorandfrancis.com).

Thermal Spray Coatings for High-Temperature Conditions
Edited by Gaurav Prashar, Hitesh Vasudev, and Lalit Thakur

Microwave Claddings
Engineering Applications
Edited by Sarbjeet Kaushal

Fabrication of Wear and Corrosion Resistant Materials
Surface Processing Techniques
Edited by Gurbhej Singh and Hitesh Vasudev

Fabrication of Wear and Corrosion Resistant Materials

Surface Processing Techniques

Edited by
Gurbhej Singh and Hitesh Vasudev

CRC Press
Taylor & Francis Group
Boca Raton London New York

CRC Press is an imprint of the
Taylor & Francis Group, an **informa** business

Designed cover image: Shutterstock

First edition published 2026
by CRC Press
2385 NW Executive Center Drive, Suite 320, Boca Raton FL 33431

and by CRC Press
4 Park Square, Milton Park, Abingdon, Oxon, OX14 4RN

CRC Press is an imprint of Taylor & Francis Group, LLC

ISBN: 978-1-032-90108-4 (hbk)
ISBN: 978-1-032-90204-3 (pbk)
ISBN: 978-1-003-54652-8 (ebk)

DOI: 10.1201/9781003546528

Typeset in Times
by Apex CoVantage, LLC

Contents

Preface

This edited book, *Fabrication of Wear and Corrosion Resistant Materials: Surface Processing Techniques*, provides researchers, engineers, and other professionals working in the material sciences with in-depth knowledge of how sophisticated surface processing technologies may improve the characteristics of materials. This book is intended to be a key resource for these individuals. The objective is not only to record the information that is already known; rather, the book aims to encourage more innovation and application in the sector.

In the next chapters, a wide range of cutting-edge surface processing methods are discussed. These techniques include, but are not limited to, plasma treatment, laser cladding, chemical vapor deposition, and nanostructuring. Each technique is evaluated to see whether it has the capability of imparting excellent durability and resistance to corrosion, hence prolonging the useful life of materials and ensuring their dependability over a wide range of applications. In addition, the book investigates the fundamental mechanisms that regulate these advances, providing readers with a solid foundation that allows them to understand the dynamic relationship that exists between surface processes and material performance.

This book is distinguished by its multidisciplinary approach. This book incorporates concepts from the fields of metallurgy, chemistry, physics, and nanotechnology. This is done in recognition of the fact that breakthroughs often make their appearance at the junction of many scientific and technical areas. Not only does this synthesis serve to extend the reader's viewpoint, but it also offers a comprehensive perspective on the difficulties and possibilities that are associated with the manufacturing of materials.

About the Editors

Dr. Gurbhej Singh is an assistant professor in the Department of Mechanical Engineering at Amritsar Group of Colleges, Punjab, India. Dr. Singh earned his Ph.D. in mechanical engineering from CT University Ludhiana, Punjab, India. He specializes in surface engineering with a focus on microwave claddings, processing of materials using microwave energy, material characterization, surface biomedical implants, biomaterial processing, polymeric composites, and nanocomposites. He has published more than 50 papers, out of which 30 are SCI-indexed papers. He was recently honored with the Top Cited Paper Awards India 2022 in the review category. Additionally, he has contributed to numerous conference papers and book chapters. Dr. Singh is actively involved in the research community, serving as a reviewer for research articles. He has over ten years of teaching experience and has guided many MTech students. Dr. Singh is guest editor of many journals. Dr. Singh has been granted a patent titled "Minimum Quantity Lubrication Device for Machining Operations".

Dr. Hitesh Vasudev is a professor in the School of Mechanical Engineering, Lovely Professional University, Phagwara, Punjab, India. He received his Ph.D. from Guru Nanak Dev Engineering College, Ludhiana, India. His research areas include surface engineering/thermal spraying; development of nanostructured, multi-modal, high-entropy alloy coatings for high-temperature oxidation and corrosion resistance; thermal barrier coatings (TBCs); and microwave processing of materials. Dr. Vasudev has contributed extensively in thermal spray coatings in reputed journals, including *Surface Coatings and Technology*, *Materials Today Communications*, *Engineering Failure Analysis*, and *Journal of Cleaner Production*, among others. He won the Research Excellence Award 2019, 2020, 2021, and 2023 at Lovely Professional University. He has written 40+ SCI-indexed papers, 100 Scopus-indexed papers, 15 conference papers, 18 book chapters, and five books. Dr. Vasudev has been granted a patent titled "High-Temperature Oxidation and Erosion Resistant Alloy-718/Al_2O_3 Composite Coatings".

About the Editors

Contributors

Atul Agnihotri
Department of Mechanical Engineering
Khalsa College of Engineering & Technology
India

Anuj Bansal
Sant Longowal Institute of Engineering and Technology
India

Jyotsna Barpute
Artificial Intelligence and Data Science Department
Dr. D.Y. Patil Institute of Technology
India

Amrut Bhosale
Department of Mechatronics Engineering
Kasegaon Education Society's Rajarambapu
Institute of Technology
India

Digvijay Bhosale
Department of Mechatronics Engineering
Rajarambapu Institute of Technology
India

Umesh Ghorpade
Mechanical Engineering Department
Dnyanshree Institute of Engineering and Technology
India

Suresh Gudala
Department of Mechanics and Mechanical Engineering
Siberian State Industrial University
Russia

Karthikeyan A.
Department of Mechanical Engineering
Sathyabama Institute of Science and Technology
India

Priyanka Kaushal
Department of Mechanical Engineering
Chandigarh Group of Colleges
India

Abhinav Kumar
Department of Nuclear and Renewable Energy
Ural Federal University
Russia

Ravi Kumar
Department of Mechanical Engineering
Lovely Professional University
India

Satish Kumar
Department of Mechanical Engineering
Chandigarh Engineering College
India

S. Maharajan
Ramco Institute of Technology
India

Omar Mbrouk
University of Sadat City
Egypt

Rahul Mehra
Department of Mechanical Engineering
Chandigarh Engineering College
India

Amrinder Mehta
Research and Development Cell
Lovely Professional University
India

F. Michael Thomas Rex
National Engineering College
India

Jeyaprakash Natarajan
School of Mechanical and Electrical Engineering
China University of Mining and Technology
China

Dhirendra Nath
Department of Mechanical Engineering
Siksha 'O' Anusandhan (Deemed to be University)
India

Gaurav Prashar
Department of Mechanical Engineering
Rayat Bahra Institute of Engineering and Nano Technology
India

Rajeev Rathi
Department of Mechanical Engineering
National Institute of Technology
India

Ranbir Singh Rooprai
Sant Longowal Institute of Engineering and Technology
India

Pankaj Sharma
O.P. Jindal Global University
India

Sunil B. Shivarudraiah
ETH Zurich
Switzerland

Bhupinder Singh
Department of Chemical and Environmental Engineering
University of Nottingham
United Kingdom

Gurbhej Singh
Department of Mechanical Engineering
Amritsar Group of Colleges
India

Harmandeep Singh
Department of Mechanical Engineering
Amritsar Group of Colleges
India

Harvinder Singh
Department of Mechanical Engineering
Chandigarh Engineering College
India

Jagdeep Singh
Transport Engineering Department
Telus Corporation
Canada

Jagtar Singh
Sant Longowal Institute of Engineering and Technology
India

Jasjeevan Singh
Department of Mechanical Engineering
Khalsa College of Engineering & Technology
India

Manpreet Singh
Biotechnology High Performance Computing Software Applications Institute
Telemedicine and Advanced Technology Research Center
USA

Shubhangi Suryawanshi
Mechanical Engineering Department
D.Y. Patil Institute of Technology
India

S. Valaiganesh
Ramco Institute of Technology
India

Hitesh Vasudev
School of Mechanical Engineering
Lovely Professional University
India

Vladimir Ivanovich Velkin
Department of Nuclear and Renewable Energy
Ural Federal University
Russia

Praveen Kumar Verma
Lovely Professional University
India

1 Perspectives and Fundamentals of Corrosion and Wear Mechanisms in Surface Engineering

Amrinder Mehta and Hitesh Vasudev

1.1 INTRODUCTION

Surface engineering is an important discipline that deals with the improvement of the properties of the surface of a material to meet specific requirements. This discipline is particularly important in combating two major challenges faced by materials in various industries: corrosion and wear. Corrosion is the gradual destruction of material through chemical action with its surroundings, which is likely to lessen the durability of parts and constructions dramatically [1]. On the other hand, wear, or wear-out, is defined as the loss or degradation of the surfaces, primarily as the result of simplified mechanical contacts. These two processes can cause great economic losses and even threats to safety in various industries. Surface treatment and modification technologies provide answers to these problems since they modify the surface properties of a material while retaining its bulk properties [2]. The role of surface engineering when it comes to changing material properties with the intention to improve performance and reliability in relation to corrosion and wear issues by applying coatings and treatments or altering materials at the molecular level is depicted in Figure 1.1 [3].

Such techniques may be provided by deposits on the surface of the materials in other forms, chemical breakdown of the surface, or alterations at microscopic levels. Through the use of these methods, engineers can improve surfaces so as to gain additional protection against corrosion, wear, and even both in order to increase the effective life of materials as well as enhance performance [4]. The scope of this chapter is to try summarizing the present state of surface engineering, mainly in the context of corrosion and abrasion protection. It discusses such questions as why corrosion and wear happen and how they develop in various materials and conditions. The chapter also examines the various surface engineering techniques used at present to address these problems; the principles, uses, and efficiency of these techniques are also explored [5]. This chapter aims at reviewing existing published articles in an attempt to provide an insight into new trends in surface engineering from current

DOI: 10.1201/9781003546528-1

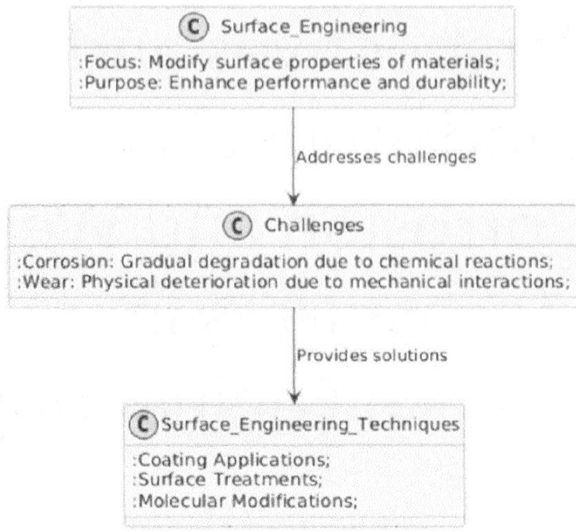

FIGURE 1.1 Role of surface engineering in mitigating corrosion and wear.

Source: Licensed under CC BY 4.0.

knowledge and advancements. This includes presently developing technologies, new types of materials, and new ways of thinking that can potentially build on improved corrosion and wear protection. The aim is to present comprehensive coverage of the presented material, which, in turn, will be highly beneficial for researchers, engineers, and other professionals in the field of study—materials science, which, as mentioned, plays an immensely important role today.

1.2 CORROSION MECHANISMS

1.2.1 Electrochemical Corrosion

Electrochemical corrosion is a universal process, which majorly determines the service life and efficiency of metallic materials in various applications, including in medical devices and engineering structures. This category of corrosion occurs due to a redox reaction, in which metal atoms lose electrons (oxidation) while other species in the medium gain electrons [6]. Electrochemical corrosion is characterized by three important criteria. The susceptibility of a metal or alloy to corrosion is influenced by its chemical composition and microstructural characteristics. Some metals are more prone to corrosion than others, while the addition of certain types of alloying products increases corrosion protection or degrades it. Features of the environment—temperature, pH, oxygen content and corrosive species—have an effect on the rate and form of corrosion [7]. Corrosion processes may be quickened by tough conditions, although they may be slowed down by favorable conditions. Corrosion starts on the material states, and its physical and chemical states, such as roughness, defectiveness, and surface energy, thus determine how corrosion begins and develops [8].

To formulate a surface engineering strategy to mitigate corrosion, these and other factors must first be appreciated. By changing the surface of the material through methods like coating application, surface treatment, or alloying, engineers develop materials that are less susceptible to corrosion [9]. Using a coating can help prevent the metal from interacting with its surroundings, while surface treatments can change the physical and chemical characteristics of the metal to increase its corrosion protection [10, 11]. Corrosion protection strategies acting on this understanding make it possible to increase the service life of materials and improve safety in respective applications, all through cutting maintenance costs. Such knowledge proves more important in areas such as biomedical engineering, where implant corrosion poses adverse health consequences, or areas involving manufacturing industries, where corrosion causes equipment failure and financial losses [12].

1.2.2 Pitting Corrosion

Pitting corrosion is a localized form of material degradation that can severely compromise structural integrity. This type of corrosion occurs when protective passive films break down, resulting in the formation of small, deep pits on the material surface. Figure 1.2 shows the application of surface engineering methods to improve corrosion resistance and pitting [13].

Pitting corrosion resistance can be improved by changing the surface features of the material through surface engineering techniques. The use of particular alloying components might enhance the corrosion protection of nanoscaled materials.

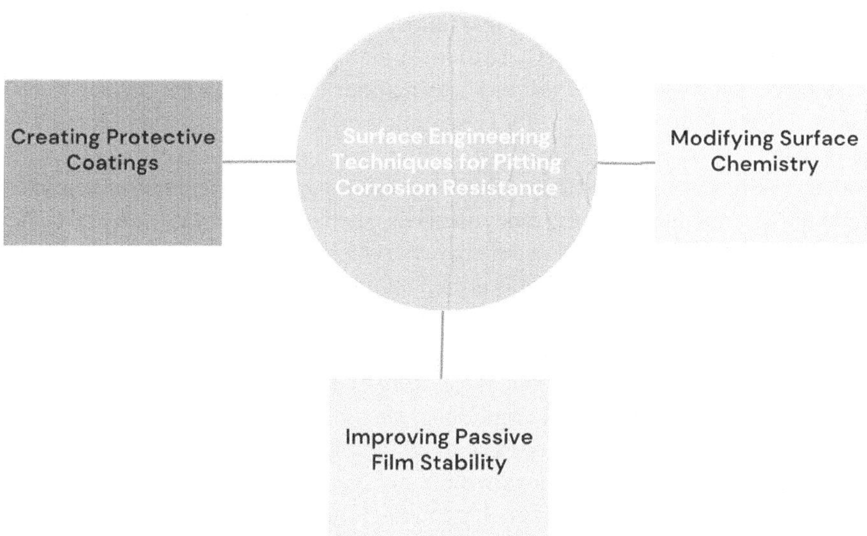

FIGURE 1.2 Pitting corrosion resistance surface engineering methodologies.

Anodic dissolution is suppressed, interface impedance is enhanced, and corrosion is substantially reduced when 4% tantalum is added to a Ti-12% Zr alloy [14]. Measures that help to stabilize passive films can, therefore, significantly increase corrosion resistance. For instance, in the anodization of aluminum alloys, a controlled oxide layer forms that protects against corrosion. However, what needs to be understood here is that under some circumstances, deep under-film pitting can still take place due to very highly occluded conditions. Speaking of nanocoatings, they are one of the popular methods to eliminate corrosion [15]. All these coatings have properties such as high adhesion strength, hardness of the surface, and excellent long-term corrosion protection. Nanocoatings are thicker and less flexible in comparison, and the main advantage is that a much thinner layer of nanocoating can be used; therefore, the maintenance costs will be significantly lower. Self-healing, eco-sustainable corrosion inhibitors are being developed for advanced materials [16]. Nanotechnology is also being used to create coatings that enhance mechanical and tribological properties. This study focuses on the design and development of technologically advanced surface coatings for biomedical applications. These techniques of material and surface modification are finding applications in the aerospace, automobile, and healthcare industries for improving the reliability and durability of components.

1.2.3 Crevice Corrosion

Crevice corrosion is a special type of corrosion in which a localized attack is found in crevices where stagnant solutions can collect; it is a major problem in implant and mechanical joint applications. This type of corrosion is dangerous because it can cause rapid and severe material loss at locations that can be very hard to inspect [17]. On the problem of crevice corrosion, some methods to address surface engineering have been established. Methods using finer finishes like electropolishing or mechanical polishing can minimize or smooth surface roughness where crevice formation can take place where corrosive solution can mitigate problems [18]. The use of other coatings is also an effective way of preventing corrosive environments from affecting the implant material; some materials, such as titanium nitride (TiN), have been found to improve the corrosion characteristics of implants [19]. An important function of design control is to reduce the amount of crevice formation at all costs, such as not having sharp edges or corners; stressing the so-called drain point, when the liquid flows out of the assembly; and most of all, avoiding bolted connections in favor of welded ones when possible. None of these strategies, implemented with the correct choice of materials and conditions, should prove detrimental to efforts of increasing resistance to crevice corrosion [20]. For instance, in biomedical implants, the material used is titanium alloys with better or enhanced surface characteristics with higher spans and fewer corrosion complications. Likewise, in industrial usage, surface engineering techniques have shown longer product life cycles and lower maintenance requirements [21]. With progress in materials science and surface engineering courses, more efficient methods and films are being discovered for enhanced barrier properties in hostile environments.

1.3 WEAR MECHANISMS

1.3.1 Adhesive Wear

Adhesive wear is a critical issue in tribological applications because the contacting surfaces create a powerful adhesive bond; this leads to material displacement and surface deformation. This process is equally deleterious to the performance and durability of mechanical parts, raising the rate of wear and tear due to increased friction, energy transfer, and abrasion [22]. To minimize adhesive wear, several surface engineering approaches have been formulated. Other forms of anti-stick coatings include the likes of diamond-like carbon (DLC) or molybdenum disulfite (MoS_2) [23], which similarly dramatically decreases adhesion between two surfaces in contact. These coatings offer a protective layer, which reduces direct contact between the metal surfaces to thus reduce the formation of very strong adhesive joints. A scientific way of lowering the contact area between surfaces involves methods such as laser texturing or micro-patterning that help in the reduction of the overall contact area [24]. Depending on the design, the true area of contact is lowered by producing features on the surface that come in the form of dimples or grooves, thus reducing adhesion and increasing wear resistance [25]. Also, micro-structuring by applying nitriding or carburization changes the chemical properties of the surface to decrease its adhesion. There are treatments that alter the surface energy and chemical reactivity and therefore reduce the formation of strong adhesive bonds. The effectiveness of these strategies has been proven in several applications ranging from automotive engines to aerospace parts, where they have led to increases in efficiency, longer working life of components, and decreased need for maintenance [26]. New methods and materials of tribology and surface engineering are under study at present with the aim to improve wear protection in specific operating conditions.

1.3.2 Abrasive Wear

Abrasion wear is one of the common types of material loss due to the contact between the harder and softer phases, which causes material deformation through ploughing, cutting, or cutting-edge fracture. This wear mechanism is quite relevant in many different industrial uses as well as interacting with the performance and durability of many different products in industries such as mining and manufacturing [27]. With the aim of improving Abrasive Wear (AW) resistance, several surface engineering treatments have been invented. The application of hardening treatments such as nitriding and carburizing is common to enhance surface hardness and wear strength. These involve the establishment of case depth by penetration of nitrogen or carbon into the material surface to produce a hard chrome-indicator layer [28]. Additional protection is delivered by hard coatings, which are layers including ceramic to guard against abrasive wear. These coatings could be materials such as aluminum oxide (Al_2O_3) or titanium nitride (TiN) and are generally much harder and more resistant to wear than the core material [29]. By using surface alloying techniques, it is desired that the hardness of the surface layer will be enhanced while at the same time ensuring that the surface has an appreciable degree of toughness to afford good wear and impact resistance. This can be done via techniques such as laser surface alloying or plasma-transferred arc welding where elements of improved properties are brought into the surface layer [30]. The applicability of these techniques has been evidenced by the effectiveness of their application in mining with earth-moving equipment and manufacturing with cutting tools and has led to increases in component durability, avoidance of frequent breakdowns, and enhanced productivity [31]. Current research in material science and surface engineering is focused on gaining better abrasion wear material properties or better ways of processing to optimize its usage and performance in extreme applications.

1.3.3 Fatigue Wear

Fatigue wear is another important problem in contact tasks, such as bearings and gears, in which cyclic loading of material surfaces and crack formation and growth occur. This mechanism can lead to earlier fatigue of the components, and in mechanical systems the damage may be complete. Several solution strategies have hence been devised and adopted in counteracting fatigue wear and other effects of surface engineering [32]. One of the ways to attempt to improve fatigue strength is by causing the formation of compressive residual stresses in the surface layer. Techniques like shot peening or laser shock peening introduce a residual compressive stress on the surface to avoid crack nucleation and extend the fatigue life of the component [33]. The next essential tactic is to enhance the surface finish since the stress increase slows down in proximity to a fatigue crack initiation site. Practices such as superfinishing or isotropic superfinishing can provide surface roughness that virtually eliminates the possibility of an RCC call initiation. Thus, the lubricant is further protected by wear-resistant coatings with fatigue characteristics [34]. Special coatings such as diamond-like carbon or certain classes of ceramic coatings, apart from providing high wear resistance, will also offer acceptable fatigue strengths, leading to improved life of the component. These surface engineering approaches have been applied in various fields such as aerospace, automotive, and power generation industries, where they have proven essential by contributing to reliable, long-lasting components with reduced maintenance requirements [35]. In aeronautics, these techniques used on fan components of aero engines have enhanced their efficiency and lifespan. With the improvement of materials science and surface engineering knowledge, new coating materials and surface treatment methods are being designed and manufactured to improve or maintain fatigue wear resistance in even more extreme service conditions [36].

1.4 SURFACE ENGINEERING TECHNIQUES FOR CORROSION AND WEAR PROTECTION

1.4.1 Thermal Spray Coatings

Thermal spraying is a variety of coating that changes the surface of different substances. This process requires the use of molten or semi-molten particulates, which are afterwards placed on a substrate to create a shield. Flame spraying, arc spraying, and plasma or high-velocity oxy-fuel spraying are the most frequently used methods of thermal spraying. These methods enable the formation of coating on the substrate with a definite chemical makeup and thickness to address one or many performance characteristics as required [37]. Despite the numerous advantages associated with thermal spraying, the coatings formed by these processes contain defects inherent in the technology. These imperfections may be porosity and material discontinuities, which may compromise the ability of the coating [38]. To this end, post-processing procedures like remelting are used as a solution to these problems. Dulles describes remelting as the process of heating the coating sprayed onto metals, and this has the added advantage of reducing porosity and increasing the hardness of the entire surface [39]. This process control enables the depth of the remelted material to be controlled and subsequently the properties of the coated material to be enhanced. However, it may be noted that several thermal sprayed coatings contain inherent porosity and weak bonding characteristics that may hinder their effectiveness in some situations [40].

1.4.2 Plasma Spray Coatings

Thermal spraying is the most progressive procedure of applying metallic coating utilizing thermal spraying. Plasma spray coating in particular is ideal for enhancing the thickness of the protective layer on different substrates. It uses a plasma gun that produces a high-temperature plasma jet, which can be between 10,000 °C and 15,000 °C. It has the ability to deposit, fuse, and solidify almost all metals, ceramics, and cermets because of high temperatures [41]. In the plasma spray process, a powder feed is injected into the plasma flame, and the material in the feedstock is immediately melted and accelerated to the substrate. When the particles make impact on the surface, they spread and solidify, producing a typical laminar structure. This technique can give a deposit thickness of a few micrometres to few millimeters [42].

The characteristic features and thus the quality of plasma spray coatings are largely determined by different factors associated with the process. These are plasma current (100 to 1000 A), stand-off distance (*t* between 60 and 250 mm), and gas flow rate (between 30 and 400 L/min).[2]. The other important process parameters include powder feed rate, substrate surface speed, and gun traverse speed. It is important to maintain control of these variables to obtain the desired characteristics of conventional coatings, such as density, adhesion strength, and microstructure [43].

Plasma spray coatings have many uses across various fields, with special application in aerospace, automotive, and industries that deal with energy. They are priceless for developing what is known as thermal barrier coatings (TBCs) for these parts, which are exposed to high heat levels in turbines and diesel engines. This can, in fact, enhance the utility of the substrate material by increasing corrosion, wear, and thermal protection of the coated substrate materials [44].

However, independent of the type of plasma, spray coatings may contain porosity, oxides, and cracks. However, improvement of studies and constant optimization of the process parameters help to raise the quality and the performance of the coating; thus, plasma spray coating is considered an effective solution for surface engineering [45]. Figure 1.3 shows the plasma-spraying process of yttria-stabilized zirconia (YSZ) coating. YSZ powder is injected through a plasma spray gun into a

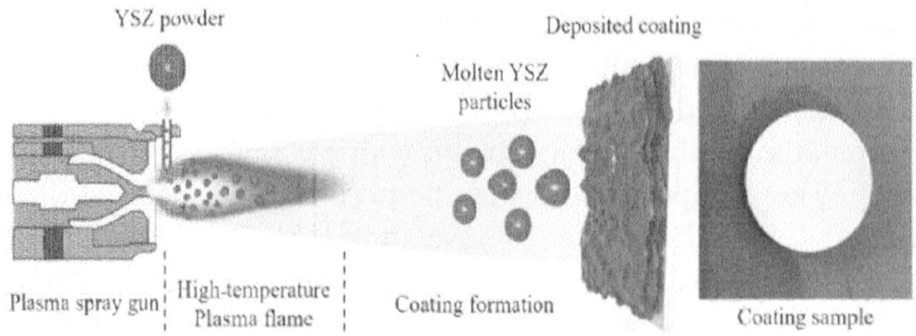

FIGURE 1.3 Schematic diagram of the YSZ coating deposited by the plasma-spraying process and images of the prepared coating samples [46].

high-temperature plasma flame, where it melts and is deposited onto the substrate. The particles melt and with increasing speed hit the substrate and start to arrange themselves in a layered structure. The work produces a final product of a white circular coating sample [46].

1.4.3 Suspension Plasma Spray Coatings

Suspension plasma spray (SPS) is a novel coating technique that has attracted considerable interest for the synthesis of coatings with specific microstructures and improved characteristics. In contrast to conventional plasma spraying, where powdery materials are employed, SPS applies a liquid suspension of fine particles. This makes it possible to deposit coatings with more refined microstructures and tailored properties, like thermally grown barrier coatings of columnar grain structures [47].

In the SPS process, a suspension of submicron or nanosized particles is introduced into a plasma jet at elevated temperature. The liquid carrier evaporates, and the particles are accelerated and then softened before meeting the substrate to eventually create a layer. It is thus possible to form a complete range of microstructures from dense coatings to coatings with porosity as desired for the needs of the process engineer, depending on the process parameters employed [48, 49].

Another advantage of SPS is the ability to fabricate columnar-structured TBCs that are effectively suitable to work in highly thermally insulated and very durable applications, such as gas turbine engines. These types of coatings are often applied at short distances, but developments have been made to extend them for more complicated shaped parts and in the process, avoid a collision with the plasma torch [50]. The parameters of SPS allow enhanced coating quality. For example, the liquid suspension flow rate, injection pressure, and stand-off distance have great impact on the coating deposition efficiency and structure [51]. A study showed that these parameters could be tuned to get a deposition efficiency of 51% and build thick columnar-structured TBCs with a porosity of approximately 22%. However, SPS is very complicated and often sensitive to the parameters to be controlled for enhanced performance. Simulation methods have been used to model SPS coating build-up, and the results offer further understanding of the processes of particle deposition and

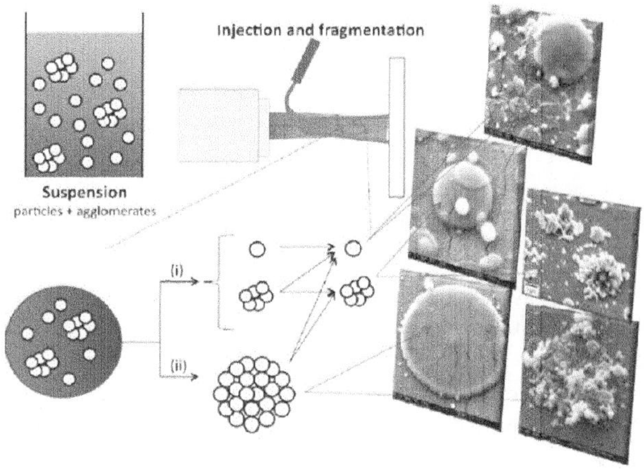

FIGURE 1.4 Schematics of the proposed deposition mechanisms in the SPS process [54].

Source: Licensed under CC BY 4.0.

improvement of the experimental techniques. As we can observe, these models are crucial for the comprehension of the SPS process and for minimizing the casework necessary to evaluate the effectiveness of its models [52, 53].

The deposition mechanisms in the SPS process are shown in Figure 1.4. It represents a suspension with particles and agglomerates that are subjected to injection and fragmentation. Two pathways are depicted: when the particles and small clusters are easily distinguished but not necessarily individual particles and when large spherical clusters can be clearly seen. The SEM images on the right show these different morphologies in the real coating structure as well as dispersed particles and consolidated agglomerates after deposition [54].

1.4.4 Solution Precursor Plasma Spray Coatings

Solution precursor plasma spray (SPPS) is an improved thermal spray coating technology that comes with several benefits enabling the development of high-performance coatings. Contrary to other plasma spray techniques, where powder feedstock is employed, SPPS makes use of liquid precursor solutions as the feedstock. This concept makes it possible to deposit functional nanostructured coatings using the methods described previously. In the SPPS process, a liquid precursor solution with metal salts or organometallic compounds is injected through a jet into a high-temperature plasma. During the precursor's travel through the plasma, there is evaporation of the precursor to form a vapor, decomposition of the precursor into smaller molecules, and chemical reactions in-flight. These processes lead to the formation of fine particles or droplets to be deposited on the substrate to form a coating with characteristic microstructures [55].

Another feature that can be regarded as one of the major benefits offered by SPPS is an opportunity to achieve enhanced high-temperature characteristics for synthesized

TBCs. For instance, SPPS-deposited YSZ has superiority in thermal cycling endurance and a lower coefficient of thermal conductivity than conventional plasma sprayed coatings. The technique also permits the introduction of dopants like ytterbium and gadolinium that will improve the thermal properties of the coating and its phase stability. SPPS provides relatively great freedom in the choice of coating composition and structure [56]. Scientists can control the microstructure of the formed coatings, from dense to highly porous, through variation of process parameters and makeup of precursors. This flexibility makes SPPS ideal for use in vigorous applications such as abradable seal coatings in gas turbine engines. Moreover, the optimization of SPPS can be used in conjunction with other materials and techniques to produce better-performing coatings [57]. The synergistic combination of yttrium aluminum garnet (YAG) with SPPS produces thermal barrier coatings that can endure higher temperatures, which extends the operating temperature range of gas turbine engines. Although SPPS has many benefits, the fabrication process's coating properties depend on the strict regulation of these steps and the choice of precursors. More studies have been carried out to develop improved precursors and process modifications on the broad application of this promising coating technique [58].

Figure 1.5 shows an epistemic diagram of the solution precursor plasma spray process that starts with a pressurized precursor tank consisting of Zn–Cu–H+/OH– complex. Five factors, solvent type, stoichiometry, concentration, pH, and feed rate, govern the first part of the process. The solution is atomized and injected into the plasma plume produced by the utilized plasma gun; other parameters determining the spraying process are distance, speed, power, and gas flow. In-situ–generated ZnO/CuO particles are the atomized droplets that successively deposit onto a substrate, and their type and pre-heat temperature determine the effect on the final film formation [59].

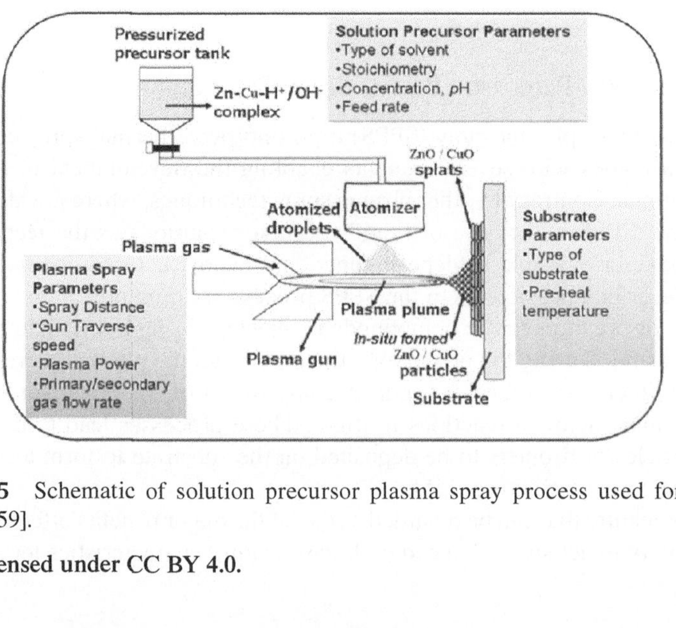

FIGURE 1.5 Schematic of solution precursor plasma spray process used for film deposition [59].

1.5 SURFACE ENGINEERING FOR SPECIFIC APPLICATIONS

1.5.1 Orthopedic Implants

Surface engineering plays an important role in orthopedic implant problems to improve performance, longevity, and biocompatibility. Orthopedic implants like joint replacements and bone plates need to fulfil various properties like biocompatibility, wear resistance, corrosion resistance, and osseointegration for them to be successful in clinical practice [60]. There is a need to avoid complex and unwanted immunologic reactions and guarantee good interaction between the implant and other tissues. The application of modifications, for instance, thin films of bioactive glass, is quite common for improving biocompatibility by stimulating tissue growth and diminishing inflammation. Tribological wear is also crucial, as some implants such as hip or knee replacements are subjected to load bearing; hence, wear of the implant surface may trigger implant fracture [61]. Nitriding and carburizing, for instance, can increase the wear resistance of the implant by hardening a surface layer of the material. Another important consideration is corrosion because implants are immersed in body fluids that are likely to corrode the material over time. Coating and surface treatment counter this problem and thus play a role in stabilizing the system in the long run. Last is osseointegration, which is a process where bone adheres directly to the implant surface and forms the basic stability of the implant [62]. HAp coatings are normally used to enhance bone formation and cell attachment and adhesion of the tissue to the surface of the implant. In the past few years, new technological developments in surface engineering have boosted orthopedic implants even further. Nanostructured coatings have been identified as a viable solution because the nanoscale topographies mimic that of the bone structure, affecting both biocompatibility and osseointegration [63]. Furthermore, while additive manufacturing of porous structures has addressed challenges like implant instability and stress shielding, it has also introduced new complexities related to bone ingrowth and mechanical interlocking between implant and bone [64]. These innovations not only solve traditional problems but also bring new possibilities for development of a new type of personalized orthopedic implant with more functions and a longer life cycle, therefore benefiting patients.

1.5.2 Aerospace Components

Surface engineering plays an important role in both automotive and aerospace industries where the functionality of vital parts is concerned. Surface engineering has a significant area of interest as far as optimizing and improving the performance of components in different industries [65]. The proposed material for turbine blades highlights that thermal barrier coatings provide effective high-temperature protection, thereby extending service life and improving performance in harsh environments. Gears and bearings benefit significantly from hard chrome plating, as it enhances their wear resistance and durability. New advances in the field have implemented nanocomposite layer protecting, with enhanced performance compared to traditional layer protecting. Further, it is possible to administer permanent dry lubricating films, which can be used in extreme conditions where normal lubricating procedures may prove ineffective [66].

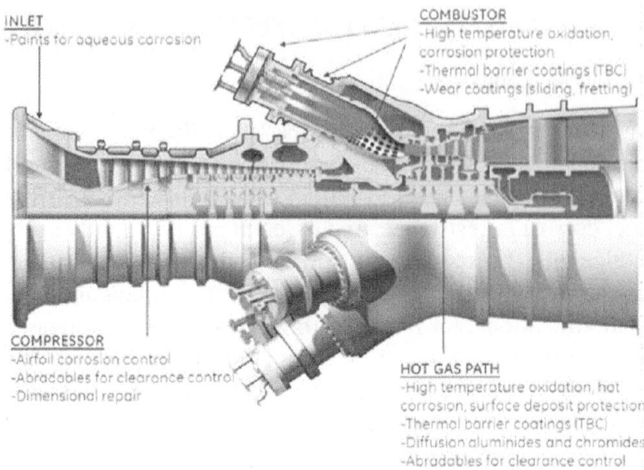

FIGURE 1.6 Thermal spray coatings in aerospace engines [67].

From Figure 1.6, one can identify the essential parts of the gas turbine engine and the protective measures one needs. The inlet is protected from aqueous corrosion treatment. The compressor incorporates airfoil corrosion control, abradable coatings to maintain clearances, and dimensional restoration of compressor blades and vanes. The combustor, in which high temperature levels are characteristic, uses thermal barrier coatings TBC, wear coatings, and protection against corrosion. Due to the high operating temperatures, the hot gas path is coated with thermal barrier materials, diffusion aluminides, and chromides and abradables for clearance control. It also needs protection against high temperature oxidation, hot corrosion, and the forming of deposits on its surface. In general, the image focuses on various specifications of coatings and materials in order to increase engine durability and the effectiveness of its parts [67].

These innovative surface engineering techniques contribute to improved component performance, increased energy efficiency, and extended service life across multiple industrial applications.

1.6 EMERGING TRENDS AND FUTURE PERSPECTIVES

1.6.1 Nanostructured Coatings

These coatings have become a revolutionary approach for the development of anti-corrosive and anti-wear coatings in different sectors. The coatings make use of the properties that the material displays at the nanoscale and are more advanced compared to traditional coating technologies [68]. Enhanced hardness and toughness are among the most attractive features of different nanostructured coatings. The right control of material structures at the nano scale results in the development of coatings with enhanced mechanical characteristics [69]. When applied to metal components, these coatings have increased hardness and enhanced fracture toughness compared

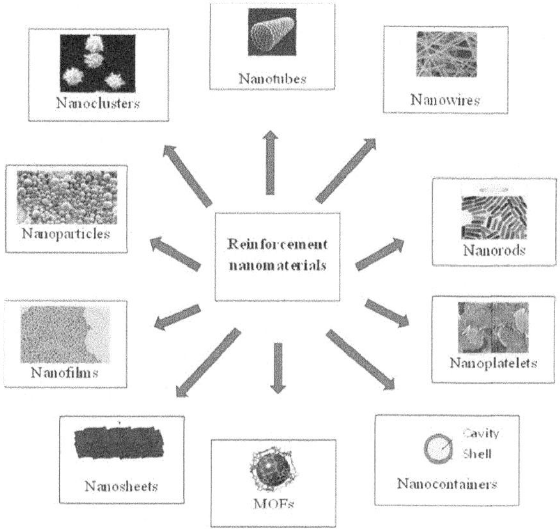

FIGURE 1.7 Materials used for nanocomposite coatings [71].

to conventional coatings, providing increased protection against wear and increased periods before the components require replacement [70].

Some of the popular reinforcement nanomaterials being utilized in construction as well as material science are shown in Figure 1.7. Nine forms of nanostructures are shown in the image. Such structures can be nanoclusters (grouped particles), nanotubes (tube-like particles), nanowires (thin elongated particles), nanorods (rod particles), nanoplatelets (platelet-like structures), nanocontainers (hollow particles with shells), metal–organic frameworks (MOFs), nanosheets (sheet-like structures), and nanofilms (materials in the form of thin films). They are illustrated with a microscope image or schematic illustration showing the structural type and morphology of each type [71].

Another important advantage of nanostructured coatings is improved corrosion protection or corrosion resistance. These coatings have features at the nanoscale level that provide the ability to design denser barriers against corrosive agents, providing enhanced protection of the base substrates [72, 73]. However, coatings with a proper dispersion of nanoparticles offer high corrosion protection because a long path forms for corrosive species to penetrate the coating, which enhances the tribological performance of basic substrate materials [74]. These coatings bring out lower friction and wear factors in sliding as well as rolling elements of mechanical aids, which makes the mechanical systems more effective and longer lasting. The size of the features at the nanoscale level yields better surface finishes and improved interfaces that provide enhanced lubrication and lower wearing rates [75].

Current research has led towards designing novel coatings that provide properties of numerous materials at the nano scale. Such coatings are found to be composed of a matrix material and nanoparticles, nanotubes, or even other nanostructured materials. Both polymer–nanoparticle composite coatings show better scratch and UV resistance, whereas ceramic-based nanocomposites exhibit superior thermal barrier properties [76]. However, there are challenges such as achieving uniform dispersion of nanomaterials within the matrix and addressing health and safety concerns associated with nanotechnology. However, enhancements of processing techniques and characterization methods have constantly improved the performance and versatility of these coatings in different sectors.

1.6.2 Smart Coatings

AM plus surface engineering is a mature step in materials science and manufacturing engineering technology. This synergistic approach has untapped potential for developing parts with very specific surface characteristics suitable for a range of further applications in a range of industries. Another advantage of integrating AM with surface engineering is the possibility of implementing elaborate surface topographies together with complex shapes [77]. AM techniques also impact structure by enabling the development of structures that are difficult, if not impossible, to generate using conventional structure manufacturing, whereas applying surface enhancement schemes integrated with AM processes enables designers to fabricate parts with desired surface features such as porosity or texturing, for instance, increased wear resistance or improved fluid flow [78]. Another advantage of this approach of process integration is the possible design and manufacture of functionally graded materials and coatings. By using AM techniques within a single part, composite structures with varying properties and compositions can be created gradually. This capability can be used to achieve designs with variations in hardness, corrosion protection, or thermal characteristics at specific areas of a part [79]. In addition, the integration of AM and surface engineering generally entails less material being wasted and fewer processing stages than standard manufacturing processes [80]. The application of AM means is certainly a more material-conservative approach because in application, most of these processes require only the quantity of material needed for the individual component. This concept can completely remove post-processing steps when applied in synergy with in-situ surface treatments or coatings, thus cutting expenses and assembly time efficiently [81].

There is still much that remains unknown for further improvement in relation to the corrosion and wear properties of AM parts. Due to the step-by-step manufacturing approaches applied in AM, the final material properties tend to have an anisotropic nature and probable flaws that compromise the surface quality [82]. Further, the contact between made substrates using AM and applied coatings might behave differently than various traditional components, which requires additional studies to demonstrate the material's permanency and stability [83]. The application of AM and surface engineering in this research will realize extensive possibilities in the future, as complex components with exemplary characteristics will be produced across multiple industries, including aerospace, automobile, and biomedical engineering [84–93]. Surface engineering has emerged as a pivotal field for mitigating corrosion and wear, enhancing the durability of materials in demanding environments. Techniques like magnetron sputtering have shown significant promise in utilizing boron nitride thin films for protecting stainless steel surfaces [94–106]. Advances in additive manufacturing and thermal spray methods underline the importance of optimizing process parameters to achieve superior hardness and wear resistance [107–115]. Moreover, microwave-assisted cladding and heat treatments further improve the mechanical and microstructural properties of coatings, offering sustainable solutions for industrial applications [116–121].

1.7 CONCLUSION

Surface engineering is a very important discipline that deals with tailoring external characteristics of materials to improve on their performance, especially in areas such as wear and corrosion. Electrochemical corrosion, pitting, and crevice corrosion are explained, and wear mechanisms like adhesive, abrasive, and fatigue wear are also explained. Several types of surface engineering coatings, such as thermal spraying, plasma spray coating, suspension plasma spray coating, and solution precursor plasma spray coating, are discussed. Included is a discussion of surface engineering as it applies to specialized fields like joint replacement components and aircraft parts, among others. Finally, future developments and directions such as nanostructured coatings together with surface engineering and additive manufacturing are proposed as future development avenues in the field.

REFERENCES

1. Abbady HE, Klinkenberg ET, de Moel L, Nicolai N, van der Stelt M, Verhulst AC, Maal TJ, Brouwers L. 3D-printed prostheses in developing countries: A systematic review. Prosthetics Orthotics International. 2022;46:19–30.
2. Abbasnezhad N, Zirak N, Champmartin S, Shirinbayan M, Bakir F. An overview of in vitro drug release methods for drug-eluting stents. Polymers. 2022;14:2751.
3. Abdulghafor MA, Mahmood MK, Tassery H, Tardivo D, Falguiere A, Lan R. Biomimetic coatings in implant dentistry: A quick update. Journal of Functional Biomaterials. 2023;15:15.
4. Abdullah JY, Abdullah AM, Hadi H, Husein A, Rajion ZA. Comparison of STL skull models produced using open-source software versus commercial software. Rapid Prototyping Journal. 2019;25:1585–91.
5. Abduo J. Fit of CAD/CAM implant frameworks: A comprehensive review. Journal of Oral Implantology. 2014;40:758–66.
6. AbuShanab WS, Moustafa EB, Taha MA, Youness RA. Synthesis and structural properties characterization of titania/zirconia/calcium silicate nanocomposites for biomedical applications. Applied Physics A. 2020;126:1–12.
7. Afzal A. Implantable zirconia bioceramics for bone repair and replacement: A chronological review. Materials Express. 2014;4:1–12.
8. Ahmad F, Al-Douri Y, Kumar D, Ahmad S. Metal-oxide powder technology in biomedicine. Metal Oxide Powder Technologies. Elsevier; 2020:121–68.
9. Mehta A, Singh G. Consequences of hydroxyapatite doping using plasma spray to implant biomaterials. Journal of Electrochemical Science Engineering Failure Analysis. 2023;13:5–23.
10. Ahmed S, Atia NN. Fast and efficient zirconia-based reversed phase chromatography for selective determination of triptans in rat plasma. Journal of Pharmaceutical Biomedical Analysis. 2017;143:241–51.

11. Liu L, Wang S, Zhang B, Jiang G, Liu H, Yang J, Wang J, Liu W. Present status and prospects of nanostructured thermal barrier coatings and their performance improvement strategies: A review. Journal of Manufacturing Processes. 2023;97:12–34.
12. Ak A, Çiçek B. Synthesis and characterization of hydrophobic glass-ceramic thin film derived from colloidal silica, zirconium (IV) propoxide and methyltrimethoxysilane via sol–gel method. Journal of Sol-Gel Science Technology. 2021;98:252–63.
13. Alifui-Segbaya F, Ituarte IF, Hasanov S, Gupta A, Fidan I. Opportunities and limitations of additive manufacturing. Springer Handbook of Additive Manufacturing. Springer; 2023:125–43.
14. Al-Naamani L, Dobretsov S, Dutta J. Chitosan-zinc oxide nanoparticle composite coating for active food packaging applications. Innovative Food Science Emerging Technologies. 2016;38:231–37.
15. Amani A, Montazer M, Mahmoudirad M. Synthesis of applicable hydrogel corn silk/ZnO nanocomposites on polyester fabric with antimicrobial properties and low cytotoxicity. International Journal of Biological Macromolecules. 2019;123:1079–90.
16. Bahremandi-Toloue E, Mohammadalizadeh Z, Mukherjee S, Karbasi S. Incorporation of inorganic bioceramics into electrospun scaffolds for tissue engineering applications: A review. Ceramics International. 2022;48:8803–37.
17. Baino F, Minguella-Canela J, Korkusuz F, Korkusuz P, Kankılıç B, Montealegre MÁ, De los Santos-López MA, Vitale-Brovarone C. In vitro assessment of bioactive glass coatings on alumina/zirconia composite implants for potential use in prosthetic applications. International Journal of Molecular Sciences. 2019;20:722.
18. Mehta A, Vasudev H, Jeyaprakash N. Role of sustainable manufacturing approach: Microwave processing of materials. International Journal on Interactive Design Manufacturing. 2023;18:5283–99.
19. Bakhsheshi-Rad H, Najafinezhad A, Hadisi Z, Iqbal N, Daroonparvar M, Sharif S, Ismail AF, Akbari M, RamaKrishna S, Berto F. Characterization and biological properties of nanostructured clinoenstatite scaffolds for bone tissue engineering applications. Materials Chemistry Physics. 2021;259:123969.
20. Bayer IS. On the durability and wear resistance of transparent superhydrophobic coatings. Coatings. 2017;7:12.
21. Berger O. The correlation between structure, multifunctional properties and application of PVD MAX phase coatings. Part III. Multifunctional applications. Surface Engineering. 2020;36:303–25.
22. Mehta A, Vasudev H, Singh S. Sustainable manufacturing approach with novel thermal barrier coatings in lowering CO_2 emissions: Performance analysis with probable solutions. International Journal on Interactive Design Manufacturing. 2023;18:5427–39.
23. Bernauer SA, Zitzmann NU, Joda T. The use and performance of artificial intelligence in prosthodontics: A systematic review. Sensors. 2021;21:6628.
24. Mehta A, Vasudev H, Singh S, Prakash C, Saxena KK, Linul E, Buddhi D, Xu J. Processing and advancements in the development of thermal barrier coatings: A review. Coatings. 2022;12:1318.
25. Bharathidasan T, Sathiyanaryanan S. Self-replenishing superhydrophobic durable polymeric nanocomposite coatings for heat exchanger channels in thermal management applications. Progress in Organic Coatings. 2020;148:105828.
26. Bhowmick A, Pramanik N, Jana P, Mitra T, Gnanamani A, Das M, Kundu PP. Development of bone-like zirconium oxide nanoceramic modified chitosan based porous nanocomposites for biomedical application. International Journal of Biological Macromolecules. 2017;95:348–56.
27. Bhowmick A, Pramanik N, Mitra T, Gnanamani A, Das M, Kundu PP. Mechanical and biological investigations of chitosan–polyvinyl alcohol based ZrO_2 doped porous hybrid composites for bone tissue engineering applications. New Journal of Chemistry. 2017;41:7524–30.

28. Bonda DJ, Manjila S, Selman WR, Dean D. The recent revolution in the design and manufacture of cranial implants: Modern advancements and future directions. Neurosurgery. 2015;77:814–24.
29. Caravaca C, Shi L, Balvay S, Rivory P, Laurenceau E, Chevolot Y, Hartmann D, Gremillard L, Chevalier J. Direct silanization of zirconia for increased biointegration. Acta Biomaterialia. 2016;46:323–35.
30. Celia E, Darmanin T, de Givenchy ET, Amigoni S, Guittard F. Recent advances in designing superhydrophobic surfaces. Journal of Colloid Interface Science. 2013;402:1–18.
31. Chatham CA, Long TE, Williams CB. A review of the process physics and material screening methods for polymer powder bed fusion additive manufacturing. Progress in Polymer Science. 2019;93:68–95.
32. Chavan S, Cha H, Orejon D, Nawaz K, Singla N, Yeung YF, Park D, Kang DH, Chang Y, Takata Y. Heat transfer through a condensate droplet on hydrophobic and nanostructured superhydrophobic surfaces. Langmuir. 2016;32:7774–87.
33. Chen S, Li X, Li Y, Sun J. Intumescent flame-retardant and self-healing superhydrophobic coatings on cotton fabric. ACS Nano. 2015;9:4070–76.
34. Cheng W, Zhang Y, Tao Y, Lu J, Liu J, Wang B, Song L, Jie G, Hu YJ. Durable electromagnetic interference (EMI) shielding ramie fabric with excellent flame retardancy and self-healing performance. Journal of Colloid Interface Science. 2021;602:810–21.
35. Choo YJ, Boudier-Revéret M, Chang MC. 3D printing technology applied to orthosis manufacturing: Narrative review. Annals of Palliative Medicine. 2020;9:4262270–4270.
36. Nasiri S, Khosravani MR. Machine learning in predicting mechanical behavior of additively manufactured parts. Journal of Materials Research Technology. 2021;14:1137–53.
37. Chung JJ, Im H, Kim SH, Park JW, Jung Y. Toward biomimetic scaffolds for tissue engineering: 3D printing techniques in regenerative medicine. Frontiers in Bioengineering Biotechnology. 2020;8:586406.
38. Prashar G, Vasudev H, Mehta A. Application of bimodal stuructured thermal sprayed coatings in power plant components. Advanced Ceramic Coatings for Energy Applications. 2024:273–87.
39. Ciasca G, Papi M, Businaro L, Campi G, Ortolani M, Palmieri V, Cedola A, De Ninno A, Gerardino A, Maulucci G. Recent advances in superhydrophobic surfaces and their relevance to biology and medicine. Bioinspiration Biomimetics. 2016;11:011001.
40. Saxena A, Mehta A, Vasudev H, Prashar G, Jasim NY. A short review on machine learning for the purpose of optimizing and predicting the properties of polymeric nanocomposites. Materials Today: Proceedings. 2023. https://www.sciencedirect.com/science/article/abs/pii/S2214785323042189
41. Croitoru C, Bode F, Calotă R, Berville C, Georgescu M. Harnessing nanomaterials for enhanced energy efficiency in transpired solar collectors: A review of their integration in phase-change materials. Energies. 2024;17:1239.
42. d'Água RB, Branquinho R, Duarte MP, Maurício E, Fernando AL, Martins R, Fortunato EJNJoC. Efficient coverage of ZnO nanoparticles on cotton fibres for antibacterial finishing using a rapid and low cost in situ synthesis. New Journal of Chemistry. 2018;42:1052–60.
43. Das I, De G. Zirconia based superhydrophobic coatings on cotton fabrics exhibiting excellent durability for versatile use. Scientific Reports. 2015;5:18503.
44. Demofonti A, Carpino G, Zollo L, Johnson MJ. Affordable robotics for upper limb stroke rehabilitation in developing countries: A systematic review. IEEE Transactions on Medical Robotics Bionics. 2021;3:11–20.
45. Depan D, Misra R. Processing–structure–functional property relationship in organic–inorganic nanostructured scaffolds for bone-tissue engineering: The response of preosteoblasts. Journal of Biomedical Materials Research Part A. 2012;100:3080–91.

46. Huang J, Sun W, Huang R, Ma W. Cracking behavior of atmospheric plasma-sprayed 8YSZ thermal barrier coatings during thermal shock test. Coatings. 2023;13:243.
47. Ataie A, Mostaghimi J, Pershin L, Xu P. Fabrication of nanostructured cobalt ferrite coatings using suspension plasma spraying (SPS) technique. Surface Coatings Technology. 2017;328:451–61.
48. Bernard B, Quet A, Bianchi L, Joulia A, Malié A, Schick V, Rémy B. Thermal insulation properties of YSZ coatings: Suspension plasma spraying (SPS) versus electron beam physical vapor deposition (EB-PVD) and atmospheric plasma spraying (APS). Surface Coatings Technology. 2017;318:122–28.
49. Sharma Y, Mehta A, Vasudev H, Jeyaprakash N, Prashar G, Prakash C. Analysis of friction stir welds using numerical modelling approach: A comprehensive review. International Journal on Interactive Design Manufacturing. 2023;18:5329–42.
50. Björklund S, Goel S, Joshi S. Function-dependent coating architectures by hybrid powder-suspension plasma spraying: Injector design, processing and concept validation. Materials Design. 2018;142:56–65.
51. Carnicer V, Orts MJ, Moreno R, Sánchez E. Microstructure assessment of suspension plasma spraying coatings from multicomponent submicronic Y-TZP/Al_2O_3/SiC particles. Ceramics International. 2018;44:12014–20.
52. Carpio P, Rayón E, Pawłowski L, Cattini A, Benavente R, Bannier E, Salvador M, Sánchez E. Microstructure and indentation mechanical properties of YSZ nanostructured coatings obtained by suspension plasma spraying. Surface Coatings Technology. 2013;220:237–43.
53. Singh H, Mehta A, Sharma Y, Vasudev H. Role of expert systems to optimize the friction stir welding process parameters using numerical modelling: A review. International Journal on Interactive Design Manufacturing. 2023;18:2609–25.
54. Joulia A, Bolelli G, Gualtieri E, Lusvarghi L, Valeri S, Vardelle M, Rossignol S, Vardelle A. Comparing the deposition mechanisms in suspension plasma spray (SPS) and solution precursor plasma spray (SPPS) deposition of yttria-stabilised zirconia (YSZ). Journal of the European Ceramic Society. 2014;34:3925–40.
55. Fan W, Bai Y. Review of suspension and solution precursor plasma sprayed thermal barrier coatings. Ceramics International. 2016;42:14299–312.
56. Mittal G, Paul S. Suspension and solution precursor plasma and HVOF spray: A review. Journal of Thermal Spray Technology. 2022;31:1443–75.
57. Von Niessen K, Gindrat M. Plasma spray-PVD: A new thermal spray process to deposit out of the vapor phase. Journal of Thermal Spray Technology. 2011;20:736–43.
58. Vaßen R, Kaßner H, Mauer G, Stöver D. Suspension plasma spraying: Process characteristics and applications. Journal of Thermal Spray Technology. 2010;19:219–25.
59. Nehe P, Sivakumar G, Kumar S. Solution precursor plasma spray (SPPS) technique of catalyst coating for hydrogen production in a single channel with cavities plate type methanol based microreformer. Chemical Engineering Journal. 2015;277:168–75.
60. Lal H, Patralekh MK. 3D printing and its applications in orthopaedic trauma: A technological marvel. Journal of Clinical Orthopaedics Trauma. 2018;9:260–68.
61. Levesque JN, Shah A, Ekhtiari S, Yan JR, Thornley P, Williams DS. Three-dimensional printing in orthopaedic surgery: A scoping review. EFORT Open Reviews. 2020;5:430–41.
62. Mehjabeen A, Song T, Xu W, Tang HP, Qian M. Zirconium alloys for orthopaedic and dental applications. Advanced Engineering Materials. 2018;20:1800207.
63. Mukherjee S, Dhara S, Saha P. Laser surface remelting of Ti and its alloys for improving surface biocompatibility of orthopaedic implants. Materials Technology. 2018;33:106–18.
64. Prakash C, Singh S, Gupta MK, Mia M, Królczyk G, Khanna N. Synthesis, characterization, corrosion resistance and in-vitro bioactivity behavior of biodegradable Mg–Zn–Mn–(Si–HA) composite for orthopaedic applications. Materials. 2018;11:1602.

65. Bhattacharya S, Agarwal AK, Prakash O, Singh S, Pandey M, Kant R. Introduction to sensors for aerospace and automotive applications. Sensors for Automotiveaerospace Applications. 2019:1–6.
66. Balaji D, Ranga J, Bhuvaneswari V, Arulmurugan B, Rajeshkumar L, Manimohan MP, Devi GR, Ramya G, Masi C. Additive manufacturing for aerospace from inception to certification. Journal of Nanomaterials. 2022;2022:7226852.
67. Hardwicke CU, Lau Y-C. Advances in thermal spray coatings for gas turbines and energy generation: A review. Journal of Thermal Spray Technology. 2013;22:564–76.
68. Farooq SA, Raina A, Mohan S, Arvind Singh R, Jayalakshmi S, Irfan UI, Haq M. Nanostructured coatings: Review on processing techniques, corrosion behaviour and tribological performance. Nanomaterials. 2022;12:1323.
69. Singh N, Mehta A, Vasudev H, Samra PS. A review on the design and analysis for the application of wear and corrosion resistance coatings. International Journal on Interactive Design Manufacturing. 2023;18:5381–405.
70. Hajizadeh-Oghaz M, Razavi RS, Ghasemi A, Valefi Z. Na_2SO_4 and V_2O_5 molten salts corrosion resistance of plasma-sprayed nanostructured ceria and yttria co-stabilized zirconia thermal barrier coatings. Ceramics International. 2016;42:5433–46.
71. Muresan LM. Nanocomposite coatings for anti-corrosion properties of metallic substrates. Materials. 2023;16:5092.
72. Singh S, Sajwan M, Singh G, Dixit AK, Mehta A. Efficient surface detection for assisting collaborative robots. Robotics Autonomous Systems. 2023;161:104339.
73. Verma PK, Mehta A, Vasudev H, Kumar VJSR. Performance of thermal spray coated metallic materials for bio-implant applications. Surface Review Letters. 2023:2250017.
74. Jamali H, Mozafarinia R, Razavi RS, Ahmadi-Pidani R. Comparison of thermal shock resistances of plasma-sprayed nanostructured and conventional yttria stabilized zirconia thermal barrier coatings. Ceramics International. 2012;38:6705–12.
75. Keyvani A, Saremi M, Sohi MH. An investigation on oxidation, hot corrosion and mechanical properties of plasma-sprayed conventional and nanostructured YSZ coatings. Surface Coatings Technology. 2011;206:208–16.
76. Keyvani A, Saremi M, Sohi MH, Valefi Z. A comparison on thermomechanical properties of plasma-sprayed conventional and nanostructured YSZ TBC coatings in thermal cycling. Journal of Alloys Compounds. 2012;541:488–94.
77. Ahmady AR, Ekhlasi A, Nouri A, Nazarpak MH, Gong P, Solouk A. High entropy alloy coatings for biomedical applications: A review. Smart Materials in Manufacturing. 2023;1:100009.
78. Sanyal S, Park S, Chelliah R, Yeon S-J, Barathikannan K, Vijayalakshmi S, Jeong Y-J, Rubab M, Oh DH. Emerging trends in smart self-healing coatings: A focus on micro/nanocontainer technologies for enhanced corrosion protection. Coatings. 2024;14:324.
79. Shamshiri M, Jafari R, Momen G. Potential use of smart coatings for icephobic applications: A review. Surface Coatings Technology. 2021;424:127656.
80. Yedida VS, Mehta A, Vasudev H, Singh S. Role of numerical modeling in predicting the oxidation behavior of thermal barrier coatings. International Journal on Interactive Design Manufacturing. 2024;18:5653–62.
81. Tejero-Martin D, Rezvani Rad M, McDonald A, Hussain T. Beyond traditional coatings: A review on thermal-sprayed functional and smart coatings. Journal of Thermal Spray Technology. 2019;28:598–644.
82. Giubilini A, Bondioli F, Messori M, Nyström G, Siqueira G. Advantages of additive manufacturing for biomedical applications of polyhydroxyalkanoates. Bioengineering. 2021;8:29.
83. Iqbal A, Zhao G, Suhaimi H, He N, Hussain G, Zhao W. Readiness of subtractive and additive manufacturing and their sustainable amalgamation from the perspective of Industry 4.0: A comprehensive review. The International Journal of Advanced Manufacturing Technology. 2020;111:2475–98.

84. Singh M, Vasudev H, Kumar R. Corrosion and tribological behaviour of BN thin films deposited using magnetron sputtering. International Journal of Surface Engineering and Interdisciplinary Materials Science (IJSEIMS). 2021 Jul 1;9(2):24–39.
85. Singh M, Vasudev H, Singh M. Surface protection of SS-316L with boron nitride based thin films using radio frequency magnetron sputtering technique. Journal of Electrochemical Science and Engineering. 2022;12(5):851–63.
86. Singh G, Singh H, Sharma Y, Vasudev H, Prakash C. Analysis and optimization of various process parameters and effect on the hardness of SS-304 stainless steel welded joints. International Journal on Interactive Design and Manufacturing (IJIDeM). 2024;18(5):2889–96.
87. Singh G, Mehta A, Vasudev H. Sustainability of additive manufacturing: A comprehensive review. Progress in Additive Manufacturing. 2024;9:2249–72.
88. Singh G, Bansal A, Vasudev H, Mishra V. Sliding wear study of the Inconel-625 clad deposits by microwave heating on SS–304. Physica Scripta. 2024;99(6):065503.
89. Singh J, Yadav A, Gupta S, Balu R, Shukla N, Singh G, Singh S. Design and computational analysis of continuous wave fiber laser cutting of flat and conformal ITO PET sheets. International Journal on Interactive Design and Manufacturing (IJIDeM). 2024:1–9.
90. Singh G, Singh R, Gul J. Machinability behavior of human implant materials. Journal of Electrochemical Science and Engineering. 2023;13(1):99–114.
91. Verma PK, Vasudev H, Singh G, Singh S. A review on the microstructural and biomedical properties of implants manufactured using additive manufacturing. Journal of the Institution of Engineers (India): Series D. 2024:1–15.
92. Mehta A, Singh G, Vasudev H. Processing of shape memory alloys research, applications and opportunities: A review. Physica Scripta. 2024. https://iopscience.iop.org/article/10.1088/1402-4896/ad48cb
93. Singh G, Bansal A, Vasudev H. Corrosion and microstructural behaviour of Inconel 625 microwave clad deposited on mild steel. Thermal Claddings for Engineering Applications. CRC Press; 2024:51–62.
94. Kapade TT, Shah SK, Shah DS, Kumar NS, Bhor JK, Panwar VS, Singh G. Thermal analysis of v-cut twisted tape fitted heat exchanger: A numerical approach. International Journal on Interactive Design and Manufacturing (IJIDeM). 2024. https://link.springer.com/article/10.1007/s12008-024-01883-2
95. Bhowmik A, Kumar R, Beemkumar N, Kumar AV, Singh G, Kulshreshta A, Santhosh AJ. Casting of particle reinforced metal matrix composite by liquid state fabrication method: A review. Results in Engineering. 2024;103152.
96. Kulkarni SD, Manjunatha CU, Manjunath KV, Prasad CD, Vasudev H. Design and optimization of polyvinyl-nitride rubber for tensile strength analysis. International Journal on Interactive Design and Manufacturing (IJIDeM). 2024;18(8):5893–908.
97. Prashar G, Vasudev H. Surface topology analysis of plasma sprayed Inconel625-Al_2O_3 composite coating. Materials Today: Proceedings. 2022;50:607–11.
98. Yedida VS, Vasudev H. A review on the development of thermal barrier coatings by using thermal spray techniques. Materials Today: Proceedings. 2022;50:1458–64.
99. Sharanabasava H, Prasad CD, Ramesh MR. Microstructure, mechanical and wear properties of SiC and Mo reinforced NiCr microwave cladding. Advances in Materials and Processing Technologies. 2023;10(4):3620–33. https://doi.org/10.1080/2374068X.2023.2257937.
100. Dutta V, Thakur L, Singh B, Vasudev H. A study of erosion–corrosion behaviour of friction stir-processed chromium-reinforced NiAl bronze composite. Materials. 2022;15(15):5401.
101. Poojari M, Hanumanthappa H, Prasad CD, Jathanna HM, Ksheerasagar AR, Shetty P, Shanmugam BK, Vasudev H. Computational modelling for the manufacturing of solar-powered multifunctional agricultural robot. International Journal on Interactive Design and Manufacturing (IJIDeM). 2024;18(8):5725–36.

102. Prashar G, Vasudev H. A review on the influence of process parameters and heat treatment on the corrosion performance of Ni-based thermal spray coatings. Surface Review and Letters. 2022;29(1):2230001.
103. Kattimani MA, Venkatesh PR, Masum H, Math MM, Bahadurdesai VN, Mustafkhadri S, Prasad CD, Vasudev H. Design and numerical analysis of tensile deformation and fracture properties of induction hardened inconel 718 superalloy for gas turbine applications. International Journal on Interactive Design and Manufacturing (IJIDeM). 2024;18(8):5939–49.
104. Sunitha K, Vasudev H. A short note on the various thermal spray coating processes and effect of post-treatment on Ni-based coatings. Materials Today: Proceedings. 2022 Jan 1;50:1452–57.
105. Majji BGR, Vasudev H, Bansal A. A review on the oxidation and wear behavior of the thermally sprayed high-entropy alloys. Materials Today: Proceedings. 2022;50:1447–51.
106. Rao GS, Mukkamala U, Hanumanthappa H, Prasad CD, Vasudev H, Shanmugam B, Kishore Kumar KC. Evaluating and optimizing surface roughness using genetic algorithm and artificial neural networks during turning of AISI 52100 steel. International Journal on Interactive Design and Manufacturing (IJIDeM). 2024;18(8):6151–60.
107. Singh J, Gill HS, Vasudev H. Computational fluid dynamics analysis on role of particulate shape and size in erosion of pipe bends. International Journal on Interactive Design and Manufacturing (IJIDeM). 2023;17(5):2631–46.
108. Anjaneya G, Sunil S, Kakkeri S, Math MM, Vaibhav MN, Solaimuthu C, Durga Prasad C, Vasudev H. 2023. Numerical simulation of microchannel heat exchanger using CFD. International Journal on Interactive Design and Manufacturing (IJIDeM). 2024;18:5847–63.
109. Nagabhushana P, Ramprasad S, Prasad CD, Vasudev H, Prakash C. Numerical investigation on heat transfer of a nano-fluid saturated vertical composite porous channel packed between two fluid layers. International Journal on Interactive Design and Manufacturing (IJIDeM). 2024;18(5):2927–44.
110. Singh P, Vasudev H, Bansal A. Effect of post-heat treatment on the microstructural, mechanical, and bioactivity behavior of the microwave-assisted alumina-reinforced hydroxyapatite cladding. Proceedings of the Institution of Mechanical Engineers, Part E: Journal of Process Mechanical Engineering. 2024;238(3):1060–71.
111. Mahantesh MM, Rao KR, Chandra AP, Vijayakumar MN, Nandini B, Prasad CD, Vasudev H. Design and modeling using finite element analysis for the sitting posture of computer users based on ergonomic perspective. International Journal on Interactive Design and Manufacturing (IJIDeM). 2024;18(8):5875–91.
112. Singh J, Singh R, Singh S, Vasudev H, Kumar S. Reducing scrap due to missed operations and machining defects in 90PS pistons. International Journal on Interactive Design and Manufacturing (IJIDeM). 2023;17(5):2527–39.
113. Arora H, Kumar V, Prakash C, Pimenov D, Singh M, Vasudev H, Singh V. Analysis of sensitization in austenitic stainless steel-welded joint. Advances in Metrology and Measurement of Engineering Surfaces: Select Proceedings of ICFMMP 2019. Springer Singapore; 2021:13–23.
114. Vasudev H. Wear characteristics of Ni-WC powder deposited by using a microwave route on mild steel: Microwave cladding of Ni-WC. International Journal of Surface Engineering and Interdisciplinary Materials Science (IJSEIMS). 2020;8(1):44–54.
115. Prashar G, Vasudev H, Thakur L, Bansal A. Performance of thermally sprayed hydroxyapatite coatings for biomedical implants: A comprehensive review. Surface Review and Letters. 2023;30(1):2241001.
116. Mehta A, Vasudev H, Jeyaprakash N. Role of sustainable manufacturing approach: Microwave processing of materials. International Journal on Interactive Design and Manufacturing (IJIDeM). 2024;18(8):5283–99.

117. Prashar G, Vasudev H. Understanding cold spray technology for hydroxyapatite deposition. Journal of Electrochemical Science and Engineering. 2023;13(1):41–62.
118. Parkash J, Saggu HS, Vasudev H. A short review on the performance of high velocity oxy-fuel coatings in boiler steel applications. Materials Today: Proceedings. 2022;50:1442–46.
119. Singh J, Vasudev H, Kumar R, Ubaidullah M. Study on wear analysis of Ni-$_{20}Al_2O_3$ HVOF micron layers using artificial neural network technique. International Journal on Interactive Design and Manufacturing (IJIDeM). 2024;18(8):5925–38.
120. Singh J, Vasudev H, Szala M, Gill HS. Neural computing for erosion assessment in Al-$_{20}TiO_2$ HVOF thermal spray coating. International Journal on Interactive Design and Manufacturing (IJIDeM). 2024;18(4):2321–32.
121. Prakash C, Gupta LR, Mehta A, Vasudev H, Tominov R, Korman E, Fedotov A, Smirnov V, Kesari KK. Computing of neuromorphic materials: An emerging approach for bioengineering solutions. Materials Advances. 2023;4(23):5882–919.

2 Advances and Challenges in Corrosion Research

Harvinder Singh, Jagdeep Singh, Priyanka Kaushal, and Gurbhej Singh

2.1 INTRODUCTION

Human civilization's materials, buildings, and systems are regularly challenged by corrosion. Corrosion takes different forms, from steel rust to biomedical implant degradation, threatening our technical and economic advancement. We assess corrosion definitions, significance, and current research in this chapter [1–3]. Corrosion is the breakdown of materials by chemical or electrical interactions with their surroundings [4, 5]. This process affects the material's characteristics, resulting in structural weakening, functional degeneration, and, in extreme circumstances, failure. The manifestations of corrosion include a broad range, comprising rust formation on iron and steel, pitting in aluminum alloys, tarnishing of metals, and the breakdown of polymers [6, 7]. Corrosion manifests in numerous forms across various sectors, highlighting its pervasive impact and far-reaching consequences on materials, structures, and systems vital to human civilization. Each industry faces unique corrosion challenges, necessitating tailored preventive and mitigation measures to protect critical assets effectively.

- **Infrastructure**: Corrosion presents a severe hazard to infrastructure, including bridges, pipelines, buildings, and transportation networks. The rusting of steel buildings, concrete deterioration owing to chloride-induced corrosion, and corrosion-induced fatigue in metallic components are among the primary symptoms recognized in this category. The economic ramifications of corrosion-related failures in infrastructure are enormous, leading to expensive repairs, service interruptions, and decreased safety.
- **Energy**: Corrosion harms power generating, transmission, and distribution systems in the energy market. Corrosion in pipelines, storage tanks, and offshore platforms may result in oil and gas leakage, environmental damage, and operational delay. Similarly, corrosion-related deterioration in nuclear power plants and renewable energy systems may threaten safety and efficiency, underscoring the necessity of corrosion-resistant materials and coatings.
- **Transportation**: Corrosion causes substantial issues in the transportation market, affecting autos, airplanes, ships, and railway infrastructure. The corrosion of automotive components, such as chassis, body panels, and exhaust systems, increases wear and compromises structural integrity,

DOI: 10.1201/9781003546528-2

leading to safety issues and shortened vehicle lifetime. Similarly, corrosion-induced fatigue in airplane structures and marine vessels may threaten passenger safety and operational dependability, demanding severe maintenance and corrosion control approaches.

- **Healthcare**: In the healthcare business, corrosion may have significant effects for biological equipment and implants. The degeneration of orthopedic implants, dental prostheses, and surgical equipment due to corrosion may lead to implant failure, tissue inflammation, and patient discomfort. Moreover, corrosion-related release of metallic ions into the body might create systemic health concerns, underlining the necessity of biocompatible materials and corrosion-resistant coatings in medical device design.

Factors that influence the characteristics of corrosion inhibitors are depicted in Figure 2.1. Corrosion research affects almost every human activity, making it crucial. Corrosion endangers bridges, pipelines, buildings, and transportation networks. Repair, replacement, and downtime costs from corrosion-related failures total billions of dollars yearly. Corrosion threatens individuals and assets in the oil and gas, aerospace, automotive, and marine sectors. Corrosion in biomedical equipment and implants compromises patient safety and healthcare results, requiring ongoing corrosion-resistant material and coating development. This critical overview covers corrosion research trends, methods, difficulties, and advances. It spans a wide range of fields, including materials science, chemistry, engineering, environmental science, and computer modeling. The chapter covers basic corrosion processes, cutting-edge

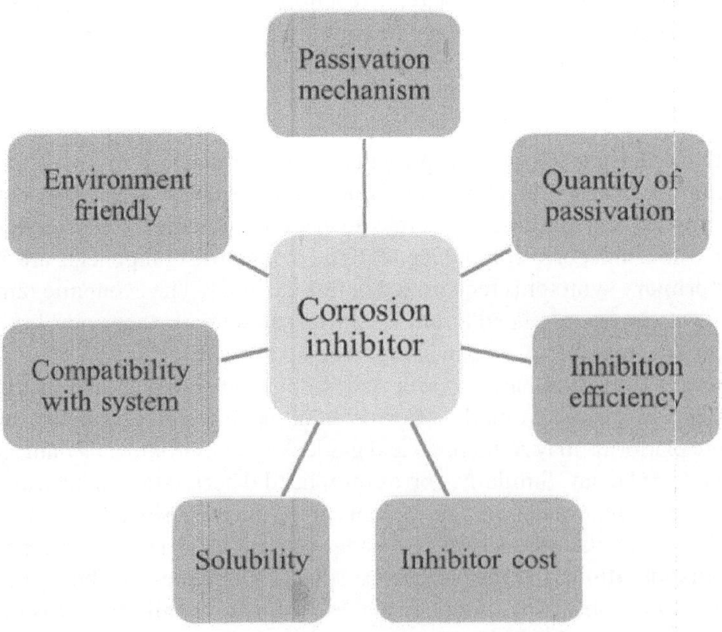

FIGURE 2.1 Factors that influence the characteristics of corrosion inhibitors [8].

experimental and computational methods, industry case studies, and future prospects in corrosion research. By critically reviewing the present status of corrosion research, we seek to stimulate improvements in corrosion prevention, mitigation, and sustainable material creation, therefore minimizing the unfavorable repercussions of corrosion on society, economy, and the environment [9–11].

2.2 FUNDAMENTALS OF CORROSION

Corrosion, an electrochemical process, represents the gradual degradation of materials due to reactions with their surrounding environment. Understanding the fundamentals of corrosion is crucial for developing effective prevention and mitigation strategies across various industries and applications. This section delves into the chemical processes involved in corrosion, the different mechanisms of corrosion, and the factors influencing corrosion susceptibility [12–14].

2.2.1 Chemical Processes Involved

Corrosion involves electrochemical reactions between a metal, an electrolyte, and environmental factors. A corrosion cell has an anode for oxidation, a cathode for reduction, an electrolyte for ion transport, and an external circuit for electron flow. Metal atoms oxidize at the anode. Metal atoms lose electrons and become positively charged ions in the electrolyte. Alternatively, iron (Fe) corrosion creates iron ions (Fe_2^+) [15, 16]:

$$Fe \rightarrow Fe2^+ + 2e^-$$

At the cathode, reduction reactions take place, often involving the reduction of oxygen or hydrogen ions. For instance, in the reduction of oxygen to form hydroxide ions:

$$1/2O_2 + H_2O + 2e^- \rightarrow 2OH^-$$

Figure 2.2 depicts the mechanism of corrosion process. The overall corrosion process involves the continuous transfer of electrons from the anode to the cathode, resulting in the deterioration of the metal [17–19].

2.2.2 Types of Corrosion Mechanisms

Depending on the specific environmental conditions, the materials that are involved, and the electrochemical processes that are taking place, corrosion may become obvious in a number of different ways. It is important to note that the following are some examples of common types of corrosion processes [17–19]:

- "Uniform corrosion" refers to a kind of corrosion in which the corrosion process is relatively consistent throughout the surface of a metal. This kind of corrosion causes a progressive reduction in the thickness of the material. Temperature, humidity, and the chemical makeup of the electrolyte are the factors that are most likely to have an impact on this kind of corrosion.

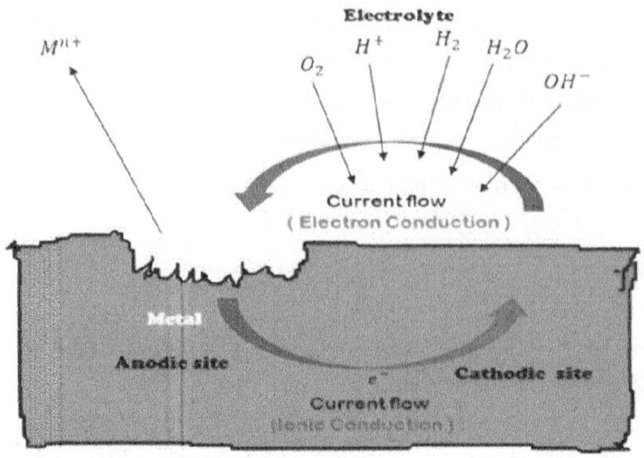

FIGURE 2.2 Schematic of corrosion process [20].

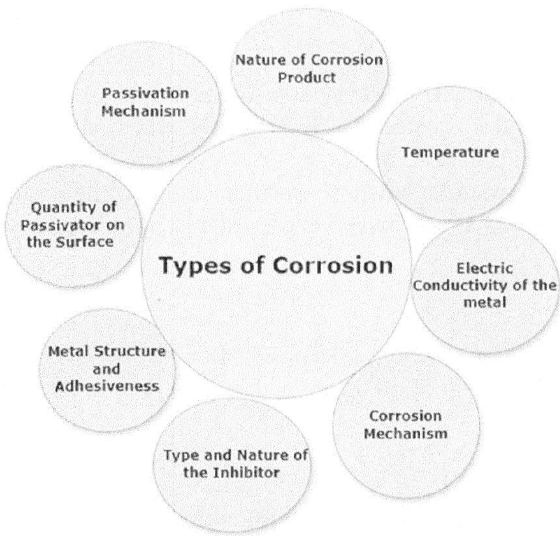

FIGURE 2.3 Parameters that determine the types of corrosion (a symmetric diagram) [8].

- Among the types of localized corrosion that may occur on metal surfaces, some examples are pitting, crevice corrosion, and galvanic corrosion. This kind of corrosion takes place in certain areas of the metal's surface. The local ambient or surface conditions are often the cause of these corrosions, which might manifest themselves in a variety of different ways.
- The formation of pits or holes on metal surfaces is a characteristic of pitting corrosion. The corrosion in question often takes place when aggressive ions or isolated flaws are present. Pitting corrosion is a destructive kind of corrosion that may quickly degrade metal and is difficult to detect and treat.
- Crevice corrosion occurs in narrow gaps or crevices between two surfaces, where stagnant electrolyte conditions promote localized corrosion. Crevice

corrosion is common in areas with poor fluid circulation, such as bolted joints, gaskets, and under-deposit corrosion.

- Galvanic corrosion occurs when two dissimilar metals are in contact in the presence of an electrolyte, leading to accelerated corrosion of the less noble (more reactive) metal. Galvanic corrosion can occur in various engineering applications, such as plumbing systems, marine environments, and electrical connections.

2.2.3 Factors Influencing Corrosion

Several factors influence the susceptibility of materials to corrosion, including [21–23]:

- **Environmental Conditions:** Environmental factors such as temperature, humidity, pH, salinity, and atmospheric pollutants can significantly impact corrosion rates. For example, high temperatures and humidity levels can accelerate corrosion reactions, while acidic or alkaline environments can promote or inhibit corrosion, depending on the material properties.
- **Material Properties:** The chemical composition, microstructure, and mechanical properties of materials play a critical role in their corrosion resistance. Materials with passive oxide layers, such as stainless steels and aluminum alloys, exhibit enhanced corrosion resistance compared to pure metals or alloys lacking protective coatings.
- **Electrochemical Potential:** The electrochemical potential of a material, determined by its position in the galvanic series, influences its tendency to corrode relative to other materials in the same electrolyte. Materials with

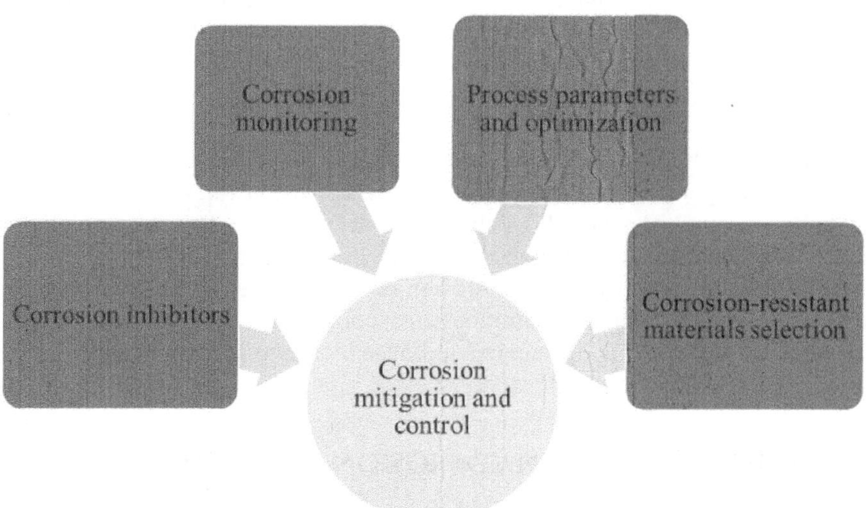

FIGURE 2.4 Parameters for corrosion mitigation and control [8].

TABLE 2.1
Factors Influencing Corrosion

Factor	Influence	Examples
Environmental	Chemical composition, temperature, humidity	High chloride concentration in marine environments
Material	Composition, microstructure, surface condition	Alloying elements in stainless steel
Electrochemical	Electrolyte conductivity, pH, oxygen concentration	Acidic environments promoting corrosion
Mechanical	Stress, strain, fatigue	Stress corrosion cracking in pipelines

more negative (less noble) potentials are more susceptible to corrosion in galvanic coupling scenarios.

- **Flow Conditions:** Fluid flow patterns and velocities can affect corrosion rates by influencing mass transport and fluid–surface interactions. Turbulent flow regimes may promote corrosion by removing protective surface films or increasing mass transfer rates, whereas laminar flow conditions may inhibit corrosion by maintaining a stable boundary layer.
- **Mechanical Stress:** Mechanical stresses, such as tensile, compressive, or cyclic loading, can exacerbate corrosion by inducing localized deformation, cracking, or surface defects. Stress corrosion cracking (SCC) is a particularly insidious form of corrosion that occurs under combined mechanical and corrosive environments.
- **Presence of Contaminants**: The presence of contaminants, such as chlorides, sulfides, oxygen, or microbial species, can accelerate corrosion processes by promoting aggressive chemical reactions or microbial-induced corrosion mechanisms. Contaminants may originate from industrial processes, environmental pollutants, or biological sources.

Table 2.1 highlights the various factors that influence corrosion processes in materials. Temperature and humidity increase corrosion rates in harsh environments like marine areas with high chloride concentrations. This is especially true in chloride-rich situations. Composition and microstructure influence corrosion susceptibility. Many alloys boost corrosion resistance. Electrochemical characteristics like pH and oxygen concentration affect corrosion kinetics, whereas mechanical variables like stress and strain may produce localized corrosion like pipe stress corrosion cracking.

2.3 METHODOLOGIES IN CORROSION RESEARCH

Research on corrosion involves a broad variety of disciplines, including experimental techniques, computer modeling, field investigations, and case studies, among others. The advancement of our knowledge of corrosion processes, the development of efficient preventative methods, and the mitigation of hazards connected to corrosion in a variety of sectors and applications are all significantly aided by these techniques. In this part, we have a discussion of the most important approaches that are used in corrosion research, as well as their contributions to the field [24–28].

2.3.1 Experimental Techniques

Corrosion research relies on experiments to study materials in different conditions. Tests used in corrosion research include [29–31]:

Electrochemical Methods: Potentiodynamic polarization, electrochemical impedance spectroscopy (EIS), and cyclic voltammetry are used to study corrosion kinetics, rates, and materials' electrochemical properties. These methodologies provide valuable insights into metal and alloy corrosion and electrochemical behavior.

Methods of Surface Analysis: For corroded surfaces, scanning electron microscopy (SEM), atomic force microscopy (AFM), X-ray photoelectron spectroscopy (XPS), and energy-dispersive X-ray spectroscopy (EDS) are used to analyze their shape, composition, and structure. These methods help identify corrosion products, surface defects, and microstructural changes caused by corrosion.

Corrosion Testing: Materials are exposed to simulated or real-world corrosive environments and monitored for degradation. Corrosion testing includes salt spray, immersion, accelerated, and electrochemical methods. Tests assess material resistance to corrosion, causes of corrosion, and performance under specific conditions. Mechanical testing methods, such as tensile, hardness, and fatigue testing, assess mechanical qualities and integrity before and during corrosion exposure. These tests may determine how corrosion affects material strength, ductility, and fracture behavior, which reveals corrosion-related mechanical failures.

2.4 COMPUTATIONAL APPROACHES

Computational techniques have been developed by researchers working in the field of corrosion research as useful tools that provide them with the ability to model corrosion processes, forecast the behavior of materials, and devise designs that are resistant to corrosion. These techniques were created as part of a study that was conducted on corrosion. Computational approaches are a useful supplement to experimental procedures for the purpose of gaining insights into complicated corrosion processes that are difficult to detect or measure via testing. This is due to the fact that computational methods provide insights on certain occurrences. The research on corrosion often makes use of the following computer techniques, which are a few examples of methodology that fall under this category [32–34]:

Molecular dynamics (MD) simulations carried out on a computer with the purpose of investigating: Researchers have the ability to explore the processes of corrosion on a molecular level via the use of molecular dynamics simulations. Modeling the atomic-scale interactions that take place between molecules and the environment in which they are situated is the purpose of these simulations. The creation of materials and coatings that are resistant to corrosion is made much easier with the use of molecular dynamics simulations. Processes such as ion transport, adsorption, and surface reactions are essential components of corrosion resistance, and these simulations give insights into these processes.

One of the capabilities of calculations that are based on density functional theory (DFT) is the capacity to predict the electronic structure and characteristics of materials. When researchers acquire this skill, they are able to undertake an examination of the thermodynamics and kinetics of corrosion processes. This capability was previously unavailable to them. Calculations that make use of density functional theory provide a substantial amount of information about the activation barriers of corrosion processes, the stability of corrosion products, and the adsorption of corrosive species. These are all aspects of corrosion that are important to consider.

The term "finite-element analysis" (FEA) refers to a technique that simulates the electrochemical and mechanical behavior of materials that have been exposed to corrosion and mechanical stress. During the process of designing corrosion-resistant structures and components, finite-element analysis simulations are utilized for the aim of providing assistance. For the purpose of assisting in the prediction of stress corrosion cracking, corrosion-induced deformation, and material deterioration under complicated loading circumstances, these simulations are employed.

In corrosive environments, simulations of computational fluid dynamics (CFD) depict fluid flow and mass transport. A better understanding of corrosion rates, corrosion product deposition, and interactions between fluids and solids may be gained via the use of these models. Simulations using computational fluid dynamics allow for the optimization of fluid flow patterns, the reduction of corrosion hotspots, and the design of equipment and pipe systems that are resistant to corrosion.

By examining massive datasets of experimental and computational findings, machine learning algorithms are designed to discover correlations, trends, and prediction models of corrosion behavior. This is accomplished via the process of using machine learning. In order to aid in the prediction of corrosion, the screening of materials, and the creation of coatings that are resistant to corrosion, several machine learning techniques are used. Some examples of these approaches include neural networks, support vector machines, and random forests.

2.5 INTERDISCIPLINARY PERSPECTIVES IN CORROSION RESEARCH

Corrosion research may benefit from insights from materials science and engineering, electrochemistry, surface science, and environmental science. This interdisciplinary approach allows researchers to study complex interactions between materials, the environment, and corrosion. This leads to unique corrosion prevention, mitigation, and material design solutions. This section examines how each field has contributed to corrosion research, including their roles in understanding corrosion processes and developing corrosion-control methods [35–38]. This section will highlight each discipline's corrosion research contributions.

- Materials scientists choose and create corrosion-resistant materials. They also make corrosion-resistant metals, composites, and coatings for specific uses and conditions. Researchers can enhance material corrosion resistance while maintaining mechanical, thermal, and functional properties [39]. When people grasp how microstructure and material characteristics relate, this is achievable. Materials scientists study atomic and microstructural corrosion processes. They study rusting. They provide information on grain boundaries, defects, and alloying components that affect corrosion susceptibility. If they study corrosion factors, researchers may develop prediction models and ways to prevent corrosion-induced deterioration. They may reduce rusting. Table 2.2 lists common corrosion processes in

TABLE 2.2
Common Corrosion Mechanisms

Mechanism	Description	Examples	Contributing Factors
Uniform corrosion	Gradual loss of material thickness over the entire surface	Rusting of steel structures, atmospheric corrosion	Presence of moisture and oxygen Acidic or alkaline environments High operating temperatures
Pitting corrosion	Formation of localized pits or craters on the surface	Pitting corrosion of stainless steel, aluminum	Halide ions (chlorides) Localized acidic environments Crevices or imperfections on the surface
Galvanic corrosion	Accelerated corrosion due to electrochemical coupling	Corrosion of metal fasteners in contact with dissimilar metals	Significant difference in electrode potentials between metals Electrolyte presence (e.g., saltwater) Large surface area ratio of active to noble metal
Crevice corrosion	Localized attack at occluded areas where stagnant conditions and concentration of corrosive species can occur	Corrosion beneath gaskets or deposits on metal surfaces	Limited oxygen supply and buildup of aggressive ions within crevices Stagnant or slow-moving electrolytes
Intergranular corrosion	Preferential attack at grain boundaries within a metal microstructure	Corrosion of sensitized stainless steel (improper heat treatment)	Chromium depletion at grain boundaries Specific microstructures and sensitization of alloys
Stress corrosion cracking (SCC)	Brittle failure due to the combined effects of tensile stress and a corrosive environment	Cracking of brass cartridge cases, stainless steel piping in high-temperature chloride environments	Susceptible material properties (e.g., some alloys) Residual or applied tensile stresses Specific corrosive media

engineered materials. Materials scientists explore surface modification strategies to increase corrosion resistance. This category includes passivation, anodization, CVD, and PVD. Surface-level treatments may affect materials' chemical composition, structure, and surface energy, resulting in corrosion-resistant barriers [37].
- Better characterization methods: Materials scientists utilize sophisticated characterization methods to study corroded surface appearance, chemistry, and structure. Gathering information is the aim. SEM, TEM, XRD, and AFM are examples of these methods. They reveal corrosion products, localized corrosion processes, and surface flaws, which guide material design and corrosion prevention [40, 41].

2.6 SURFACE SCIENCE

Surface science focuses on the characterization, manipulation, and control of material surfaces and interfaces, providing insights into surface reactions, adsorption phenomena, and surface-mediated processes relevant to corrosion research. Surface science techniques are used to study surface morphology, chemistry, and reactivity, enabling researchers to understand the interactions between materials and corrosive environments. Key contributions of surface science to corrosion research include [41–43]:

- **Surface Analysis Techniques:** Surface science techniques, such as XPS, Auger electron spectroscopy (AES), secondary ion mass spectrometry (SIMS), and scanning probe microscopy (SPM), are employed to analyze the composition, structure, and properties of surfaces and interfaces. These techniques provide detailed information on surface chemistry, corrosion products, and adsorbed species, aiding in the characterization of corroded surfaces and the identification of surface reactions.
- **Surface Modification and Functionalization:** Surface science techniques enable the modification and functionalization of material surfaces to enhance their corrosion resistance and performance. Surface treatments, such as plasma etching, ion implantation, and self-assembled monolayers (SAMs), modify surface chemistry, wettability, and adhesion properties, creating barriers against corrosive attack and improving material durability.
- **Corrosion Kinetics at Interfaces:** Surface science studies investigate the kinetics of corrosion reactions at interfaces, including metal–electrolyte interfaces, metal–oxide interfaces, and metal–polymer interfaces. By quantifying reaction rates, activation energies, and interfacial processes, researchers elucidate the mechanisms of corrosion initiation, propagation, and inhibition at the nanoscale, guiding the design of corrosion-resistant materials and coatings.

2.7 ENVIRONMENTAL SCIENCE

Environmental science examines the interactions between natural and anthropogenic factors and their impact on ecosystems, climate, and environmental quality. In the context of corrosion research, environmental science provides insights into the effects of environmental conditions, pollutants, and climate change on corrosion processes and material degradation. Key contributions of environmental science to corrosion research include [8, 42]:

- **Corrosive Environments:** Environmental science studies characterize corrosive environments, such as marine environments, industrial atmospheres, and polluted waters, assessing their chemical composition, temperature, humidity, and exposure conditions. By understanding the corrosive agents and mechanisms prevalent in different environments, researchers develop strategies to mitigate corrosion risks and protect infrastructure.
- **Atmospheric Corrosion:** Environmental science investigates atmospheric pollutants, such as sulfur dioxide (SO_2), nitrogen oxides (NO_x), ozone (O_3), and particulate matter, and their role in atmospheric corrosion processes. By monitoring air quality and studying corrosion rates on outdoor surfaces, researchers assess the impact of air pollution on material degradation and urban infrastructure.
- **Water Chemistry and Corrosion:** Environmental science examines water chemistry, pH, alkalinity, conductivity, and dissolved ions (e.g., chlorides, sulfates, carbonates) in natural and industrial waters, evaluating their corrosivity and potential for inducing corrosion in metals and alloys. By analyzing water quality parameters, researchers assess the risk of localized corrosion, microbiologically influenced corrosion (MIC), and corrosion-related water contamination.
- **Climate Change and Corrosion:** Environmental science investigates the effects of climate change, including temperature fluctuations, extreme weather events, and sea level rise, on corrosion processes and infrastructure resilience. By modeling climate projections and assessing vulnerability to corrosion-induced damage, researchers develop adaptation strategies and sustainable infrastructure designs to mitigate the impacts of climate change on corrosion.

2.8 CHALLENGES AND LIMITATIONS IN CORROSION RESEARCH

Corrosion research faces numerous challenges and limitations, ranging from technical constraints to economic and environmental considerations. Overcoming these challenges is essential for advancing corrosion science, developing effective prevention strategies, and mitigating the economic and environmental impacts of corrosion. This section explores some of the key challenges and limitations in corrosion research and their implications for corrosion monitoring, prevention, and control, as well as the economic and environmental impacts of corrosion [42, 43].

2.8.1 Corrosion Monitoring and Detection

Due to their lack of sensitivity and resolution, conventional corrosion monitoring systems may miss early-stage corrosion or restricted corrosion processes. This is because standard corrosion monitoring techniques lack sensitivity and resolution.

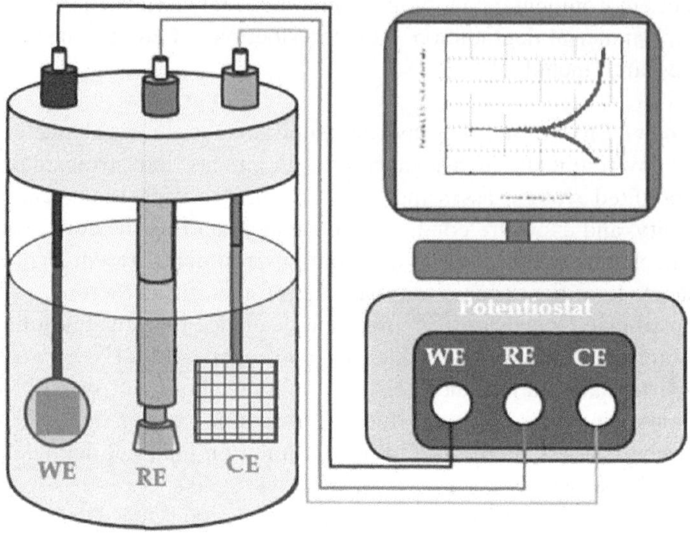

FIGURE 2.5 Schematic diagram of three-electrode cell with WE, RE, and CE and potentiostat for electrochemical experiments [44].

Increased detection limits and spatial resolution are needed to locate corrosion hotspots and start remediation quickly. Environments in the real world are very complicated: Monitoring corrosion in marine, industrial, and aerospace settings is difficult due to environmental complexity. The aerospace sector has these complex conditions, including temperature changes, fluid dynamics, and chemical variability. To accurately measure corrosion for evaluation, monitoring devices must be resilient and able to perform in adverse settings. Advanced corrosion monitoring technologies like remote sensing, wireless sensor networks, and corrosion sensors may be too costly to operate and maintain, preventing their broad adoption. This is especially true for resource-constrained enterprises and infrastructure. Financial obstacles must be removed and corrosion monitoring technology made more accessible to maximize corrosion control efficiency [43]. Figure 2.5 shows a schematic diagram of a three-electrode cell with WE, RE, and CE and potentiostat for electrochemical experiments.

2.8.2 Corrosion Prevention and Control

Table 2.3 outlines various corrosion prevention strategies employed to mitigate material degradation. Protective coatings act as a barrier between the material and corrosive environment, preventing direct contact and reducing corrosion rates. Cathodic protection systems, such as sacrificial anodes or impressed current systems, provide an electrochemical shield to inhibit corrosion. Material selection plays a crucial role, with the use of corrosion-resistant materials or alloys tailored to specific applications, such as stainless steel in marine environments. Implementing these strategies

TABLE 2.3
Corrosion Prevention Strategies

Strategy	Description	Examples
Protective Coatings	Barrier between material and corrosive environment	Paint coatings on metal surfaces
Cathodic Protection	Sacrificial anode or impressed current systems	Galvanic cathodic protection in buried pipelines
Material Selection	Corrosion-resistant materials or alloys	Stainless steel for marine applications

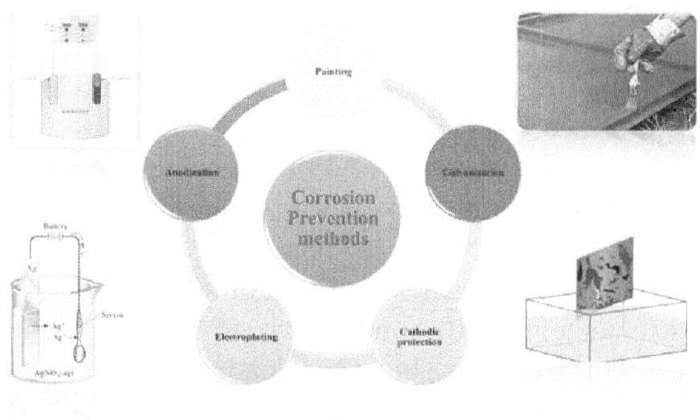

FIGURE 2.6 Corrosion prevention methods [20].

effectively can significantly extend the lifespan of materials and infrastructure [43]. Figure 2.6 depicts corrosion prevention methods.

- **Multifaceted Nature of Corrosion:** Corrosion is a multifaceted phenomenon influenced by various factors, including material properties, environmental conditions, and operational parameters. Developing comprehensive corrosion prevention strategies requires integrating diverse approaches, such as material selection, design optimization, surface treatments, and corrosion inhibitors.
- **Effectiveness of Preventive Measures:** The effectiveness of corrosion prevention measures, such as protective coatings, inhibitors, and cathodic protection systems, may vary depending on the specific application, environmental conditions, and operational requirements. Continuous monitoring and evaluation of preventive measures are necessary to ensure long-term corrosion protection and performance.
- **Compatibility and Interactions:** Corrosion prevention methods must be compatible with other system components and materials to avoid adverse interactions or unintended consequences. Compatibility issues may arise from differences in material properties, chemical compatibility, or electrochemical interactions, requiring careful selection and integration of corrosion control measures.

2.9 FUTURE DIRECTIONS AND RECOMMENDATIONS

As corrosion continues to pose significant challenges across various industries, future research efforts should focus on innovative solutions, interdisciplinary collaborations, and policy interventions to address these challenges effectively. Bridging the gap between research and industry, fostering collaborative initiatives, and implementing robust regulatory frameworks are essential for advancing corrosion science and mitigating its economic and environmental impacts.

2.9.1 Bridging the Gap between Research and Industry

- **Technology Transfer and Commercialization:** Facilitating the transfer of corrosion research findings and technologies from academia to industry is critical for translating scientific discoveries into practical applications. Establishing technology transfer offices, incubators, and industry-academic partnerships can accelerate the commercialization of corrosion-resistant materials, coatings, and corrosion monitoring technologies.
- **Industry Engagement and Stakeholder Collaboration:** Engaging industry stakeholders, including manufacturers, engineers, asset managers, and regulatory agencies, in collaborative research projects fosters knowledge exchange, identifies industry needs, and validates research outcomes in real-world settings. Industry–academic consortia, joint research programs, and technology road mapping initiatives promote synergistic collaborations and mutual benefits for academia and industry.

2.9.2 Collaborative Initiatives and Knowledge Exchange

- **Interdisciplinary Research Consortia:** Establishing interdisciplinary research consortia and centers of excellence dedicated to corrosion science promotes collaboration among materials scientists, chemists, engineers, and environmental scientists. These consortia facilitate interdisciplinary research, share resources and expertise, and address complex corrosion challenges through collaborative projects and joint funding opportunities.
- **Knowledge Exchange Platforms:** Developing knowledge exchange platforms, such as conferences, workshops, webinars, and online forums, fosters communication, networking, and dissemination of best practices in corrosion research and technology. These platforms provide opportunities for researchers, practitioners, policymakers, and industry stakeholders to share insights, showcase innovations, and build partnerships for collaborative research and technology transfer.

  Advancements in contemporary corrosion research have been significantly influenced by innovative surface modification techniques and the growing importance of sustainable manufacturing practices [45–55]. The development of boron nitride thin films using magnetron sputtering has emerged as a key area of study, as these films offer enhanced corrosion and tribological resistance [56–63]. These coatings provide long-lasting protection in harsh environments, particularly for stainless steel and other alloys. Additionally, the exploration of microwave heating for the deposition of Inconel 625 clad layers on stainless steel has shown promising results in terms of corrosion resistance and microstructural stability [64–72]. A significant challenge in corrosion research remains the optimization of process parameters to achieve consistent and reliable coating properties. The influence of heat treatments and the development of new coating materials, such as hydroxyapatite coatings in biomedical applications, also play a crucial role in enhancing corrosion resistance [73–77]. Moreover, the integration of additive manufacturing and computational modeling techniques is helping to optimize the design and performance of corrosion-resistant materials [78–83]. These advancements are essential for addressing the growing demand for durable materials in industries like the aerospace, biomedical, and energy sectors while promoting environmentally sustainable practices.

2.10 CONCLUSION

In conclusion, corrosion science plays a crucial role in addressing the economic, environmental, and societal challenges posed by corrosion. By bridging the gap between research and industry, fostering collaborative initiatives, and implementing robust regulatory frameworks, we can advance corrosion science, promote technology transfer, and mitigate the economic and environmental impacts of corrosion. Sustained research efforts, interdisciplinary collaborations, and policy interventions are essential for safeguarding critical infrastructure, protecting natural resources, and ensuring sustainable development in the face of corrosion challenges.

- Corrosion poses significant economic, environmental, and safety risks across various industries and sectors.
- Bridging the gap between research and industry is essential for translating scientific discoveries into practical applications and technologies.
- Collaborative initiatives and knowledge exchange platforms facilitate interdisciplinary research, technology transfer, and best practice dissemination in corrosion science.
- Policy implications and regulatory frameworks promote standardization, certification, and funding support for corrosion research, development, and implementation.
- Sustained research efforts and continued investments in corrosion science are crucial for addressing emerging challenges and advancing sustainable corrosion management practices.

REFERENCES

1. Mahdi, B. S., Jawad, Q., Zinad, D., Salim, R. D., Al-Amiery, A. A., Gaaz, T. S., Takriff, M. S., & Kadhum, A. (2022). Corrosion inhibition of mild steel: Gravimetric, thermodynamic, and computational studies. *Molecules*, *27*(15), 4857. https://doi.org/10.3390/molecules27154857

2. Jawad, Q., Zinad, D., Salim, R. D., Al-Amiery, A. A., Gaaz, T. S., Takriff, M. S., & Kadhum, A. (2019). Synthesis, characterization, and corrosion inhibition potential of novel thiosemicarbazone on mild steel in sulfuric acid environment. *Coatings*, *9*, 729. https://doi.org/10.3390/coatings9110729
3. Al-Baghdadi, S., Noori, F., Ahmed, W. K., & Al-Amiery, A. A. (2016). Thiadiazole as a potential corrosion inhibitor for mild steel in 1 M HCl. *Journal of Advanced Electrochemistry*, *2*, 67–69.
4. Resen, A. M., Hanoon, M., Salim, R. D., Al-Amiery, A. A., Shaker, L. M., & Kadhum, A. A. H. (2020). Gravimetrical, theoretical, and surface morphological investigations of corrosion inhibition effect of 4-(benzoimidazole-2-yl) pyridine on mild steel in hydrochloric acid. *Koroze a Ochrana Materiálů*, *64*, 122–130. https://doi.org/10.2478/kom-2020–0018
5. Junaedi, S., Al-Amiery, A., Kadihum, A., Kadhum, A., & Mohamad, A. (2013). Inhibition effects of a synthesized novel 4-aminoantipyrine derivative on the corrosion of mild steel in hydrochloric acid solution together with quantum chemical studies. *International Journal of Molecular Sciences*, *14*, 11915–11928. https://doi.org/10.3390/ijms140611915
6. Solomon, M. M., Uzoma, I. E., Olugbuyiro, J. A. O., & Ademosun, O. T. (2022). A censorious appraisal of the oil well acidizing corrosion inhibitors. *Journal of Petroleum Science and Engineering*, *215*, 110711. https://doi.org/10.1016/j.petrol.2022.110711
7. Verma, P. K., Kumar, V., & Vasudev, H. (2024). Bioactivity and corrosion analysis of thermally sprayed hydroxyapatite based coatings. *Journal of Electrochemical Science and Engineering*, *14*(4), 419–439.
8. Abbas, A., Adesina, A. Y., & Suleiman, R. K. (2023). Influence of organic acids and related organic compounds on corrosion behavior of stainless steel: A critical review. *Metals*, *13*(8), 1479.
9. El-Haitout, B., Selatnia, I., Lgaz, H., Al-Hadeethi, M. R., Lee, H.-S., Chaouiki, A., Ko, Y. G., Ali, I. H., & Salghi, R. (2023). Exploring the feasibility of new eco-friendly heterocyclic compounds for establishing efficient corrosion protection for N80 steel in a simulated oil well acidizing environment: From molecular-level prediction to experimental validation. *Colloids and Surfaces A*, *656*, 130372. https://doi.org/10.1016/j.colsurfa.2022.130372
10. Tiu, B. D. B., & Advincula, R. C. (2015). Polymeric corrosion inhibitors for the oil and gas industry: Design principles and mechanism. *Reactive and Functional Polymers*, *95*, 25–45. https://doi.org/10.1016/j.reactfunctpolym.2015.08.006
11. Singh, H., Kumar, M., Singh, R., & Kumar, S. (2024). Microstructure and cavitation erosion performance of cold-sprayed WC-12Co and WC-17Co coatings on hydraulic turbine steels. *Journal of Materials Engineering and Performance*. Advance online publication. https://doi.org/10.1007/s11665-024-10097-2
12. Singh, H., Kumar, M., Singh, R., & Kumar, S. (2024). Deposition behaviour of WC-17Co cold-sprayed composite coating on hydro turbine steels. *Surface Review and Letters*. Advance online publication. https://doi.org/10.1142/S0218625X25500209
13. Singh, H., Kumar, M., & Singh, R. (2024). Cavitation erosion behaviour of WC-Co coatings on turbine steel deposited by cold spray. *Surface Review and Letters*. Advance online publication. https://doi.org/10.1142/S0218625X25500052
14. Singh, H., Kumar, M., & Singh, R. (2024). Slurry erosion performance of cold-sprayed WC-12Co composite coatings. *Surface Review and Letters*. Advance online publication. https://doi.org/10.1142/S0218625X2450050
15. Farsak, M., Keleş, H., & Keleş, M. A. (2015). New corrosion inhibitor for protection of low carbon steel in HCl solution. *Corrosion Science*, *98*, 223–232. https://doi.org/10.1016/j.corsci.2015.05.036

16. Alamiery, W. N. R. W., Isahak, H. S. S., Aljibori, H. A., Al-Asadi, H. A., & Kadhum, A. A. H. (2021). Effect of the structure, immersion time and temperature on the corrosion inhibition of 4-pyrrol-1-yl-N-(2,5-dimethyl-pyrrol-1-yl)benzoylamine in 1.0 M HCl solution. *International Journal of Corrosion and Scale Inhibition*, *10*(2), 700–713. https://doi.org/10.17675/2305-6894-2021-10-2-14
17. Al-Baghdadi, S., Hashim, F., Salam, A., Abed, T., Gaaz, T., Al-Amiery, A., Kadhum, A. H., Reda, K., & Ahmed, W. (2018). Synthesis and corrosion inhibition application of NATN on mild steel surface in acidic media complemented with DFT studies. *Results in Physics*, *8*, 1178–1184. https://doi.org/10.1016/j.rinp.2018.02.007
18. Al-Azzawi, W. K., Al Adily, A. J., Sayyid, F. F., Al-Azzawi, R. K., Kzar, M. H., Jawoosh, H. N., Al-Amiery, A. A., Kadhum, A. A. H., Isahak, W. N. R. W., & Takriff, M. S. (2022). Evaluation of corrosion inhibition characteristics of an N-propionanilide derivative for mild steel in 1 M HCl: Gravimetrical and computational studies. *International Journal of Corrosion and Scale Inhibition*, *11*(3), 1100–1114. https://doi.org/10.17675/2305-6894-2022-11-3-12
19. Mustafa, A., Sayyid, F., Betti, N., Hanoon, M., Al-Amiery, A., Kadhum, A., & Takriff, M. (2021). Inhibition evaluation of 5-(4-(1H-pyrrol-1-yl)phenyl)-2-mercapto-1,3,4-oxadiazole for the corrosion of mild steel in an acid environment: Thermodynamic and DFT aspects. *Tribologia*, *38*, 39–47. https://doi.org/10.30678/fjt.105330
20. Bijapur, K., Molahalli, V., Shetty, A., Toghan, A., De Padova, P., & Hegde, G. (2023). Recent trends and progress in corrosion inhibitors and electrochemical evaluation. *Applied Sciences*, *13*(18), 10107.
21. Abdulsahib, Y. M., Eltmimi, A. J. M., Alhabeeb, S. A., Hanoon, M. M., Al-Amiery, A. A., Allami, T., & Kadhum, A. A. H. (2021). Experimental and theoretical investigations on the inhibition efficiency of N-(2,4-dihydroxytoluene-ylidene)-4-methylpyridin-2-amine for the corrosion of mild steel in hydrochloric acid. *International Journal of Corrosion and Scale Inhibition*, *10*(3), 885–899. https://doi.org/10.17675/2305-6894-2021-10-3-3
22. Khudhair, A. K., Mustafa, A. M., Hanoon, M. M., Al-Amiery, A. A., Shaker, L. M., Gaaz, T., Mohamad, A. B., Kadhum, A. H., & Takriff, M. S. (2022). Experimental and theoretical investigation on the corrosion inhibitor potential of N-MEH for mild steel in HCl. *Progress in Color, Colorants and Coatings*, *15*, 111–122. https://doi.org/10.30509/pccc.2021.166815.1111
23. Zinad, D. S., Salim, R. D., Betti, N., Shaker, L. M., & Al-Amiery, A. A. (2022). Comparative investigations of the corrosion inhibition efficiency of a 1-phenyl-2-(1-phenylethylidene)hydrazine and its analog against mild steel corrosion in hydrochloric acid solution. *Progress in Color, Colorants and Coatings*, *15*, 53–63. https://doi.org/10.30509/pccc.2021.166786.1108
24. Salim, R. D., Betti, N., Hanoon, M., & Al-Amiery, A. A. (2021). 2-(2,4-Dimethoxybenzylidene)-N-phenylhydrazinecarbothioamide as an efficient corrosion inhibitor for mild steel in acidic environment. *Progress in Color, Colorants and Coatings*, *15*, 45–52. https://doi.org/10.30509/pccc.2021.166775.1105
25. Al-Amiery, A. A., Shaker, L. M., Kadhum, A. H., & Takriff, M. S. (2021). Exploration of furan derivative for application as a corrosion inhibitor for mild steel in hydrochloric acid solution: Effect of immersion time and temperature on efficiency. *Materials Today: Proceedings*, *42*, 2968–2973. https://doi.org/10.1016/j.matpr.2020.12.807
26. Resen, A. M., Hanoon, M. M., Alani, W. K., Kadhim, A., Mohammed, A. A., Gaaz, T. S., Kadhum, A. A. H., Al-Amiery, A. A., & Takriff, M. S. (2021). Exploration of 8-piperazine-1-ylmethylumbelliferone for application as a corrosion inhibitor for mild steel in hydrochloric acid solution. *International Journal of Corrosion and Scale Inhibition*, *10*(1), 368–387. https://doi.org/10.17675/2305-6894-2021-10-1-21

27. Hanoon, M. M., Resen, A. M., Al-Amiery, A. A., Kadhum, A. A. H., & Takriff, M. S. (2021). Theoretical and experimental studies on the corrosion inhibition potentials of 2-((6-Methyl-2-ketoquinolin-3-yl)methylene) hydrazinecarbothioamide for mild steel in 1 M HCl. *Progress in Color, Colorants and Coatings*, *15*, 21–33. https://doi.org/10.30509/pccc.2020.166739.1095
28. Hashim, F. G., Salman, T. A., Al-Baghdadi, S. B., Gaaz, T., & Al-Amiery, A. A. (2020). Inhibition effect of hydrazine-derived coumarin on a mild steel surface in hydrochloric acid. *Tribologia*, *37*, 45–53. https://doi.org/10.30678/fjt.95510
29. Al-Amiery, A. A., Shaker, L. M., Allami, T., Kadhum, A. H., & Takriff, M. S. (2021). A study of acidic corrosion behavior of furan-derived Schiff base for mild steel in hydrochloric acid environment: Experimental, and surface investigation. *Materials Today: Proceedings*, *44*, 2337–2341. https://doi.org/10.1016/j.matpr.2020.12.431
30. Al-Baghdadi, S., Al-Amiery, A., Gaaz, T., & Kadhum, A. (2021). Terephthalohydrazide and isophthalo-hydrazide as new corrosion inhibitors for mild steel in hydrochloric acid: Experimental and theoretical approaches. *Koroze a Ochrana Materiálu*, *65*, 12–22. https://doi.org/10.2478/kom-2021–0002
31. Hanoon, M. M., Resen, A. M., Shaker, L. M., Kadhum, A., & Al-Amiery, A. A. (2021). Corrosion investigation of mild steel in aqueous hydrochloric acid environment using N-(naphthalen-1yl)-1–(4-pyridinyl)methanimine complemented with antibacterial studies. *Biointerface Research in Applied Chemistry*, *11*, 9735–9743. https://doi.org/10.33263/BRIAC112.97359743
32. Al-Baghdadi, S., Gaaz, T. S., Al-Adili, A., Al-Amiery, A., & Takriff, M. (2021). Experimental studies on corrosion inhibition performance of acetylthiophene thiosemicarbazone for mild steel in HCl complemented with DFT investigation. *International Journal of Low-Carbon Technologies*, *16*, 181–188. https://doi.org/10.1093/ijlct/ctaa050
33. Al-Amiery, A. (2021). Anti-corrosion performance of 2-isonicotinoyl-N-phenylhydrazinecarbothioamide for mild steel hydrochloric acid solution: Insights from experimental measurements and quantum chemical calculations. *Surface Review and Letters*, *28*, 2050058. https://doi.org/10.1142/S0218625X20500584
34. Al-Amiery, A. (2022). Investigations on the corrosion inhibitory effect of newly quinoline derivative on mild steel in HCl solution complemented with antibacterial studies. *Biointerface Research in Applied Chemistry*, *12*, 1561–1568. https://doi.org/10.33263/BRIAC122.15611568
35. Al-Amiery, A. (2021). Short report of mild steel corrosion in 0.5 M H2SO4 by 4-ethyl-1–(4-oxo-4-phenylbutanoyl)thiosemicarbazide. *Tribologia*, *30*, 90–99.
36. Al-Amiery, A. A., Isahak, W. N. R. W., & Takriff, M. S. (2021). Inhibition of mild steel corrosion by 4-benzyl-1–(4-oxo-4-phenylbutanoyl)thiosemicarbazide: Gravimetrical, adsorption and theoretical studies. *Lubricants*, *9*, 93. https://doi.org/10.3390/lubricants9090093
37. Dawood, M. A., Alasady, Z. M. K., Abdulazeez, M. S., Ahmed, D. S., Sulaiman, G. M., Kadhum, A. A. H., Shaker, L. M., & Al-Amiery, A. A. (2021). The corrosion inhibition effect of a pyridine derivative for low carbon steel in 1 M HCl medium: Complemented with antibacterial studies. *International Journal of Corrosion and Scale Inhibition*, *10*, 1766–1782. https://doi.org/10.17675/2305-6894-2021-10-4-25
38. Al-Amiery, A. (2021). Corrosion inhibition effect of 2-N-phenylamino-5–(3-phenyl-3-oxo-1-propyl)-1,3,4-oxadiazole on mild steel in 1 M hydrochloric acid medium: Insight from gravimetric and DFT investigations. *Materials Science for Energy Technologies*, *4*, 398–406. https://doi.org/10.1016/j.mset.2021.09.002
39. Al-Amiery, A. (2021). Anticorrosion effect of thiosemicarbazide derivative on mild steel in 1 M hydrochloric acid and 0.5 M sulfuric acid: Gravimetrical and theoretical studies. *Materials Science for Energy Technologies*, *4*, 263–273. https://doi.org/10.1016/j.mset.2021.07.004

40. Bannwarth, C., Caldeweyher, E., Ehlert, S., Hansen, A., Pracht, P., Seibert, J., Spicher, S., & Grimme, S. (2021). Extended tight-binding quantum chemistry methods. *WIREs Computational Molecular Science*, *11*(2). https://doi.org/10.1002/wcms.1493
41. Chermette, H. (1999). Chemical reactivity indexes in density functional theory. *Journal of Computational Chemistry*, *20*(1), 129–154. https://doi.org/10.1002/(SICI)1096–987X(19990115)20:13.0.CO;2-A
42. Parr, R. G., & Chattaraj, P. K. (1991). Principle of maximum hardness. *Journal of the American Chemical Society*, *113*(5), 1854–1855. https://doi.org/10.1021/ja00005a072
43. Iczkowski, R. P., & Margrave, J. L. (1961). Electronegativity. *Journal of the American Chemical Society*, *83*(17), 3547–3551. https://doi.org/10.1021/ja01478a001
44. Shin, D. H., Hwang, H. K., Kim, H. H., & Lee, J. H. (2022). Evaluation of commercial corrosion sensors for real-time monitoring of pipe wall thickness under various operational conditions. *Sensors*, *22*(19), 7562.
45. Singh, G., Singh, H., Sharma, Y., Vasudev, H., & Prakash, C. (2024). Analysis and optimization of various process parameters and effect on the hardness of SS-304 stainless steel welded joints. *International Journal on Interactive Design and Manufacturing (IJIDeM)*, *18*(5), 2889–2896.
46. Parkash, J., Saggu, H. S., & Vasudev, H. (2022). A short review on the performance of high velocity oxy-fuel coatings in boiler steel applications. *Materials Today: Proceedings*, *50*, 1442–1446.
47. Singh, J., Vasudev, H., Kumar, R., & Ubaidullah, M. (2024). Study on wear analysis of Ni-$_{20}Al_2O_3$ HVOF micron layers using artificial neural network technique. *International Journal on Interactive Design and Manufacturing (IJIDeM)*, *18*(8), 5925–5938.
48. Singh, J., Vasudev, H., Szala, M., & Gill, H. S. (2024). Neural computing for erosion assessment in Al-$_{20}TiO_2$ HVOF thermal spray coating. *International Journal on Interactive Design and Manufacturing (IJIDeM)*, *18*(4), 2321–2332.
49. Mehta, A., Singh, G., & Vasudev, H. (2024). Processing of shape memory alloys research, applications and opportunities: A review. *Physica Scripta*. https://iopscience.iop.org/article/10.1088/1402-4896/ad48cb
50. Singh, G., Bansal, A., & Vasudev, H. (2024). Corrosion and microstructural behaviour of Inconel 625 microwave clad deposited on mild steel. In *Thermal Claddings for Engineering Applications* (pp. 51–62). CRC Press.
51. Singh, M., Vasudev, H., & Kumar, R. (2021). Corrosion and tribological behaviour of BN thin films deposited using magnetron sputtering. *International Journal of Surface Engineering and Interdisciplinary Materials Science (IJSEIMS)*, *9*(2), 24–39.
52. Singh, G., Mehta, A., & Vasudev, H. (2024). Sustainability of additive manufacturing: A comprehensive review. *Progress in Additive Manufacturing*, *9*, 2249–2272.
53. Singh, G., Bansal, A., Vasudev, H., & Mishra, V. (2024). Sliding wear study of the Inconel-625 clad deposits by microwave heating on SS–304. *Physica Scripta*, *99*(6), 065503.
54. Singh, J., Yadav, A., Gupta, S., Balu, R., Shukla, N., Singh, G., & Singh, S. (2024). Design and computational analysis of continuous wave fiber laser cutting of flat and conformal ITO PET sheets. *International Journal on Interactive Design and Manufacturing (IJIDeM)*, 1–9. https://link.springer.com/article/10.1007/s12008-024-02171-9
55. Singh, G., Singh, R., & Gul, J. (2023). Machinability behavior of human implant materials. *Journal of Electrochemical Science and Engineering*, *13*(1), 99–114.
56. Verma, P. K., Vasudev, H., Singh, G., & Singh, S. (2024). A review on the microstructural and biomedical properties of implants manufactured using additive manufacturing. *Journal of the Institution of Engineers (India): Series D*, 1–15.
57. Kulkarni, S. D., Manjunatha, C. U., Manjunath, K. V., Prasad, C. D., & Vasudev, H. (2024). Design and optimization of polyvinyl-nitride rubber for tensile strength analysis. *International Journal on Interactive Design and Manufacturing (IJIDeM)*, *18*(8), 5893–5908.

58. Prashar, G., & Vasudev, H. (2022). Surface topology analysis of plasma sprayed Inconel625-Al_2O_3 composite coating. *Materials Today: Proceedings*, *50*, 607–611.
59. Prashar, G., & Vasudev, H. (2022). A review on the influence of process parameters and heat treatment on the corrosion performance of Ni-based thermal spray coatings. *Surface Review and Letters*, *29*(1), 2230001.
60. Kattimani, M. A., Venkatesh, P. R., Masum, H., Math, M. M., Bahadurdesai, V. N., Mustafkhadri, S., Prasad, C. D., & Vasudev, H. (2024). Design and numerical analysis of tensile deformation and fracture properties of induction hardened inconel 718 superalloy for gas turbine applications. *International Journal on Interactive Design and Manufacturing (IJIDeM)*, *18*(8), 5939–5949.
61. Singh, M., Vasudev, H., & Singh, M. (2022). Surface protection of SS-316L with boron nitride based thin films using radio frequency magnetron sputtering technique. *Journal of Electrochemical Science and Engineering*, *12*(5), 851–863.
62. Prashar, G., Vasudev, H., Thakur, L., & Bansal, A. (2023). Performance of thermally sprayed hydroxyapatite coatings for biomedical implants: A comprehensive review. *Surface Review and Letters*, *30*(1), 2241001.
63. Mehta, A., Vasudev, H., & Jeyaprakash, N. (2024). Role of sustainable manufacturing approach: Microwave processing of materials. *International Journal on Interactive Design and Manufacturing (IJIDeM)*, *18*(8), 5283–5299.
64. Prashar, G., & Vasudev, H. (2023). Understanding cold spray technology for hydroxyapatite deposition. *Journal of Electrochemical Science and Engineering*, *13*(1), 41–62.
65. Yedida, V. S., & Vasudev, H. (2022). A review on the development of thermal barrier coatings by using thermal spray techniques. *Materials Today: Proceedings*, *50*, 1458–1464.
66. Sharanabasava, H., Raviprakash, M., Prasad, C. D., Ramesh, M. R., Phanibhushana, M. V., Vasudev, H., & Kumar, S. (2023). Microstructure, mechanical and wear properties of SiC and Mo reinforced NiCr microwave cladding. *Advances in Materials and Processing Technologies*, *10*(4), 3620–3633. https://doi.org/10.1080/2374068X.2023.2257937
67. Dutta, V., Thakur, L., Singh, B., & Vasudev, H. (2022). A study of erosion–corrosion behaviour of friction stir-processed chromium-reinforced NiAl bronze composite. *Materials*, *15*(15), 5401.
68. Poojari, M., Hanumanthappa, H., Prasad, C. D., Jathanna, H. M., Ksheerasagar, A. R., Shetty, P., Shanmugam, B. K., & Vasudev, H. (2024). Computational modelling for the manufacturing of solar-powered multifunctional agricultural robot. *International Journal on Interactive Design and Manufacturing (IJIDeM)*, *18*(8), 5725–5736.
69. Prashar, G., & Vasudev, H. (2023). Understanding cold spray technology for hydroxyapatite deposition. *Journal of Electrochemical Science and Engineering*, *13*(1), 41–62.
70. Prakash, C., Gupta, L. R., Mehta, A., Vasudev, H., Tominov, R., Korman, E., Fedotov, A., Smirnov, V., & Kesari, K. K. (2023). Computing of neuromorphic materials: An emerging approach for bioengineering solutions. *Materials Advances*, *4*(23), 5882–5919.
71. Kapade, T. T., Shah, S. K., Shah, D. S., Kumar, N. S., Bhor, J. K., Panwar, V. S., & Singh, G. (2024). Thermal analysis of v-cut twisted tape fitted heat exchanger: A numerical approach. *International Journal on Interactive Design and Manufacturing (IJIDeM)*. https://link.springer.com/article/10.1007/s12008-024-01883-2
72. Bhowmik, A., Kumar, R., Beemkumar, N., Kumar, A. V., Singh, G., Kulshreshta, A., & Santhosh, A. J. (2024). Casting of particle reinforced metal matrix composite by liquid state fabrication method: A review. *Results in Engineering*, 103152.
73. Sunitha, K., & Vasudev, H. (2022). A short note on the various thermal spray coating processes and effect of post-treatment on Ni-based coatings. *Materials Today: Proceedings*, *50*, 1452–1457.

74. Majji, B. G. R., Vasudev, H., & Bansal, A. (2022). A review on the oxidation and wear behavior of the thermally sprayed high-entropy alloys. *Materials Today: Proceedings, 50*, 1447–1451.
75. Rao, G. S., Mukkamala, U., Hanumanthappa, H., Prasad, C. D., Vasudev, H., Shanmugam, B., and Kishore Kumar, K. C. (2024). Evaluating and optimizing surface roughness using genetic algorithm and artificial neural networks during turning of AISI 52100 steel. *International Journal on Interactive Design and Manufacturing (IJIDeM), 18*(8), 6151–6160.
76. Singh, J., Gill, H. S., & Vasudev, H. (2023). Computational fluid dynamics analysis on role of particulate shape and size in erosion of pipe bends. *International Journal on Interactive Design and Manufacturing (IJIDeM), 17*(5), 2631–2646.
77. Anjaneya, G., Sunil, S., Kakkeri, S., Math, M. M., Vaibhav, M. N., Solaimuthu, C., Durga Prasad, C., & Vasudev, H. (2023). Numerical simulation of microchannel heat exchanger using CFD. *International Journal on Interactive Design and Manufacturing (IJIDeM), 18*, 5847–5863.
78. Nagabhushana, P., Ramprasad, S., Prasad, C. D., Vasudev, H., & Prakash, C. (2024). Numerical investigation on heat transfer of a nano-fluid saturated vertical composite porous channel packed between two fluid layers. *International Journal on Interactive Design and Manufacturing (IJIDeM), 18*(5), 2927–2944.
79. Singh, P., Vasudev, H., & Bansal, A. (2024). Effect of post-heat treatment on the microstructural, mechanical, and bioactivity behavior of the microwave-assisted alumina-reinforced hydroxyapatite cladding. *Proceedings of the Institution of Mechanical Engineers, Part E: Journal of Process Mechanical Engineering, 238*(3), 1060–1071.
80. Mahantesh, M. M., Rao, K. R., Chandra, A. P., Vijayakumar, M. N., Nandini, B., Prasad, C. D., & Vasudev, H. (2024). Design and modeling using finite element analysis for the sitting posture of computer users based on ergonomic perspective. *International Journal on Interactive Design and Manufacturing (IJIDeM), 18*(8), 5875–5891.
81. Singh, J., Singh, R., Singh, S., Vasudev, H., & Kumar, S. (2023). Reducing scrap due to missed operations and machining defects in 90PS pistons. *International Journal on Interactive Design and Manufacturing (IJIDeM), 17*(5), 2527–2539.
82. Arora, H., Kumar, V., Prakash, C., Pimenov, D., Singh, M., Vasudev, H., & Singh, V. (2021). Analysis of sensitization in austenitic stainless steel-welded joint. In *Advances in Metrology and Measurement of Engineering Surfaces: Select Proceedings of ICFMMP 2019* (pp. 13–23). Springer Singapore.
83. Vasudev, H. (2020). Wear characteristics of Ni-WC powder deposited by using a microwave route on mild steel: Microwave cladding of Ni-WC. *International Journal of Surface Engineering and Interdisciplinary Materials Science (IJSEIMS), 8*(1), 44–54.

3 Fabrication of Surfaces with Laser Cladding, Microwave Cladding, and Surface Alloying Techniques

Amrinder Mehta, Hitesh Vasudev, and Jeyaprakash Natarajan

3.1 INTRODUCTION

Surface engineering is a critical process that improves the performance and wear resistance of industrial materials. The need for high-performance mechanical properties has encouraged researchers to work on new surface modification methods. Laser cladding, microwave cladding, and surface alloying have gained attention for their ability to generate surfaces with improved wear and corrosion-resistant features [1]. Laser cladding has been lauded for its applicability in developing strong surface coatings that enhance wear behavior and extend component service life. This technique entails laser welding of a filler material to the substrate, creating a metallurgically bonded layer that greatly improves the surface characteristics. Microwave cladding is another process of surface modification, where microwave energy is used to impart change on the material surface and creates benefits like shorter processing times and suitability of difficult shapes [2]. In contrast, surface alloying refers to the ability to atomically dissolve supposed alloying elements in the surface layer that can result in enhanced hardness and wear and corrosion resistance. Numerous works have been written that provide evidence for the reliability of these techniques on mechanical and tribological characteristics [3]. Previous investigations indicated that laser-treated surfaces have lower wear scars and enhanced load-carrying capacities, which are important characteristics for use in abrasive settings. Moreover, the incorporation of state-of-the-art surface modification techniques can result in the emergence of environmentally benign ones that will suit the current global trends of a green economy [4]. Surface engineering will certainly remain crucial as industries keep looking for new ways to improve material properties and extend the longevity of various products. This chapter intends to give the reader an understanding of those three surface fabrication techniques and their working principles, methods, applications, and any updates.

DOI: 10.1201/9781003546528-3

3.1.1 LASER CLADDING

Laser cladding is a novel surface modification process that makes use of a high-energy density laser to melt a coating material to become a metallurgical bond with a substrate to improve the performance of the latter substantially. It affords wear resistance, corrosion resistance, and oxidation resistance in coatings during the service of components in harsh conditions [5]. In comparison to traditional coating technologies, the laser cladding method can be characterized by shorter production time, better thermal management, and limited heat-affected zone (HAZ), which in turn predetermine the formation of higher maturity for coatings [6]. The dynamic action of laser cladding includes a process where possibly metallic powder material is fed directly into a molten pool formed by laser irradiation. During the laser scan, the material solidifies and leaves a clad bead profile at the substrate surface. Several factors affect the microstructure of the cladding layer, such as laser power, scanning speed, and rate of powder feeding, to name but a few, and all these must be controlled to obtain the right mechanical properties for the layer. Further, the design tools, including the finite-element method, have been used widely in assessing the thermal and mechanical stresses that occur in the cladding process that may in one way or another influence the performance of the coated components [7, 8].

As can be seen in Figure 3.1, a direct energy deposition (DED) additive manufacturing system is illustrated. The setup consists of several key components: a fiber laser source, a powder feeder system used with an argon gas, and an industrial robotic arm containing a laser head part shifted at 45° angle. The system spreads metallic powder on a substrate lying on the table at the same time as it melts the

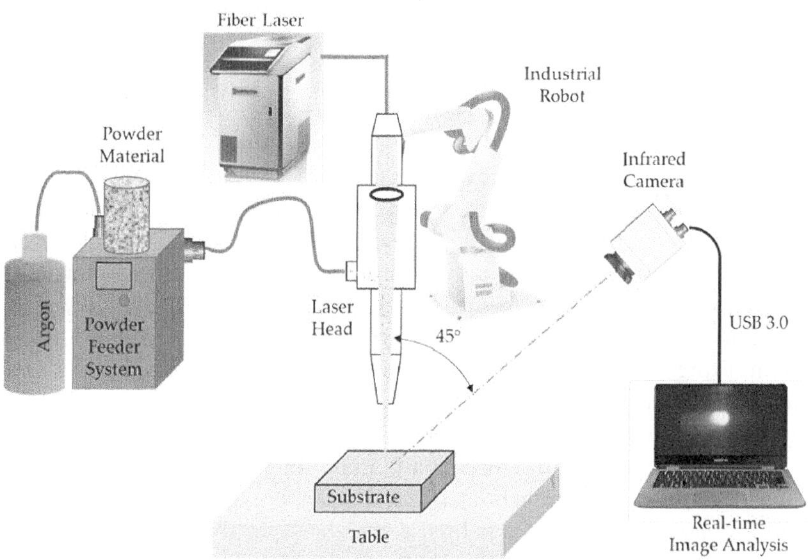

FIGURE 3.1 Process monitoring of laser cladding [9].

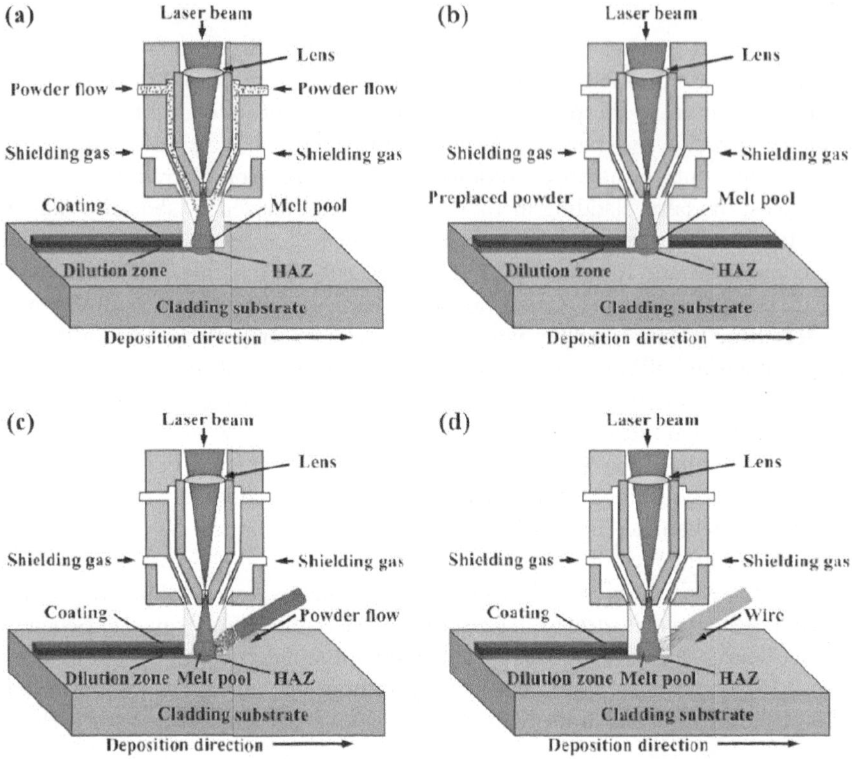

FIGURE 3.2 LC process: (a) coaxial powder feeding, (b) preplaced powder, (c) off-axis powder feeding, (d) wire feeding.

powder using the laser beam. Nearby, there is an infrared camera that records the process and transfers thermal images in real time through USB 3.0 to a computer. This integrated system provides the flexibility of independent control plus the effectiveness of integrated shuttling and monitoring of the metal deposition process [9].

Four types of laser cladding processes depicted in Figure 3.2 are shown as (a), (b), (c), and (d). Both configurations demonstrate a laser beam passing through a lens system, using shielding gas for protection. The general features of the various setups are laser beam, focusing lens, shielding gas supply, and clad substrate, with a dilution zone and HAZ. All configurations generate a melt pool on the cladding substrate which translates in the direction marked for deposition. Molten material is also shielded from further reactions in the atmosphere by a shielding gas, which effectively isolates it. This leaves behind a coating layer that has several regions, such as the dilution zone, where the coating material is blended with the substrate, and the HAZ, where dependence of the substrate on heat is changed [10].

Figure 3.3 gives a cross-sectional representation of a laser cladding process. In the process, a semi-elliptical cross-section is developed depending on the molten

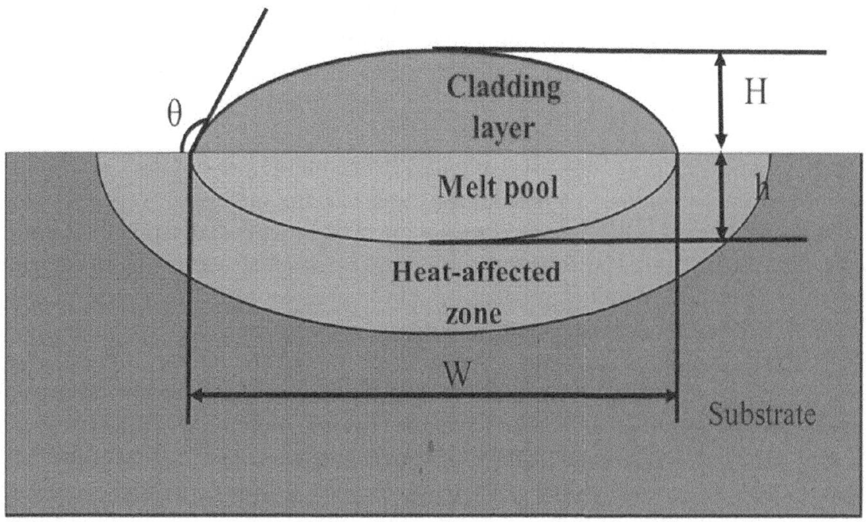

FIGURE 3.3 Typical cross-section of single cladding [11]. The diagram shows multiple distinct layers: It consists of two discrete layers of color: a red cladding layer at the top, which is covered by a layer of green melt pool from the exposure and irradiation process, and a layer of blue of heat-affected zone beneath it resting on a purple matrix substrate material. The geometric parameters of interest include dimensions of width (W) and height (H), and an angle θ at the interface.

material provided, which can cover or strengthen the substrate. This schematic can be useful in visualizing how material is deposited by laser cladding and how the substrate is influenced from a heat transfer perspective [11].

The use of advanced laser cladding techniques, for instance, with high-power diode lasers boosting the production of the cladded layers, has been investigated recently. This has popularized this technique in many industrial uses across industries but most especially the aerospace and automobile industries, where mechanical stressing is a significant factor due to the importance of using the strongest and highest-performing materials out there. In this light, laser cladding is promising for the future of this subject and its extensive applications in different fields of surface engineering [12].

3.2 TYPES OF LASER CLADDING

The process of laser cladding can be carried out in a number of different ways.

3.2.1 Powder-Fed Laser Cladding

Powder-fed laser cladding (PFLC) is a state-of-the-art surface engineering technology that prepares a laser beam to consolidate a metallic powder into a molten pool formed at the substrate surface. This method has gained attention because it can improve the mechanical characteristics of materials like wear resistance and fatigue life by producing a uniform cladding layer on the surface [13]. The process works in a directed energy deposition fashion in which the energy focus of the laser beam concentrates heat density sufficient to melt the powdered material and allow it to weld metallurgically with the substrate in the subsequent cooling phase. Specific aspects of the cladding head—particularly its configuration—are important as they determine the pattern of powder injection into the melt pool. In most cases, a Gaussian laser beam is used, and powder is injected either co-currently or in parallel to the laser beam [14]. The feeding method has been identified as the most critical determinant of the thermal characteristics of the process, affecting everything from the powder flow characteristics to the quality of the resulting clad. In particular, the dynamic behavior of the laser beam, powder stream, and molten pool is multifaceted, and control of these factors is critical for the formation of dense and metallurgically sound cladding layers [15]. The enhancements in the technology of laser cladding are as follows: practically implemented improvements create higher efficiency in the procedure of laser cladding. Advanced wide-band laser cladding technologies are under investigation to improve the existing limitations of this method through improving the laser–powder interaction and decreasing the powder dispersal [16]. In addition, numerical simulations have been used to investigate the powder transport phenomena and how this influences the changes in the melt pool and thus enhances the process parameters. On balance, the application of powder-fed laser cladding has vast potential as a tool for surface engineering for numerous fields of industry [17].

3.2.2 Wire-Fed Laser Cladding

Wire-fed laser cladding is a particular type of additive manufacturing that differs from powder-bed techniques and uses wire materials instead. This method also confers certain benefits, such as deposition rates which are always higher than those for powder particle settling systems [18]. Figure 3.4 shows a complex configuration for a laser wire additive manufacturing system component. At the center is a KUKA robot which is directly integrated with a direct diode laser and a wire feeding system. The system includes several key components: a high-speed camera for observing profile changes, a spectrometer that enables the determination of material composition, and a shield gas system for the protection of the weld pool. The latter gives the feeding wire material, and the hot-wire power source regulates the process. The scanning direction goes across the substrate as the laser helps melt the wire as it builds up layers. This setup is a complex configuration of additive manufacturing for manufacturing metals with a high degree of accuracy [19].

As pointed out, the efficiency of wire-fed laser cladding is significantly higher than that for other cladding methods because it directly supplies material to the melt pool formed by the laser beam. However, where the wire-fed approach gives these benefits, it also has drawbacks with respect to the layout of the material feedstock. While powder systems are more versatile, in the sense that they can accommodate a range of materials and blends, wire-fed systems are somewhat more limited by the alloy or composite used in the wire. It will also limit the possibilities for choosing the cladding material to respond to required performance parameters or specific

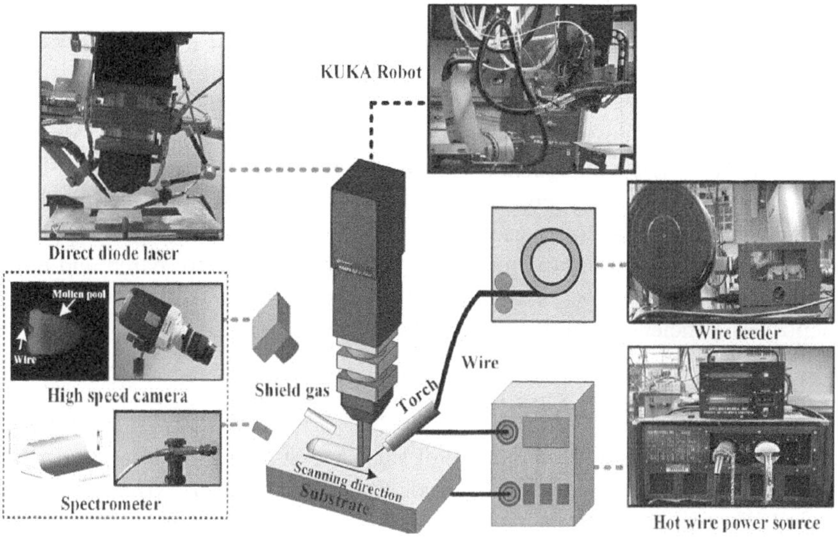

FIGURE 3.4 Schematic illustration of a laser hot-wire cladding system [19].

microstructure [20]. The introduction of reinforcements like tungsten carbide is difficult with wire as compared to powder metallurgy, in which mixing and layering mechanisms are easy. More importantly, within the wire-fed laser cladding process, laser power, wire feed rate, and traverse speed act as important controlling factors in defining the characteristics of the formed layer. The findings of research have also revealed that by tuning these parameters, it is possible to enhance the mechanical properties and grain structure of the cladding for use in repair and surface enhancement technologies of structural components [21]. As such, wire-fed laser cladding has its benefits, but wire and laser choices and other operational parameters need to be considered optimally to achieve these advantages.

3.2.3 Preplaced Powder Laser Cladding

Preplaced powder laser cladding is an elaborate manufacturing process in which metal powders are placed on the substrate before the melting of the metal layer is conducted by a high-energy laser beam. This method proves especially beneficial in small areas where complicated geometries do not allow for the application of coating techniques. The process starts with the selective application of powder that, upon melting with laser, forms a metallurgically bonded layer on the substrate [22]. This technique offers the best opportunity in responding to the thickness and composition of the coating, in accordance with the acceptable specifications for use. Another advantage of the preplaced powder laser cladding technique is the restraint of the dilution between the cladding material and the substrate, factors which retain the properties of the protection layer [23]. Research has shown that the dilution effect of the laser can be controlled by several factors such as traverse speed of the laser and type of powder used. Further, the process offers very high heating and solidification rates, helping to produce a high-quality cladding layer with improved mechanical characteristics. Studies have shown that this technique is now widely applied for such purposes as reclamation of worn components and modification of the surface features of substrates, which include Inconel alloy [24]. This flexibility in developing surface coatings with enhanced wear and corrosion characteristics can make pre-placed powder laser cladding a useful technique across a broad range of applications from aerospace and automotive industries to others. Further research and development works are focused on enhancing the process parameters for the enhancement of deposition efficiency and the quality of the clad layers [25].

3.3 MICROWAVE CLADDING

Microwave cladding is a novel method of surface engineering that uses microwaves to heat and fuse ceramic coating materials to base matrixes at extremely high rates of speed. There is growing interest in using this method because it offers a straightforward way of achieving better characteristics in the resulting coatings, including improved wear and corrosion resistance [26]. Microwave cladding has been defined by its distinct volumetric heating mode, which promotes molecular, uniform heating and therefore a fine-grained structure and reduced treatment times compared to conventional techniques. Current investigations have shown that the technique of microwave cladding is efficient in the formation of composite layers like Ni–40Cr_3C_2 and Ni-Si-La_2O_3 on a stainless steel base with improved mechanical characteristics and better wear resistance [27].

Figure 3.5 depicts a microwave-assisted material processing experimental arrangement in a schematic manner. At its top, there exists microwave radiation, which is pointing down towards the susceptor layer, made of charcoal powder for this instance. Below this is a separator made of graphite plate. The La_2O_3 is the outer

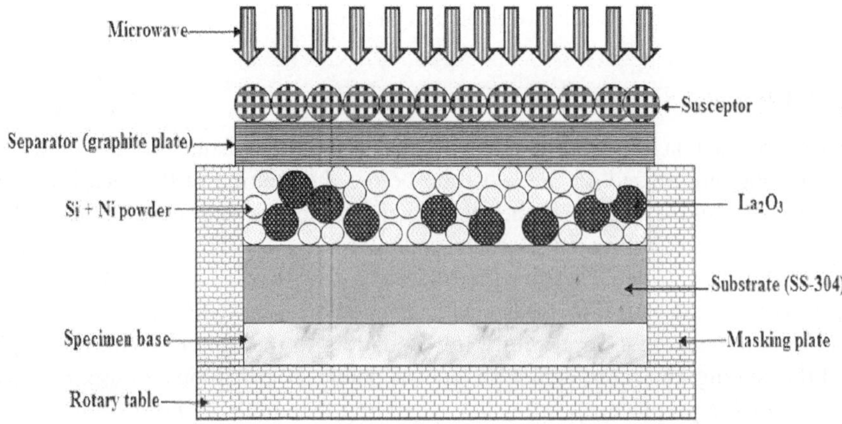

FIGURE 3.5 Microwave cladding procedure [28].

layer and in the middle layer the Si + Ni powder mixture is developed. The substrate layer is also fashioned from stainless steel (SS-304) and is located on the specimen base. The whole assembly is located on a rotary table and housed in masking plates. This setup seems to be initiated for the synthesis or treatment of the material in question under the method of microwave heating technology [28].

The process reduces dimensional distortions on the cladding layer and sharpens the hardness gradient on the covered surface, which has implications for the performance of the coated material. In addition, the microwave cladding technique has revealed a strong metallurgical bond between the cladding and the substrate that enhances the durability and service life of the coating. The incorporation of microwave cladding to different types of materials has been done effectively, particularly to nickel-based alloys and functionally graded materials, where tribological improvement has been realized [29]. Further, the use of microwave energy in the cladding process has been associated with the development of uniform microstructures that do not show porosity and interfacial crack defects, which are evident in traditional cladding procedures. Furthermore, microwave cladding as a subject of ongoing research is underpinning the significance of the technology for surface engineering of materials and their applications to more innovative and practical material processing techniques [30].

Figure 3.6 shows the temperature profile and the process of microwave heating of charcoal and clad powder through time. The temperature increases from room temperature, reaches approximately 850 °C within 200 seconds, and then decreases slowly. A1, A2, A3 below the graph represents the following: initial skin depth of

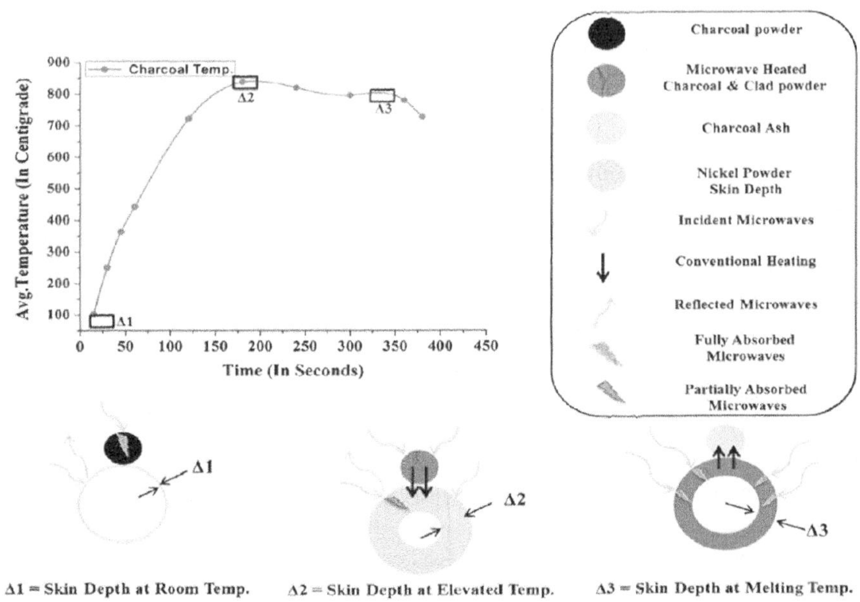

FIGURE 3.6 Schematic representation of possible material interaction of microwave radiation with materials at different temperatures [31].

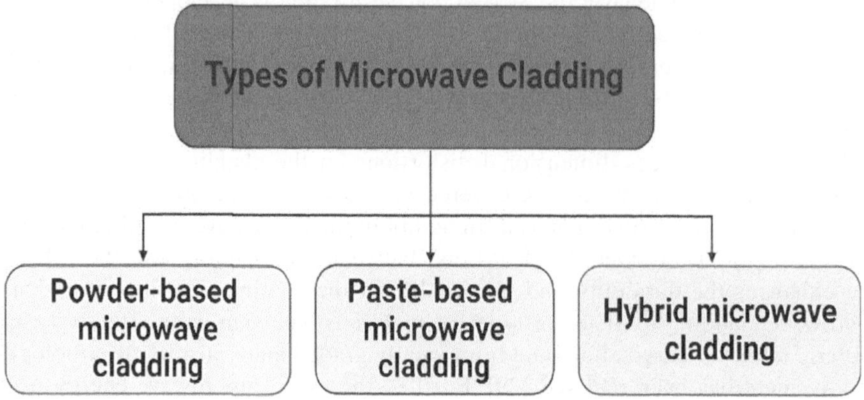

FIGURE 3.7 Microwave cladding can be classified based on the coating material application method.

the charcoal at the room temperature, second phase characterizing the contact of microwave heated charcoal with the nickel powder, the final phase characterized by the formation of the charcoal ash at melting temperature. The legend indicates various components using color coding: charcoal black for the charcoal powder, red for microwave-heated mixture, grey for other parts of the charcoal ashes, and yellow is assigned to nickel powder skin depth. In addition to that, microwave interactions such as incident, reflected fully absorbed, and partially absorbed are depicted by unique symbols [31].

Figure 3.7 also depict a simple flowchart showing the three major categories of microwave cladding processes. The diagram branches from a central header "Types of Microwave Cladding" into three categories, such as powder-based microwave cladding, paste-based microwave cladding, and hybrid microwave cladding, which are shown in three corresponding boxes with rounded corners in different colors [32].

Microwave cladding from powder is a novel method of material surface modification that makes use of microwave energy to bond metal or ceramic particles to the base material. This process consists of applying a layer of powder to the target surface as well as irradiating this layer with microwaves [33]. Microwave energy heats the particles of the powder at very high rates and melts them into the substrate. The advantages of this method include but are not limited to thickness control, heating uniformity, and formation of complex geometrical structures. It is especially used in improvement of surface features like wear, corrosion, and thermal features, including heat resistance [34]. The technique is not limited to but enables deposition of a range of metallic and ceramic materials to various substrates. Microwave cladding using paste is a modern technique of surface coating that is usually achieved after powders mixing with binders to develop a paste-like consistency. Compared to powder-based techniques, this method provides enhanced applicability convenience. This paste is normally constituted of metal or ceramic powders and organic or inorganic binders, which are spread on the substrate surface [35]. As the paste absorbs microwave radiation, such as during the curing process, it quickly heats, and the binders in the

TABLE 3.1
Process Parameters and Their Effects

Parameter	Laser Cladding	Microwave Cladding	Surface Alloying
Energy Input	Laser power	Microwave power	Varies by technique [41]
Scanning Speed	Critical for layer thickness	N/A	Affects alloying depth [42]
Material Feed Rate	Determines deposition rate	N/A	Influences alloy composition [43]
Shielding Gas	Prevents oxidation	Optional (in controlled atmosphere)	Depends on technique [44]
Substrate Temperature	Preheating may be required	Affects coupling efficiency	Influences diffusion and mixing [45]

mixture become dry while the working powder melts and coalesces into a conglomerate on the substrate. The maximized benefit accrued by using this technique is that it is easier to adjust the thickness and densities of the applied coatings to match the geometrical shapes. It can be applied very effectively for improving surface characteristics, including wear and corrosion resistance and heat insulation in industrial uses [36]. Microwave hybrid cladding is a novel method of creating a surface coating that combines the use of microwaves as the primary heating source with other supplementary energy sources for better operational control and reduced energy consumption [37]. This method is usually incorporated with other heating methods like induction, laser, or infrared radiation to complement microwave energy. Remote control of both energy sources facilitates better control of temperature and general curing across the entire length of the coated pieces at the same time [38]. Exploring the advantages of two or more heating sources of the heat treatment process, hybrid microwave cladding can provide solutions that enable faster processing times, better energy efficiency, and improved coating quality. This technique is especially helpful for products where it is difficult to heat evenly through microwaves alone, to broaden the range of substrates and coating materials that can be clad successfully [39].

Table 3.1 compares three surface modification techniques: laser cladding, microwave cladding, and surface alloying, which is the method chosen for presentation here. It defines some essential characteristics of each method such as the input energy, the rate of table movement, feed rate of material, shielding gas, and substrate temperature [40].

3.4 CURRENT CHALLENGES

Shown as a flow chart in Figure 3.8, it is labeled as four current issues of what seems to be a manufacturing or engineering discipline. The main node "Current Challenges" branches out into four distinct areas: These are "Process control and optimization" (orange), "Material compatibility" (pink/magenta). "Cost reduction" (purple) and "Residual stress management" (red). The diagram is relatively uncluttered and

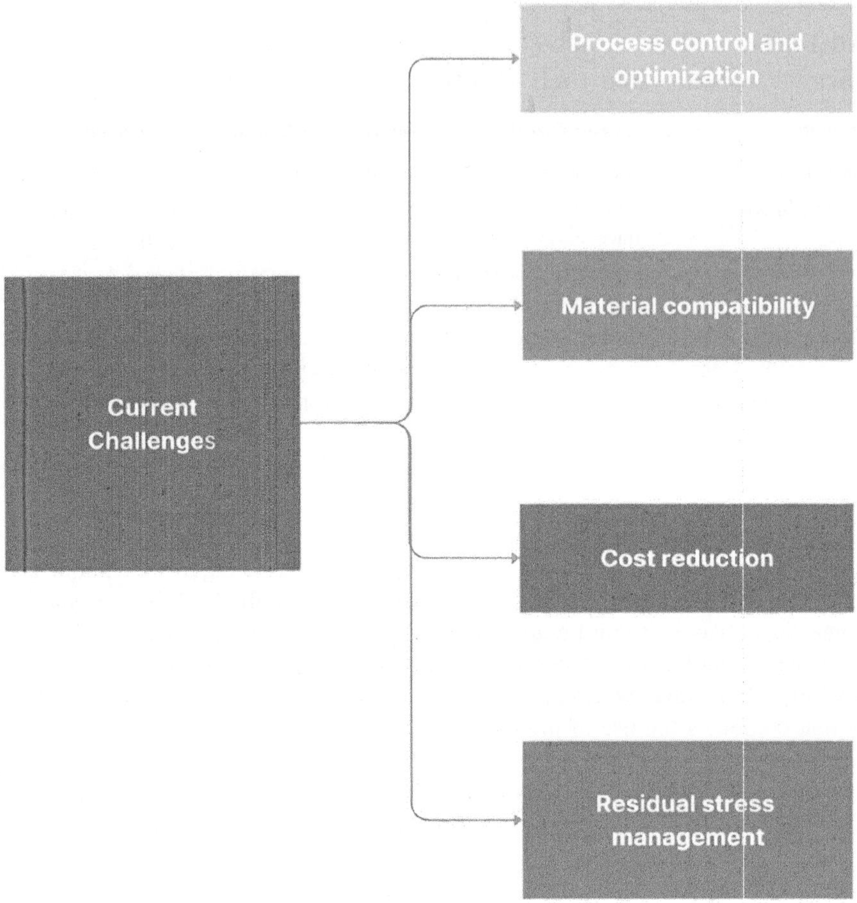

FIGURE 3.8 Current challenges in additive manufacturing.

employs straight connection bar and rectangles of different colors to highlight the orientation of these challenges. Every one of them is a vital component that has to be implemented in a particular process or system [46]. Several factors have been identified when it comes to the process control and optimization of quality in laser cladding on large zones and geometrically large structures. There are various parameters which are involved in coating processes, such as laser power, scanning speed, powder feeding rate, and overlap rate which affect coating quality [47]. Neural networks and genetic algorithms, for example, can used to model and predict parameters within a process to achieve the goals of bonding strength and microhardness [48]. There are other quality evaluation techniques such as AHP-FCE, which provide forming quality, microstructure, and surface performance [49]. Basically, control of defects is realized by selecting appropriate materials, improving manufacturing processes, heat treatment, and numerical modeling. In the case of method A, improving the closed-loop control of the molten pool temperature also improves the consistency [50].

In the case of laser cladding and its related technique of microwave cladding, then material compatibility expansion by adding more feedstock material selections must examine feeds that can bond metallurgically to a wide range of substrates with as little interface mixing as possible. This has implications on melting behaviors and the nature of interaction during cladding process as well as the properties of the materials involved. The main cost-cutting measures targeted at laser-based techniques concern the enhancement of the processes used in manufacturing [51]. It is with high-power laser cladding that researchers have been able to find the best solution for using rotor coating in automotive applications. In general, efficient catchment of powders, as shown in the ABA cladding method, can greatly minimize wastage and the time taken for processing the powders. Also, optimization of corporate activities and human resource management may imply cost reduction plans [52].

High-volume cladding applications require larger pores than those used in smaller components. Additionally, stacked bar configurations must be designed to accommodate industrial-scale cladding requirements. These will involve the enhancement of high processing rates and volume, bigger workspace, and the integration of technology in the working processes [53]. ABA cladding, which is used to fill the space between large intervals of tracks, appears to hold potential for higher density in large-scale use. Stress management, particularly after treatment, is recommended to reduce distortion and cracking in treated parts. This entails working on the parameters of a process, creating methods of relieving stress, and placing special instrumental facilities [54]. A lot of studies have pointed to the use of machine learning models in modeling and, perhaps more importantly, controlling residual stresses by examining bigger data sets and controlling process parameters.

3.5 FUTURE RESEARCH DIRECTIONS

Remote sensing devices can play a critical role in ensuring quality in surface fabrication methods due to in-situ monitoring and control of processes. Monitoring systems which track activity progression in real time and adaptive control means allow maximizing parameters and keeping the corresponding results in the necessary range during the production flow. In laser cladding, a feedback system for precise regulation of temperature of the molten pool surface has been added to improve part repeatability [55]. Figure 3.9 represents the schematic of a laser-based additive manufacturing system. The system consists of several key components; power regulation is accomplished by using the laser conductor, control is managed through the help of the PLC actuator, while monitoring is done through usage of the CMOS camera. Material supply is facilitated with the powder feeder, and the temperature is measured using the pyrometer. This process occurs on the previously deposited material layer, where the process builds up coating layers. A circular inset with considerable detail illustrates the melting as well as the solidification of the material records the possible flow characteristics of the molten material. There is a PC-based operations interface which tracks various process parameters and controls the setting of the entire system [56].

Table 3.2 presents a summary of a wide range of aspects related to the laser cladding process and its monitoring and control. Processing signals are the records of the

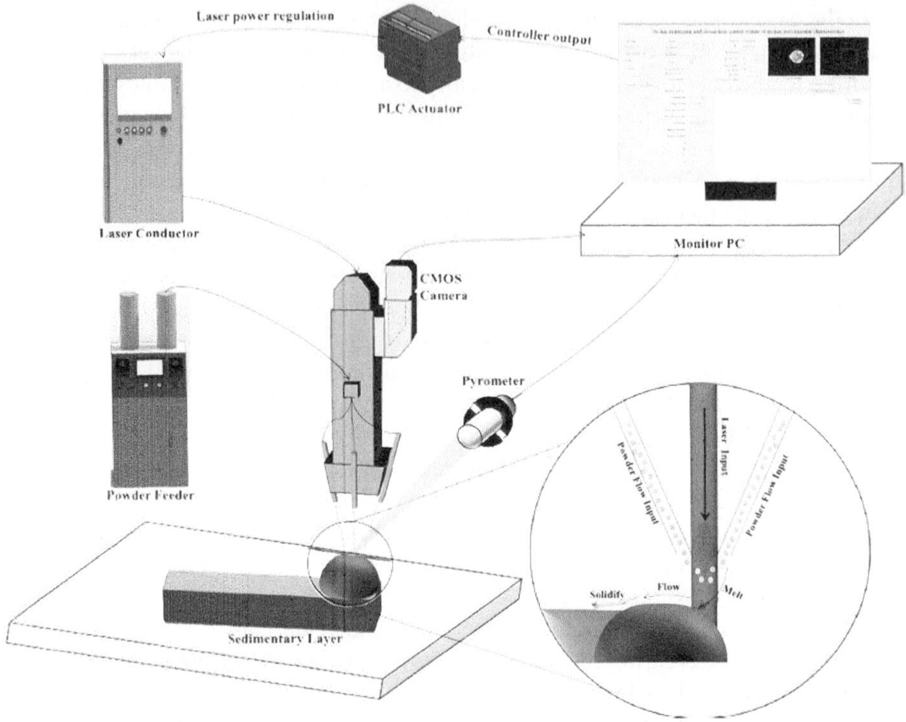

FIGURE 3.9 Real-time closed-loop DLD processing system [56].

TABLE 3.2
Important Parameters of Signal Process Monitoring and Control

Processing Signal	Monitoring Object	Control Object	Adjusted Parameter
Photodiode signal	Melt pool	Coating quality	Laser power [58]
Acoustic emission	Track morphology	Melt pool size	Scanning speed [59]
Optical spectra	Temperature profile	Dilution ratio	Powder feed rate [60]
Hyperspectral imaging	Coating geometry	Track dimensions	Focal position [61]
Machine vision	Melt pool area	Forming quality	Nozzle tilt angle [62]
Light emission	Element composition	Bonding strength	Overlap rate [63]
Thermal radiation	Microstructure	Microhardness	Shielding gas flow rate [64]
Plume emission	Surface performance	Porosity	Layer thickness [65]
Molten pool temperature	Defects	Crack formation	Spot size [66]
Acoustic signal	Residual stress	Spatter	Standoff distance [67]
Infrared imaging	Dilution	Thermal gradients	Preheating temperature [68]
High-speed camera	Coating thickness	Cooling rate	Powder catchment efficiency [69]

data acquired during the cladding process, while monitoring objects are the corresponding features or characteristics under observation. Control objects are the parts of the process with regard to which regulation or optimization is taking place, and the adjusted parameter is the factor that can be altered with a view to a particular result [57].

Neural networks are used together with genetic algorithms to carry out advanced modeling to achieve process parameters that would lead to bonding strength and microhardness. Water loss from surface irrigation has led to the development of real-time adaptive control frameworks—such as surface irrigation simulation, calibration, and optimization (SISCO), Auto Furrow, and VARIwise—that use in-field sensors, automated processors, and actuators to improve irrigation management automated computer processors, infield actuators [70, 71]. Multifunctional, functionally graded coatings provide surfaces with spatially varying compositions and properties, enabling them to meet diverse performance demands. One method for producing these kinds of material coatings has recently received much attention, namely laser-directed energy deposition (L-DED). Dual-carriage plasma transferred arc (PTA) welding systems can form layers of up to four different powders in one layer, appropriate for high mechanical loads and wear and temperature resistance. In this technique, the mechanical chemical properties can be fully controlled by simple variation of the powder feeding positions [72].

Combining surface fabrication technology and AM allows the creation of surface properties on the product to be improved. Specifically, L-DED is suitable for this integration because it can fabricate both metal matrix composites and FGMs in a single step [73]. These two parameters cumulatively provide fresh opportunities for alloying concepts along with the creation of wear-resistant coatings. However, problems persist with the uniform distribution of the various components of multiphase powder blends and the impact of thermal history on microstructure characteristics [74]. Nanostructured and nanocomposite coatings purposely incorporate nanoscale effects to improve the characteristics of the surface. For example, hierarchical micro/nanostructured ZnO/Cu-ZnMOFs@SA superhydrophobic composite coatings have been synthesized which possess multipurpose application such as anticorrosive, blood-flying, and antibacterial properties. These coatings offer robust protective characteristics such as water resistance, abrasion endurance, durability, and thermal stability that make the medical instruments usable in severe conditions [75].

Environmental concerns arising from use of surface fabrication techniques are currently being countered through the development of more sustainable means of processing. *Piper longum* extract has been used in the recyclable synthesis of PtPd alloy nanoclusters to enhance electrolytic performance in the oxidation of formic acid and methanol. This approach supports eco-friendly fabrication. Additionally, ongoing research aims to maximize biomass pyrolysis yields for potential applications in surface engineering and environmental management [76].

Machine learning is growing in application for process optimization and determining surface characteristics. For biochar derived from biomass pyrolysis, machine learning models have been trained to predict and optimize yield, total N content, and specific surface area. In laser cladding, reinforcement learning has been integrated with artificial microswimmers to perform dynamics of learning in stochastic conditions, and suitable algorithms to solve navigation challenges involving Brownian motion have been found [77]. New classes of both crystalline and amorphous alloys like high-entropy alloys (HEAs) have been introduced for SE applications. Surface segregation has been investigated employing molecular simulations in AgAuCuPdPt HEAs that are potential catalysts for CO_2 electrochemical reduction. These last simulations show that doping enhances the number of different catalytically active sites on the surface and indicate that HEAs can develop a wide variety of surface structures that can be advantageous for catalysis [78]. The specific area of surface engineering is expanding dynamically in recent years with the trends in real-time surface inspection, multifunctional coating techniques, surface engineering in combination with AM, nanoscale surface topography, and environmentally friendly surface treatment as well as AI-driven methodologies and novel alloy systems [79]. These advancements are being used to enhance coating quality, processes, and the opportunity to make structures possessing desirable characteristics to be used over structures that are required for a particular purpose or use [80]. Nevertheless, some issues like controlling the composition of multiple materials at the same time, analyzing relationships between microstructures, and introducing these innovative techniques for production in volume remain unresolved at the present stage [81]. Further investigations on these topics will be imperative for proper exploitation of several surface engineering techniques in several sectors, including production and healthcare equipment manufacturing.

3.6 CONCLUSION

This chapter provides a comprehensive overview of three prominent surface fabrication techniques: laser cladding, microwave cladding, and surface alloying and surface cladding. Laser cladding defines a process wherein a material is melted using a laser and applied on a substrate to create a metallurgically bonded wear- or corrosion-resistant layer. In microwave cladding, microwave energy is used to heat the coating material at a fast rate and then melt it to form a clad layer with better characteristics. The surface alloying process is done through the diffusion of alloying elements in the surface layer of the material, increasing its hardness as well as applying wear and corrosion resistance. However, these techniques have limitations which include process control and/or optimization, compatibility of material to be treated, cost involved in the treatment, and stress remaining in the component. Possible research areas for future work include in-situ diagnostics and control, multi-layer and graded coatings, uptake of additive technologies, nanoscale and nanocomposite coatings, coatings manufactured through sustainable processes, and the application of AI and machine learning for coatings.

REFERENCES

1 Ali SRA-S, Hussein AHA, Nofal A, Elnaby SIH, Elgazzar H. A contribution to laser cladding of Ti-6Al-4V titanium alloy. Metallurgical Research Technology. 2019;116:634.
2. Alizadeh-Sh M, Marashi S, Ranjbarnodeh E, Shoja-Razavi R. Laser cladding of Inconel 718 powder on a non-weldable substrate: Clad bead geometry-solidification cracking relationship. Journal of Manufacturing Processes. 2020;56:54–62.

3. Aravazhi S, Geskus D, van Dalfsen K, Vázquez-Córdova SA, Grivas C, Griebner U, García-Blanco SM, Pollnau M. Engineering lattice matching, doping level, and optical properties of KY (WO 4) 2: Gd, Lu, Yb layers for a cladding-side-pumped channel waveguide laser. Applied Physics B. 2013;111:433–46.
4. Bhatnagar S, Mullick S. A study on the influence of reinforcement particle size in laser cladding of TiC/Inconel 625 metal matrix composite. Optics Laser Technology. 2023;161:109115.
5. Liu L, Wang S, Zhang B, Jiang G, Liu H, Yang J, Wang J, Liu W. Present status and prospects of nanostructured thermal barrier coatings and their performance improvement strategies: A review. Journal of Manufacturing Processes. 2023;97:12–34.
6. Cai Y, Chen Y, Luo Z, Gao F, Li L. Manufacturing of FeCoCrNiCux medium-entropy alloy coating using laser cladding technology. Materials Design. 2017;133:91–108.
7. Candel J, Jiménez JA, Franconetti P, Amigó V. Effect of laser irradiation on failure mechanism of TiCp reinforced titanium composite coating produced by laser cladding. Journal of Materials Processing Technology. 2014;214:2325–32.
8. Cao Y-b, Ren H-t, Hu C-s, Meng Q-x, Liu Q. In-situ formation behavior of NbC-reinforced Fe-based laser cladding coatings. Materials Letters. 2015;147:61–63.
9. Srisungsitthisunti P, Kaewprachum B, Yang Z, Gao G. Real-time quality monitoring of laser cladding process on rail steel by an infrared camera. Metals. 2022;12:825.
10. Liu Y, Ding Y, Yang L, Sun R, Zhang T, Yang X. Research and progress of laser cladding on engineering alloys: A review. Journal of Manufacturing Processes. 2021;66:341–63.
11. Li T, Long H, Qiu C, Wang M, Li D, Dong Z, Gui Y. Multi-objective optimization of process parameters of 45 steel laser cladding Ni60PTA alloy powder. Coatings. 2022;12:939.
12. Chen G-D, Liu X-B, Zhang F-Z, Liu Q-S, Ou H-Z, Zhang S-H. Fabrication and tribological properties of laser cladding WC-Cu/Co-based composite coatings. Surface Coatings Technology. 2023;472:129930.
13. Chen J, Li J, Song R, Bai L, Shao J, Qu C. Effect of the scanning speed on microstructural evolution and wear behaviors of laser cladding NiCrBSi composite coatings. Optics Laser Technology. 2015;72:86–99.
14. Chen L, Zhao Y, Song B, Yu T, Liu Z. Modeling and simulation of 3D geometry prediction and dynamic solidification behavior of Fe-based coatings by laser cladding. Optics Laser Technology. 2021;139:107009.
15. Chen T, Li W, Liu D, Xiong Y, Zhu X. Effects of heat treatment on microstructure and mechanical properties of TiC/TiB composite bioinert ceramic coatings in-situ synthesized by laser cladding on Ti6Al4V. Ceramics International. 2021;47:755–68.
16. Deng D, Li T, Huang Z, Jiang H, Yang S, Zhang Y. Multi-response optimization of laser cladding for TiC particle reinforced Fe matrix composite based on Taguchi method and grey relational analysis. Optics Laser Technology. 2022;153:108259.
17. Mehta A, Singh G. Consequences of hydroxyapatite doping using plasma spray to implant biomaterials. Journal of Electrochemical Science Engineering failure analysis. 2023;13:5–23.
18. Deng Z, Liu D, Liu G, Liu J, Li C, Li S, Chen T. Numerical simulation study on the formation mechanism of hydroxyapatite-silver gradient bioactive ceramic coatings under wide-band laser cladding. Optics Laser Technology. 2023;163:109412.
19. Chen X, Yao M, Kong F, Fu Y, Wu J, Zhang H. In-situ quality monitoring of laser hot wire cladding process based on multi-sensing diagnosis and machine learning model. Journal of Manufacturing Processes. 2023;87:183–98.
20. Deng Z, Liu D, Liu G, Xiong Y, Xin S, Li S, Li C, Chen T. Study on the corrosion resistance of SiC particle-reinforced hydroxyapatite-silver gradient bioactive ceramic coatings prepared by laser cladding. Surface Coatings Technology. 2023:129734.
21. Ding M, Lu C, Cao L, Huang W, Ni Y, Xu Z. Transparent glass coatings incorporated with upconversion nanocrystals by laser cladding method. Applied Surface Science. 2013;277:176–81.

22. Ding Z, Fang Z, Wang G, Wu C, Yang S. Microstructure and properties of Mo-Si-B-Y2O3 composite coating on nickel-based alloy by laser cladding. Optics Laser Technology. 2023;164:109473.
23. Fan L, Chen H, Dong Y, Li X, Dong L, Yin Y. Corrosion behavior of Fe-based laser cladding coating in hydrochloric acid solutions. Acta Metallurgica Sinica. 2018;54:1019–30.
24. Feng K, Chen Y, Deng P, Li Y, Zhao H, Lu F, Li R, Huang J, Li Z. Improved high-temperature hardness and wear resistance of Inconel 625 coatings fabricated by laser cladding. Journal of Materials Processing Technology. 2017;243:82–91.
25. Findik F. Laser cladding and applications. Sustainable Engineering Innovation. 2023;5:1–14.
26. Sharanabasava H, Prasad CD, Ramesh M. Effect of Mo-and SiC-reinforced NiCr microwave cladding on microstructure, mechanical and wear properties. Journal of the Institution of Engineers: Series D. 2023;104:539–51.
27. Sharma A, Singh A, Chawla V, Grewal J, Bansal A. Microwave processing and characterization of alumina reinforced HA cladding for biomedical applications. Materials Today: Proceedings. 2022;57:650–56.
28. Dwivedi SP, Sharma S, Sharma KP, Kumar A, Agrawal A, Singh R, Eldin SM. The microstructure and properties of Ni-Si-La_2O_3 coatings deposited on 304 stainless steel by microwave cladding. Materials. 2023;16:2209.
29. Singh G, Mehta A, Bansal A. Electrochemical behaviour and biocompatibility of claddings developed using microwave route. Journal of Electrochemical Science Engineering. 2023;13:173–92.
30. Verma PK, Vasudev H, Singh G, Singh S. A review on the microstructural and biomedical properties of implants manufactured using additive manufacturing. Journal of the Institution of Engineers (India): Series D. 2024:1–15.
31. Mago J, Bansal S, Gupta D, Jain V. Investigation of microwave processing parameters on development of Ni-40Cr 3 C 2 composite clad and their characterization. Metallurgical Materials Transactions A. 2020;51:4288–300.
32. Bansal A, Zafar S, Sharma AK. Microstructure and abrasive wear performance of Ni-WC composite microwave clad. Journal of Materials Engineering Performance. 2015;24:3708–16.
33. Sharanabasava H, Prasad CD, Ramesh M. Characterization and wear behavior of NiCrMoSiC microwave cladding. Journal of Materials Engineering Performance. 2024;33:763–75.
34. Dwivedi SP, Sharma S, Srivastava AP, Sethi VA, Mohammed KA, Kumar A, Khan MI, Abbas M, Tag-Eldin EM. Homogeneity, metallurgical, mechanical, wear, and corrosion behavior of Ni and B4C coatings deposited on 304 stainless steels developed by microwave cladding technique. Journal of Materials Research Technology. 2023;27:5854–67.
35. Mehta A, Vasudev H, Jeyaprakash N. Role of sustainable manufacturing approach: Microwave processing of materials. International Journal on Interactive Design Manufacturing. 2024;18:5283–99.
36. Dwivedi SP, Sharma S. Metallic cladding through microwave energy of the mixture of Ni and 15% SiC powder on AISI 304: A green approach in surface engineering. Proceedings of the Institution of Mechanical Engineers, Part E: Journal of Process Mechanical Engineering. 2024;238:2545–56.
37. Mehta A, Vasudev H, Jeyaprakash N. Role of sustainable manufacturing approach: Microwave processing of materials. International Journal on Interactive Design Manufacturing. 2023:1–17.
38. Singh B, Zafar S. Understanding temperature characteristics during microwave cladding through process modeling and experimental investigation. CIRP Journal of Manufacturing Science Technology. 2022;37:401–13.

39. Sharma AK, Gupta D. On microstructure and flexural strength of metal–ceramic composite cladding developed through microwave heating. Applied Surface Science. 2012;258:5583–92.
40. Gupta D, Sharma AK. Investigation on sliding wear performance of WC10Co2Ni cladding developed through microwave irradiation. Wear. 2011;271:1642–50.
41. Fu Y, Guo N, Zhou C, Wang G, Feng J. Investigation on in-situ laser cladding coating of the 304 stainless steel in water environment. Journal of Materials Processing Technology. 2021;289:116949.
42. Ganjali M, Ganjali M, Sadrnezhaad S, Pakzad Y. Laser cladding of Ti alloys for biomedical applications. *Laser Cladding of Metals*. Cham: Springer International Publishing; 2021:265–92.
43. Gao M, Li S, Guan W, Xie H, Wang X, Liu J, Wang H. Excellent thermal shock resistance of NiCrAlY coatings on copper substrate via laser cladding. Journal of Materials Science Technology. 2022;130:93–102.
44. Garcia-Herrera J, Henao J, Espinosa-Arbelaez D, Gonzalez-Carmona J, Felix-Martinez C, Santos-Fernandez R, Corona-Castuera J, Poblano-Salas C, Alvarado-Orozco J. Laser cladding deposition of a Fe-based metallic glass on 304 stainless steel substrates. Journal of Thermal Spray Technology. 2022;31:968–79.
45. Gouge MF, Heigel JC, Michaleris P, Palmer TA. Modeling forced convection in the thermal simulation of laser cladding processes. The International Journal of Advanced Manufacturing Technology. 2015;79:307–20.
46. Guo C, He S, Yue H, Li Q, Hao G. Prediction modelling and process optimization for forming multi-layer cladding structures with laser directed energy deposition. Optics Laser Technology. 2021;134:106607.
47. Mehta A, Vasudev H, Singh S, Prakash C, Saxena KK, Linul E, Buddhi D, Xu J. Processing and advancements in the development of thermal barrier coatings: A review. Coatings. 2022;12:1318.
48. Han B, Zhang M, Qi C, Cui N, Wang Y. Characterization and friction-reduction performances of composite coating produced by laser cladding and ion sulfurizing. Materials Letters. 2015;150:35–38.
49. Nasiri S, Khosravani MR. Machine learning in predicting mechanical behavior of additively manufactured parts. Journal of Materials Research Technology. 2021;14:1137–53.
50. Hao J, Yang S, Le X, Królczyk G, Sulowicz M, Glowacz A, Li Z. Bead morphology prediction of coaxial laser cladding on inclined substrate using machine learning. Journal of Manufacturing Processes. 2023;98:159–72.
51. He G, Du Y, Liang Q, Zhou Z, Shu L. Modeling and optimization method of laser cladding based on GA-ACO-RFR and GNSGA-II. International Journal of Precision Engineering Manufacturing-Green Technology. 2022;10:1207–1222.
52. Huang J, Zhu Z, Li K, Shi W, Zhao Y, He M. Microstructures and mechanical properties of an AlCoCrNiFe HEA/WC reinforcing particle composite coating prepared by laser cladding. Materials. 2022;15:8020.
53. Ibrahim MZ, Sarhan AA, Shaikh M, Kuo T, Yusuf F, Hamdi M. Investigate the effects of the laser cladding parameters on the microstructure, phases formation, mechanical and corrosion properties of metallic glasses coatings for biomedical implant application. Additive Manufacturing of Emerging Materials. 2019:299–323.
54. Imam HZ, Zheng Y, Martinez P, Ahmad R. Vision-based damage localization method for an autonomous robotic laser cladding process. Procedia CIRP. 2021;104:827–32.
55. Jeyaprakash N, Karuppasamy SS, Yang C-H. Application of wear-resistant laser claddings. *Handbook of Laser-Based Sustainable Surface Modification and Manufacturing Techniques*. CRC Press; 2023:1–26.

56. Liu B, Liu W, Xia Y, Huang Y, Liu Y, Zhao Y, Liu H, Wang Z, Li W, Song J. Real-time closed-loop control of molten pool transient area in direct laser deposition via PID algorithm with enhanced robustness. The International Journal of Advanced Manufacturing Technology. 2024;130:4529–42.
57. Jeyaprakash N, Yang C-H. Comparative study of NiCrFeMoNb/FeCrMoVC laser cladding process on nickel-based superalloy. Materials Manufacturing Processes. 2020;35:1383–91.
58. Jeyaprakash N, Yang C-H, Sivasankaran S. Laser cladding process of Cobalt and Nickel based hard-micron-layers on 316L-stainless-steel-substrate. Materials Manufacturing Processes. 2020;35:142–51.
59. Khamidullin B, Tsivilskiy I, Gorunov A, Gilmutdinov AK. Modeling of the effect of powder parameters on laser cladding using coaxial nozzle. Surface Coatings Technology. 2019;364:430–43.
60. Kim H-G, Kim I-H, Jung Y-I, Park D-J, Park J-Y, Koo Y-H. Adhesion property and high-temperature oxidation behavior of Cr-coated Zircaloy-4 cladding tube prepared by 3D laser coating. Journal of Nuclear Materials. 2015;465:531–39.
61. Kumar S, Mandal A, Das AK. The effect of process parameters and characterization for the laser cladding of cBN based composite clad over the Ti6Al4V alloy. Materials Chemistry Physics. 2022;288:126410.
62. Li J, Chen C, He Q. Influence of Cu on microstructure and wear resistance of TiC/TiB/TiN reinforced composite coating fabricated by laser cladding. Materials Chemistry Physics. 2012;133:741–45.
63. Li J, Luo X, Li G. Effect of Y_2O_3 on the sliding wear resistance of TiB/TiC-reinforced composite coatings fabricated by laser cladding. Wear. 2014;310:72–82.
64. Li K, Li D, Liu D, Pei G, Sun L. Microstructure evolution and mechanical properties of multiple-layer laser cladding coating of 308L stainless steel. Applied Surface Science. 2015;340:143–50.
65. Li R, Feng A, Zhao J, Pan X, Zhang G, Zhu Y, Chen C. Study on process optimization of WC-Ni60A cermet composite coating by laser cladding. Materials Today Communications. 2023;37:107400.
66. Li S, Huang K, Zhang Z, Zheng C, Li M, Wang L, Wu K, Tan H, Yi X. Wear mechanisms and micro-evaluation of WC+ TiC particle-reinforced Ni-based composite coatings fabricated by laser cladding. Materials Characterization. 2023;197:112699.
67. Li W, Xu P, Wang Y, Zou Y, Gong H, Lu F. Laser synthesis and microstructure of micro-and nano-structured WC reinforced Co-based cladding layers on titanium alloy. Journal of Alloys Compounds. 2018;749:10–22.
68. Li X, Feng Y, Liu B, Yi D, Yang X, Zhang W, Chen G, Liu Y, Bai P. Influence of NbC particles on microstructure and mechanical properties of AlCoCrFeNi high-entropy alloy coatings prepared by laser cladding. Journal of Alloysn Compounds. 2019;788:485–94.
69. Lim WYS, Cao J, Suwardi A, Meng TL, Tan CKI, Liu H. Recent advances in laser-cladding of metal alloys for protective coating and additive manufacturing. Journal of Adhesion Science Technology. 2022;36:2482–504.
70. Lin C-M. Parameter optimization of laser cladding process and resulting microstructure for the repair of tenon on steam turbine blade. Vacuum. 2015;115:117–23.
71. Prashar G, Vasudev H, Mehta A. Application of bimodal stuructured thermal sprayed coatings in power plant components. *Advanced Ceramic Coatings for Energy Applications*. 2024:273–87.
72. Lin Y, Lei Y, Fu H, Lin J. Mechanical properties and toughening mechanism of TiB2/NiTi reinforced titanium matrix composite coating by laser cladding. Materials Design. 2015;80:82–88.
73. Singh G, Bansal A, Vasudev H, Mishra V. Sliding wear study of the Inconel-625 clad deposits by microwave heating on SS–304. Physica Scripta. 2024;99(6):065503.

74. Liu F, Mao Y, Lin X, Zhou B, Qian T. Microstructure and high temperature oxidation resistance of Ti-Ni gradient coating on TA2 titanium alloy fabricated by laser cladding. Optics Laser Technology. 2016;83:140–47.
75. Liu J. Online measurement based tool path generation toward integrated robotic laser cladding. Industrial Robot: An International Journal. 2013;40:494–501.
76. Liu J, Liu D, Li S, Deng Z, Pan Z, Li C, Chen T. The effects of graphene oxide doping on the friction and wear properties of TiN bioinert ceramic coatings prepared using wide-band laser cladding. Surface Coatings Technology. 2023;458:129354.
77. Mocanu A-C, Miculescu F, Stan GE, Pasuk I, Tite T, Pascu A, Butte TM, Ciocan L-T. Modulated laser cladding of implant-type coatings by bovine-bone-derived hydroxyapatite powder injection on Ti6Al4V substrates—Part I: Fabrication and physico-chemical characterization. Materials. 2022;15:7971.
78. Mocanu A-C, Miculescu F, Stan GE, Tite T, Miculescu M, Ţierean MH, Pascu A, Ciocoiu R-C, Butte TM, Ciocan L-T. Development of ceramic coatings on titanium alloy substrate by laser cladding with pre-placed natural derived-slurry: Influence of hydroxyapatite ratio and beam power. Ceramics International. 2023;49:10445–54.
79. Singh G, Bansal A, Vasudev H. Corrosion and microstructural behaviour of Inconel 625 microwave clad deposited on mild steel. *Thermal Claddings for Engineering Applications*. CRC Press; 2024:51–62.
80. Sharma Y, Mehta A, Vasudev H, Jeyaprakash N, Prashar G, Prakash C. Analysis of friction stir welds using numerical modelling approach: A comprehensive review. International Journal on Interactive Design Manufacturing. 2024;18:5329–5342.
81. Nie M, Zhang S, Wang Z, Zhang H, Zhang C, Chen H. Development of a novel method for measuring the interfacial bonding strength of laser cladding coatings. Optics Laser Technology. 2022;148:107699.

4 Environmental Impact and Sustainability of Advanced Manufacturing Techniques, Materials, and Coatings

Amrinder Mehta, Hitesh Vasudev, and Bhupinder Singh

4.1 INTRODUCTION

Advanced manufacturing technologies, materials, and coating technologies are under immense pressure because of heightened global environmental concerns. New production technologies, for example, additive manufacturing (3DP) and lean manufacturing, are seen to provide a wide scope for enhancing sustainability across different industries. Most of these techniques improve efficiency, preventing waste, and therefore have a positive impact on sustainability [1].

In Figure 4.1, a schematic of a bubble diagram is used to give an idea of sustainable coatings and their characteristics. The left side shows three overlapping circles representing the core sustainability pillars: Environmental, Economic and Social. The center features "Sustainable coatings" as the focus, connected to five essential characteristics: Cost concerns, versatility, environmentally sensitive factors, safe production for customers, and healthy product. Such connections are illustrated by using dotted and dashed lines which elucidate that sustainable coatings encompass a number of positive features and at the same time provides a solution in terms of environmental, economic and societal impacts [2].

Lean manufacturing is a management strategy which aims at improving organizational efficiency through successful elimination of waste and its sources in production. By increasing the quality of the product and reducing waste, lean manufacturing is also able to greatly decrease the use of raw materials and energy for production. By improving product quality and reducing waste, lean manufacturing optimizes resource consumption and utilization, thereby minimizing ecological impact, a critical concern in resource-sensitive sectors such as the automotive industry [3]. AM technologies are also similar because they create the required goods layer by layer, and as a result, they use only the required amount of material

 DOI: 10.1201/9781003546528-4

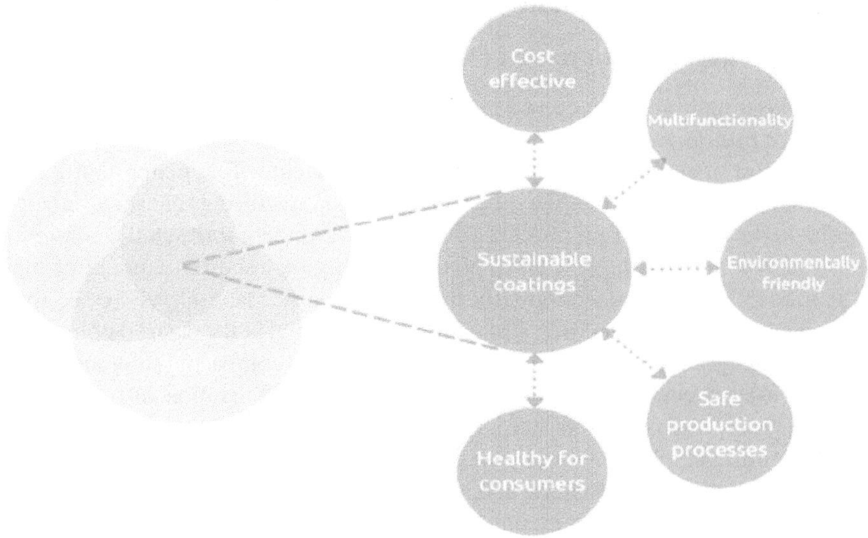

FIGURE 4.1 Schematic depiction of sustainable coatings based on the three pillars of sustainability [2].

Source: Licensed under CC BY 4.0.

to construct the product, so the percentage of material waste is significantly lower compared to SM. Besides these approaches, environmentally sustainable product development in production systems is critical [4]. Businesses are advised to embark on eco-innovative activities to go beyond meeting the proposed environmental standards while improving environmental performance during the different stages of a product life cycle. This approach forms part of sustainable manufacturing since it seeks to assess the effects of the products from acquisition of raw materials to their disposal. Life cycle assessments (LCAs) are crucial in this case as they enable companies to track sources of waste, energy usage, and emission during the life of the product [5].

The contribution of advanced materials and coatings constitutes another important element of sustainable efforts. There are better material solutions, lighter composites, and high-performance coatings that have the potential to create longer-lasting products and thus lower emission of waste [6]. However, making sure that the production processes of these advanced materials can be sustained is important. This includes the use of renewable energy sources, minimizing material waste, and developing recycling techniques to be used. New manufacturing processes, materials, and coatings are beneficial from the point of view of performance and productivity, but they present new problems in the field of ecology [7]. The efficiency promoted by business regulation and corporate programs should mitigate adverse impacts of production on the environment and sustain long-term efficiency in manufacturing. Thus, a symbiosis of innovation and environmental studies allows industries to take a more active part in striving for a sustainable future.

4.2 ADVANCED MANUFACTURING TECHNIQUES

4.2.1 Additive Manufacturing (3D Printing)

Selective laser melting has received much attention for potential uses in applications that require minimization of material usage and the development of complex structures. Additive manufacturing is an optimal advancement within material utilization during manufacturing [8]. Unlike other forms of manufacturing that involve cutting material to develop parts and products, AM constructs them through the continuous deposit of one material layer after another, and only the required amount of material is used [9]. This approach greatly minimizes waste material, as we see in the image, where the powder feeding system delivers material at the desired point. Local production comes easily to the technology thanks to the compact system installation involving a laser conductor and powder feeder. This localization also minimizes bulky transport channels and storage facilities as well as costs, including those related to emission of greenhouse gases [10]. The usage of the material is well controlled through the PLC actuators, and actual monitoring is done through the CMOS camera, whereas the pyrometer helps in proper fusion of the material. The degree of control achieved at this level leads to wide production of parts with reduced material content, making AM a sustainable manufacturing method [11]. Dependent upon the technology and material used in AM, energy can be consumed in relatively small amounts or large quantities. Compared to metal material deposition-based manufacturing processes, metal 3D printing can be a quite energy-demanding process in some cases. The energy efficiency of AM is still questionable, mostly due to the energy efficiency that depends on the type of technology used, type of material used, and complexity of the part being manufactured [12].

Some AM processes use more energy at the print stage, though they could recover this through less material usage and no post-processing, which are energy intensive. For example, based on some research findings, it was found that the energy consumption of AM is lower than that of the conventional manufacturing paradigm across the whole process if the batch size is small or when producing complex geometric parts and components [13]. AM 3DP processes involving thermoplastic materials or metal can generate ultrafine particles (UFPs) and volatile organic compounds (VOCs) in the atmosphere. Such emissions are dangerous to human health since UFPs can travel deep in the respiratory system, and VOCs produce various effects. To avoid such risks, the right form of ventilation and filtration is mandatory [14]. Solutions are enclosures incorporating air filters, HEPA particle filters, and activated carbon filters for VOC absorption. However, forced air flow as well as the location of filters within close reach of the emission source will go a long way in improving air quality. Other materials such as an ABS material, which emits fewer toxic particles, may require complex filtering methods [15]. Multi-material 3D printed parts have troubling issues as far as their end-of-life disposal is concerned. The primary challenge with utilizing recycled thermoplastics is their degradation. Sequencing layers of different material in multiple material print is usually very costly and makes recycling for economic practicality impossible [16].

Figure 4.2 presents a detailed description of the process of the 3D printing, where four major steps are outlined from left to right. The process of 3D printing starts with

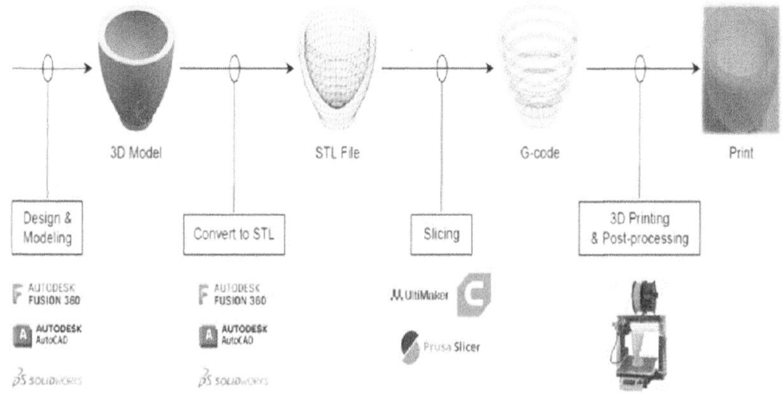

FIGURE 4.2 Overall workflow of the additive manufacturing process [17].

Source: Licensed under CC BY 4.0.

preparation of a 3D map with an aid of special programs such as Autodesk Fusion 360, AutoCAD, or SolidWorks. It is then exported to STL, a stereolithography file format. After that, the STL file is sliced using software such as UltiMaker Cura or Prusa Slicer to slice the full STL file into layers and convert it into G-code—instructions for the 3D printer. The last step demonstrates the real functional and aesthetic 3D printed object along with other printing procedures. Every stage is described with the help of the icons of the applicable software and an illustration of a 3D printer [17].

Additive manufacturing creates options for easier parts repair and replacement. The technology also allows for the near real-time manufacturing of spare parts, thus minimizing spare stock and enhancing autonomy against supply disruptions. This is particularly important for end-of-life products, especially when it comes to repairing obsolete or discontinued parts on older equipment that may otherwise be considered for scrapping [18]. By enabling repair operations, 3D printing—particularly through methods such as direct energy deposition—can extend component service life by supporting direct remanufacturing of damaged parts. For 3D printing to accomplish its recycling goals, more work must be done on recycling 3D-printed objects. By applying sustainable design in 3D printing that aims at the efficient use of materials and designing for ease of repair or recycling, the 3D printing industry can greatly decrease the rate at which it generates waste [19].

4.2.2 Advanced CNC Machining

Computer numeric control (CNC) machining has advanced to offer high accuracy and speed; however, concerns remain regarding its negative environmental impact. One of the major challenges in implementing 3D part production through additive techniques is the issue of material waste, especially in metal 3D printing. Such technologies tend to produce large quantities of scrap material, which can become quite expensive and/or damaging to the planet [20]. In powder bed fusion processes, up to 95% of the powder may remain unused after a print job is complete. However, there is considerable opportunity for recycling the metal chips and coolant after such processes [21]. Metal powders can generally be sorted using this method and sometimes can also be recycled or utilized in other subsequent printings, although this is only possible for a limited number of times because of the changes in the properties of the powders over time [22]. There are also current practices by some companies that embrace closed-loop systems for metal powder recycling. Another important area is the reuse of coolant. Sophisticated filters which include absorption CO_2 technology techniques are capable of purifying used coolants; hence, they can be recycled several times [23]. This not only helps to bring down waste but also operation costs. Work is still being done to enhance material utilization within 3D printing, such as using part orientation to come up with the fewest support structures required alongside using enhanced powder-collecting techniques. Concerns about energy consumption and high-power requirements remain major issues concerning the operation of machines such as cloud computing data centers and electric propulsion systems, among others [24]. These problems arise from the meteoric rise in the need for computational, storage, and data transmission resources. To this end, researchers are investigating several energy optimization prospects applying sophisticated control systems. One initiative includes virtualization strategies such as virtual machine computing and server virtualization to enhance resource optimality while diminishing power demands. New techniques like the grasshopper optimized heuristic technique for different types of workloads found in virtual machine migration also hold the potential for increasing energy efficacy in cloud computing [25]. In electric propulsion, there appear to be opportunities for energy savings in permanent magnet-free technologies such as switched reluctance machines. Furthermore, strategic models for multi-objective optimization based on the Taguchi and non-dominated sorting genetic algorithm II (NSGA-II) are also being integrated for improving efficiency in designing and operating machines. These constitute new control systems for meeting the performance criteria while at the same time being energy efficient, thus making the machine or system more efficient and economical [26]. Conventional coolants as well as lubrication fluids that are widely applied in industries, especially using metals, give rise to various environmental as well as health problems. Most of these fluids are petroleum based and contain additives with dangerous characteristics and are likely to pollute the environment and humans in cases of spillages and leakage. Notably, due to the increasing problems facing the world and the emerging sustainable requirements, more effort has been made towards environmentally friendly and biodegradable products [27]. Biodegradable lubricants based on renewable biomass feedstock have recently found a place as potential replacements for traditional oil-metal lubricants. For instance, cutting fluids based on groundnut-enhanced oil has displayed high potential in metal machining processes and enhanced performance characteristics in corrosion protection, friction, and drilling [28]. Another area of advanced application is the use of hot vapor in green cutting technology, where overheated water vapor serves as both a coolant and lubricant, offering benefits in reducing cutting forces, controlling temperature, and minimizing tool wear. These green solutions that meet ecological challenges also tend to improve the process and the quality of the surface in machining processes, resulting in better productivity in manufacturing [29].

4.2.3 Laser-Based Manufacturing

Other applications of lasers are cutting, welding, and surface treatments, and though the processes are highly effective, they have environmental constraints. Even though LMSs provide high accuracy in manufacturing and offer the shortest throughput time, these systems can consume a lot of power. This is mainly because of laser requirements and accessories like cooling and extraction of energy from the laser system [30]. However, these systems also have the prospect of overall energy saving because of high process efficiency. Laser technology allows parts of an object to be treated selectively, often with high rates of processing, which can reduce the time required to process an item and cut down on waste [31]. Some laser systems such as high-power VCSELs are comparatively compact and thus can be installed in current manufacturing equipment, indicating the potential to replace energy-consuming batch operations and large furnaces. This shows that laser-based manufacturing consumes a high amount of power, but the process is fine-tuned to generate maximum output and hence requires strict control and efficient optimization [32].

Figure 4.3 is a schematic diagram of an LPBF 3D printing system. The setup involves an optical system combined with a laser source that fires a beam through a beam expander. It is used to melt powder while constructing the object layer by layer, mainly on a powder bed. The system includes several key components. While SCSSs have a powder delivery system on the left side and a powder collector at the right side, the entire SS with the build container and the moving build platform is in the middle. There is always a coater tasked with applying new particles of powder on the build zone of a powder dispenser. The whole process occurs in a build chamber with particular parameters maintained within the chamber [33].

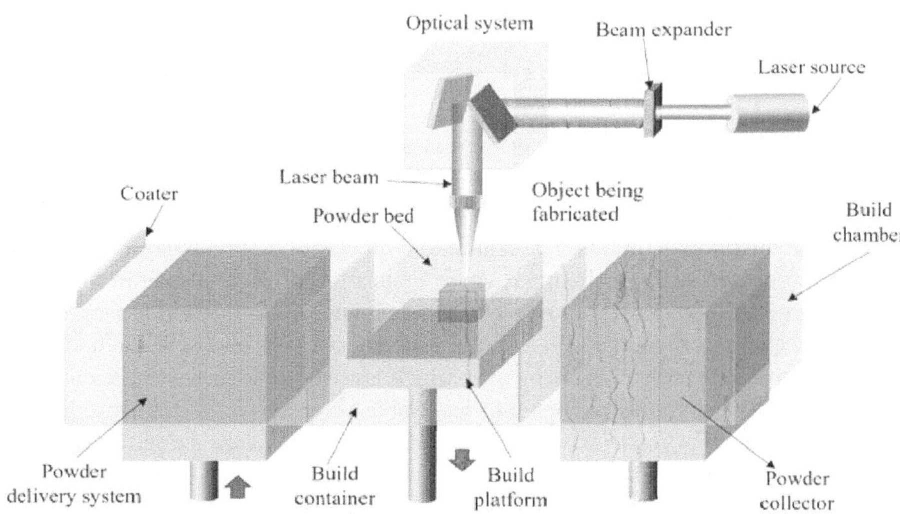

FIGURE 4.3 Schematic diagram of laser powder bed fusion [33].

Despite advances in manufacturing technologies based on lasers within the last couple of years, improvements have also been observed in the way manufacturing industries deal with material integrity. The sophisticated computer-aided design systems preferred in laser cutting apply strategies like division and optimization to material sheets to minimize utilization and optimize cutting [34]. These methods ensure that intricate design patterns are separated into segments and these segments rearranged in the least wasteful manner. For clamp applications and laser welding, new advanced techniques such as the IMF-LCD advanced laser cladding deposition technique come with flexibility in forming and high levels of powder reclaim rates [35]. This approach allows for controlled material to be deposited or shaped in the required places, which is very useful in manufacturing and repairing structures such as TRT blades. Controlled material deposition also minimizes waste, optimizes material flow and part formation, and achieves a part surface finish ranging from 2 to 10 µm, with some processes attaining a surface roughness of 1.323 µm. Laser processing produces many by-products, with particles and fumes being a health and environmental hazard [36]. Laser cutting and engraving generate airborne emissions that are toxic, such as part per million (ppm) and respirable particles, volatile organic compounds, and toxic emissions. These emissions change with the type of material being lased and power levels during lasing. Cardboard cutting releases large amounts of respirable particulate matter and ultrafine particles [37]. To minimize these risks, a well-installed fume hood and extraction and filtration systems play a very important role. However, even with fume extractors, gases and particulate matter can be released whenever the lid of a laser cutter is opened after operation. Research has also established that at least for 20 minutes after cutting, concentrations of nanoplastic and microplastic particles rise [38]. Thus, in laser-based manufacturing, it is crucial to provide efficient extraction systems together with the correct kind of ventilation to safeguard employees and contribute to environmental well-being.

4.3 ADVANCED MATERIALS

4.3.1 Nanomaterials

Nanomaterials have specific characteristics but may pose threats to the environment. The production of nanomaterials involves high energy costs because of the rigorous production conditions. For example, synthesis of carbon nanotubes (CNTs) requires processes such as chemical vapor deposition at high temperatures or an arc discharge method. However, optimizing nanomaterial production processes can bring high energy efficiency [39]. Efforts are underway to find new methods like the microwave-assisted synthesis method which are labor and energy efficient. One more obvious line is the employment of renewable energy sources in the production of nanomaterials. For instance, solar thermal reactors for synthesizing nanomaterials exist as a far more sustainable solution. Furthermore, practical ways such as the use of continuous flow reactors and increased scale-up process are being explored to improve energy densities for production of nanomaterials and possibly lower the overall energy requirements for the process as well as maintaining product value [40]. Currently, the consequences of nanoparticles for the environment in the long term are unknown, with recent concerns about bioaccumulation and disturbance of ecosystems. Research also has revealed that materials such as silver and titanium oxide can become deposited within plants and water animals. This bioaccumulation is a source of concern in issues of food safety, especially for agriculturists and marine biologists [41]. The behavior of nanoparticles in ecosystems depends on several factors, including surface charge, organic matter, and biological interactions. For example, rice root exudates can change the behavior of aggregated dissolutions and bioaccumulation of silver nanoparticles depending on their surface characteristics. Also, there is evidence that nanoparticles also tend to settle in bottom-dwelling invertebrates, with implications for aquatic food webs. The results of the present study underscore the importance of conducting further studies to assess the ecological consequences of nanoparticles released into any environment [42].

The various components and effects of phase change material (PCM) nano-additives are clearly depicted in Figure 4.4. There are three main categories of additives shown: metal oxides (alumina, copper oxide, titanium dioxide), metals (copper, aluminum, gold, silver), and nanocarbons (carbon nanotubes). The diagram indicates three primary effects of these nano-additives: microscopic modifications of latent heat, coefficient of thermal conductivity, and viscosity. These additives are employed widely in the enhancement of the thermal proprieties of phase change materials. The circular design of the diagram demonstrates how each of these components plays a role in the overall PCM nano-additive system [40].

There are great difficulties with the properties of recycling nanomaterials from composites because of the problem inherent to these structures and compositions, namely the impossibility of separating and recovering the components. The situation

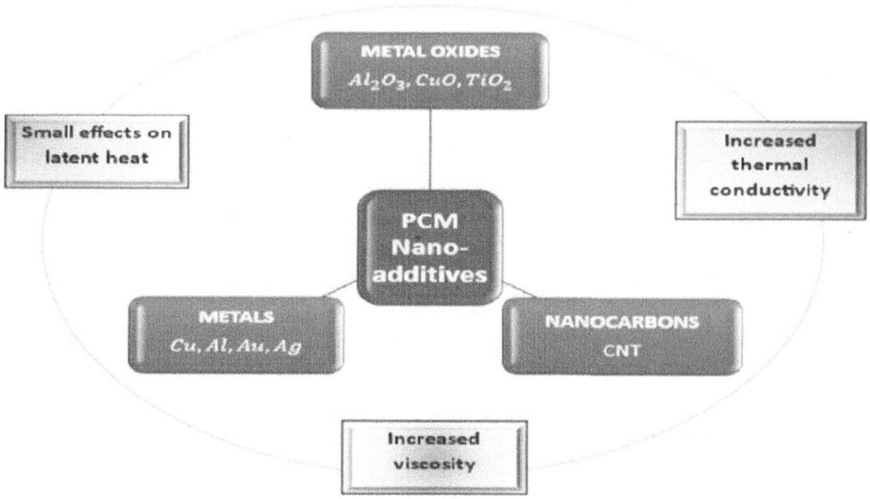

FIGURE 4.4 Nanomaterials with high conductivity for increasing the thermal performance of latent heat thermal energy storage systems [40].

is aggravated by the presence of mixed fibers and contaminants in documents as well as in textiles. For instance, integrating elastane in stretch fabrics is a problem within recycling process as well [43]. That's why specialized recycling processes are required to solve these problems. Chemical and enzymatic recycling strategies are proposed to address fossil polymer recycling in textile materials. Furthermore, the application of advanced methods such as the production of nanomaterials from agricultural waste also provides a viable option [44]. However, the idea of nanocomposites and the vastly differing types of nanomaterials encompassed in various utility categories call for the creation of specific methods for recycling nanomaterials to maximize the efficiency of reusing these valuable components.

4.3.2 Composite Materials

Fiber-reinforced polymers (FRPs), as a class of composite materials, are not fully sustainable materials. Figure 4.5 presents a basic graphic demonstration of three principal threats related to sustainability. From a central box labeled "Challenges for sustainability," three arrows branch out to different aspects: They are "Light weight design," "End of life management," and "production energy". This diagram illustrates core issues that have to be addressed for sustainable development in orange color [45].

One potential rationale for lightweighting in transportation is the need to reduce fuel consumption while improving product durability and reliability. Through the application of the next generation of materials, including high-strength steel, aluminum, and composites, vehicles can reach an optimal level of reduction of weight while attaining enhanced safety and performance. A reduction in weight has a direct impact on fuel economy, as research proves that a drop of 10% in vehicle weight easily equates to a fall in fuel usage of between 6% and 8% [46]. Lightweight materials are also more durable and offer better corrosion resistance than traditional materials,

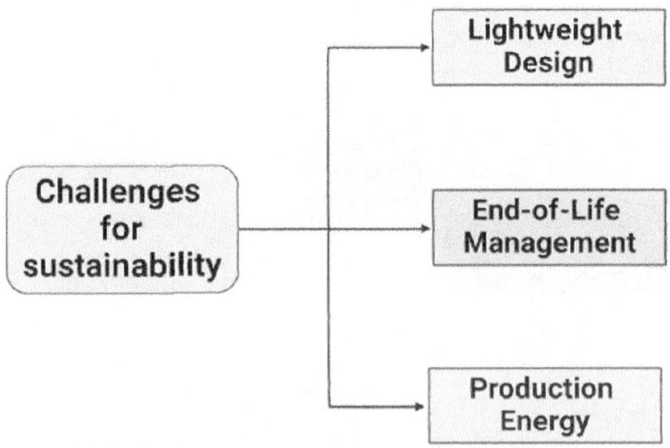

FIGURE 4.5 Fiber-reinforced polymers present challenges for sustainability.

Source: Licensed under CC BY 4.0.

leading to longer product lifespans. For example, in aerospace applications, carbon fiber reinforced polymers not only reduce weight but also provide high fatigue endurance, potentially doubling the life cycle of some parts [47].

Recycling mixed-material composites is a major issue because the various polymers used in most applications such as footwear, carpets, and multilayer packaging are chemically bonded, but solutions to these problems are emerging. Polydiketoenamines (PDKs) may well be used for controlled depolymerization rates, which are needed for sequential dismantling of MP composites. This chemospecific approach also enables the recycling of monomers, fillers, and additives in their as-new state [48]. Also, there is a need for fabricating bio-based and recyclable polymer composites for their options in several applications. For example, there are propositions of lignin- and tannin-based non-isocyanate polyurethane (NIPU) adhesives for wood composites. Still, using bio-based materials for preparing these NIPUs results in improved mechanical characterizations and reduced curing time. Another approach involves reusing electrochemical sensors fabricated by additive manufacturing to produce new conductive and nonconductive 3D-printing filaments, supporting the principles of a circular economy [49, 50].

The energy consumption associated with some of the processes used to create some composites is very high, especially for thermoset polymer composites. Most of the conventional thermal curing techniques have been known to consume large amounts of energy and hence are not suitable for large-scale curing. However, there are great opportunities for the further decrease of energy consumption through optimization of the curing technology [51]. A few other processes such as UV-induced curing, electron beam curing, and infrared curing seem to have greater potential in making composite manufacturing a more energy-efficient process. It should also be noted that these methods can help to shorten curing time and energy consumption many times. For instance, curing processes using UV have been tested for LBM of composite layers with surprisingly low cure level variation across layers [52]. Also, active research is ongoing for direct curing methodology during the additive manufacturing process for forming thermoset polymer composites at higher rates with complete scalability issues of post-curing.

4.3.3 Advanced Alloys

There are cost effective and high-performance alloys which are used in aircraft because of the resistance and better efficiency, but these have environmental impacts. While high-performance alloys contain scarce and critical materials to improve their characteristics, this creates a problem of resource availability and shortage. Significant relevance has been observed in connection with the use of these elements in high-performance alloys for several industries, including aerospace and energy, which has led to growth in demand and supply risks [53]. In response to these issues, a SCM approach to sustainable sourcing and recycling is what we see emerging. Scholars are now trying to come up with various techniques to offset resource constraints by coming up with new alloys that contain minimal quantities of the critical elements. Furthermore, research is being conducted on efficient methods of extracting and reusing resources from end-of-life products by applying circular economy principles to alloy manufacturing and use, thereby ensuring future supply and reducing environmental impact [54]. Many manufacturing processes used in creating advanced alloys require energy to be applied, especially in the melting and processing steps. But there are some effective opportunities for energy saving in near-net-shape manufacturing techniques. Selective laser melting (SLM) in powder bed systems can also provide a new method for creating permanent magnets made of particles of rare metals, bypassing the need for prolonged heat treatment in high energy–consumption oven sintering. This method also optimizes the use of material, and the rate of waste during the post-processing stage is low [55]. Likewise, HIP near-net-shape processing of superalloy parts for sCO_2 turbomachinery witnesses a similar possibility of cost optimization and energy saving. This technique makes it possible to fabricate detailed parts such as turbine nozzle rings and casings to accurate dimensions, thus possibly cutting down the total manufacturing cycle and hence reducing the overall energy utilization when compared with other methods [56].

Most metals form alloys, and these are usually very easy to recycle, which is a bonus under the provision of sustainable resources. Many base metals used to create alloys already have recycling programs in place requiring the melting and recasting of the wasted material, and the recycling process can be repeated over and over without too much of a quality loss. This characteristic makes metal alloys useful in a circular economy model. However, selectively extracting and concentrating alloying elements in multicomponent systems remains a challenge [57]. The incorporation of more than one element in the packaging can, at times, make the recycling process difficult and may need special technology for separation. The recycling of aluminum alloys entails dealing with diverse contaminants and other aluminum alloys. The introduced sorting and refining processes, including X-ray fluorescence sorting technology, are in the process of further enhancements to increase the efficiency of alloy separation and element recovery to provide higher-quality recycled materials [58].

4.4 ENVIRONMENTAL IMPACT ASSESSMENT

4.4.1 Life Cycle Assessment

Life cycle assessment is an important means for assessing the environmental characteristics of innovative manufacturing technologies, materials, and coatings. End of pipe or end of life thinking is good for the overall assessment since sustainability should be carried out through the whole life cycle of products and processes. It is an approach to the evaluation of environmental impacts in all processes of a product's life from its conception to its disposal, commonly known as life cycle assessment [59]. However, identifying those system boundaries is problematic when targeting C/P products and processes. The complexity of the integrated supply chains and new manufacturing modes complicates the task of drawing a parsimonious cut-off point. The emission savings of electric vehicles must consider the battery production, electricity generation, and recycling end of life phase. Industry and researchers are working to standardize methods and tools that make LCA results less variable and comparable with different industries and product types [60].

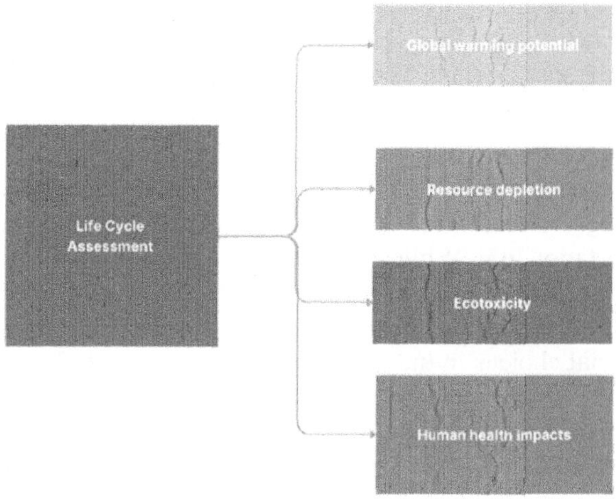

FIGURE 4.6 Life cycle assessment evaluates multiple environmental impact categories.

Source: Licensed under CC BY 4.0.

Figure 4.6 shows the life cycle assessment model, which encompasses a complex methodology for assessing impacts on the environment. The diagram shows four main impact categories branching from LCA; Climate change effect (measured by global warming potential in orange), Natural capital depletion (measured by resource depletion in pink), Impact on ecosystems (measured by ecotoxicity in purple) and Impact on human health (measured in red). Such a basic flowchart shows why LCA can consider all the positive and negative impacts on the environment and health due to a given product or process all through its life cycle [61].

This gives a broad perspective of the environmental impact that is best understood when looked at from a systems perspective. However, two major concerns exist: establishment of quality data and data availability, particularly for new technologies. Insufficient data that can be collected often result in incomplete or superficial assessment [62]. The creation of standard databases, and even more importantly, standard methodologies, ensures that the results obtained can be made as directly comparable as possible. There are continuing initiatives for enhancing the credibility of data that are being collected, as well as public databases, with the intention of refining and increasing the credibility of life cycle assessments [63].

4.4.2 Carbon Footprint Analysis

Carbon auditing is much narrower and aims at evaluating greenhouse gas emissions. Life cycle assessments that map carbon footprint, or rather greenhouse gas emissions, mainly address direct emissions of the manufacturing processes. This includes direct quantification of emissions produced at production sites throughout the creation of energy and other processes in manufacturing. One major factor that the analysis considers is the carbon factor in the various energy sources that are used in production, since this plays a big role [64]. Factories with renewable energy sources tend to have lower direct emissions because of operating power compared to factories that use fossil energy. These emissions must then be measured and monitored to determine regions issuing the most carbon output to effectively reduce it. This assists manufacturers in understanding their direct environmental influence and provides prerequisite analysis and planning for reduction of carbon dioxide emissions. These are emissions of greenhouse gases acting as precursors during raw material production and transportation [65]. While comparing a car manufacturing company's indirect emissions, the carbon footprint of steel manufacturing is also incorporated. This is especially relevant for energy-intensive products, highlighting the importance of future raw material choices in the generation of electricity [66]. That is, the energy used by electronics all through their cycle of usage produces more emissions than the power used in manufacturing them. Estimation of these indirect emissions offers a better understanding of a product's external environmental footprint. It can pinpoint areas at the supply chain end and when the product is used to allow proper planning on matters concerning materials, transportation, and how to minimize carbon footprint [67].

Carbon offsetting and mitigation are techniques used to decrease and balance carbon output in production lines. Carbon capture and utilization (CCU) technologies are being seen as effective approaches to reducing emissions in the manufacturing sector. CCU processes can generate useful chemicals and at the same time minimize environmental greenhouse gases and lower the exploitation of fossil resources [68]. CO_2 can be produced in fuels, carbonates, polymers, and chemicals. The assessments also suggest that select sectors, such as steel and cement manufacturing, could have relatively low capture costs for CO_2 for utilization. Also, there are innovations from raw materials such as calcium carbonate cement, which employs CO_2 containing flue gas in addition to industrial waste in construction, which are carbon negative [69].

4.5 SUSTAINABILITY STRATEGIES

4.5.1 Circular Economy Approaches

Utilizing circular economy principles can bring about a range of improvement in the sustainability performance of advanced manufacturing. Design for circularity is a major concept in achieving circular economy solutions aiming at the conception of products that can be recirculated faster. The items are built in a modular system that offers ease of repair and replacement of parts in a product. This saves on product costs, and the company can market the same product for longer durations [70]. Consequently, this reduces waste. Products are designed with the concept of easy disassembly, enabling easy disassembly and disposal for recycling them, including coatings that do not hinder recycling. These principles elevate the concept of waste reduction, material optimization, and the formation of objects that fit into a circular economy approach [71].

Another successful method in circular economy business models is remanufacturing and refurbishment strategies for given products through the application of advanced manufacturing technology. Further, laser applications for engineering materials include reclamation through laser cladding to restore damaged parts like turbine blades that could be 75% cheaper [72]. Technologies including plasma electrolytic oxidation, physical vapor deposition, and laser surface modification improve corrosion and mechanical characteristics of alloys like aluminum. Moreover, self-healing coatings are still in the process of production to add additional years to the lifespan of equipment. This allows the manufacture of spare and auxiliary parts, thus promoting product updates. It also enables the minimization of waste of resources and the enhancement of product quality and longevity, thereby achieving sustainable manufacturing [73].

In closed-loop material systems, recycling technology for high-performance materials and composites and take-back and reverse logistics activities are pursued. These systems are meant to cut operational costs by reducing the waste of resources. This includes techniques for recycling composite and other technically challenging materials [74]. Manufacturers put measures in place to ensure that consumers return products at the end of their lifetimes. These programs can mix positive net-margin remanufacturing activities with negative net-margin material recycling functions, improving return management capabilities for reclaiming and transporting used products back to manufacturers. This also includes establishing most favorable cost structures regarding buy-back and transport so that the program remains sustainable [75]. These strategies can be useful for lengthening product lifetimes, getting back valuable substances, and decreasing environmental footprint, in accordance with the circular economy.

4.5.2 Energy Efficiency and Renewable Energy Integration

Concerning energy, its conservation and the use of kinetic energy sources are two principal goals for sustainability. In manufacturing process optimization, there is an increased use of improved control systems and artificial intelligence in the form of energy use, heat recovery systems, and waste energy utilization. AI-based approaches are being investigated at the device-level controller up to the individual machine controller and the controller of an entire manufacturing plant [76]. This seeks to enhance capabilities for self-learning, organizational performance, and the efficient utilization of all resources. Three functions of energy saving and emission reduction are central to the use of heat recovery and waste energy utilization. In the case of cotton manufacturing, waste heat recovery has led to annual reductions in coal and electricity consumption and greenhouse gas emissions such as CO_2 and SO_2. Innovations such as Tesla turbines are being developed as possibly optimal for waste heat recovery in many industries due to simplicity, reliability, and low manufacturing costs [77]. Generation is also occurring through distributed generation with solar panels or wind turbines installed within business premises. This approach reduces dependence on grid electricity and helps minimize carbon emissions. Companies are engaged in purchasing renewable power from the utility under some of the plans. These initiatives help companies contribute to financing clean energy; together, these two approaches help reduce the carbon footprint of manufacturing operations even if they are unable to generate power on premises. For instance, a cross-sectional study of green energy purchase intentions showed that environmental attitudes, self-identity, and perceived behavioral control act as antecedents on the amount of program participation. Also, to regulate consumer energy usage, some energy providers are employing such things as remotely controlled energy use systems, which could improve the use of renewable energy [79]. Energy storage systems or the integration into the smart grid has emerged as important to levelize energy loads in manufacturing and improve renewable energy acquisition. These systems enable factories to store energy during periods of low usage while using it during higher-demand times, thus minimizing energy costs and pressures on the grid [80]. It is found that manufacturers have started adopting demand response programs in which energy consumption depends on the requirements of the grid. It helps to strengthen grid stability and can offer efficiency to companies by way of financial remunerations. For instance, more flexibility of the grid using vehicle-to-grid (V2G) systems is currently under development. For instance, electric vehicle storage at production centers could be useful for power demands during peak hours by acting as an energy storage system [81]. These strategies contribute positively to energy conservation while also supporting the development of stable structures and robust electrical networks.

4.6 FUTURE PERSPECTIVES

Advanced manufacturing has the potential to include improved sustainability through the application of emerging technologies. AI and ML are involved in the process of enhancing manufacturing procedures, and they are highly effective means of lowering environmental effects. When it comes to the application of data analyses, AI allows optimizing the distribution of inspection activities and directing regulators' attention to critical enterprises as an approach to preventing violations of the surrounding environment [82].

Figure 4.7 shows a simple network of artificial intelligence using concentric circles to show the relative importance of each field of study. At the periphery, artificial intelligence is defined by broad capabilities such as natural language processing, common vision, and automaton reasoning. Within this is formed a subfield, machine learning, consisting of various algorithms such as k-means, support vector machines, and linear regression, among others. The second inner circle shows neural networks, including the Boltzmann neural network and MLPs. Fundamentally, deep learning is addressed here, illustrated in detail about different architectures such as convolutional neural networks, recurrent neural networks, generative adversarial networks, and deep belief networks. Each successive inner circle represents a narrower field than the one that contains it [83].

Further, AI technologies such as dry machining and MQL improve machining operations by lowering the use of cutting fluids, thus promoting conservation of environment. Also, the ML-driven approach to predictive maintenance increases the equipment efficiency and utilization and prolongs the life of the manufacturing equipment by using data from its operation to make a prognosis of the faults that may occur in the future to prevent sustainable manufacturing [84]. The IoT and DT highly increase the efficiency of monitoring and controlling energy and resources used in different industries in real time. Through the IoT, always acquiring info is made convenient, allowing organizations to monitor formalized energy and resources in an efficient way, therefore enhancing operations. Moreover, integrated DT technology

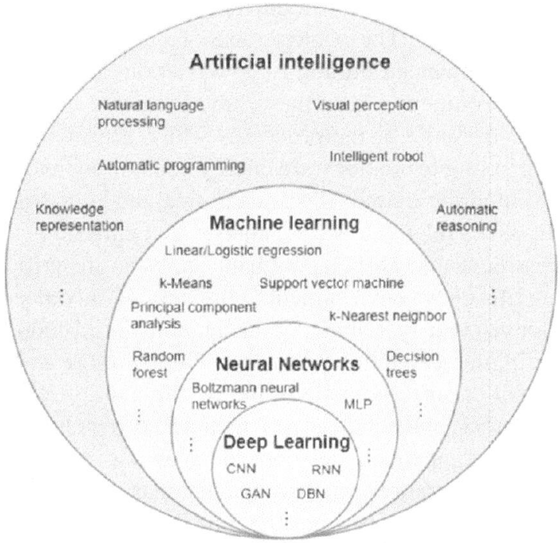

FIGURE 4.7 Relationship between artificial intelligence, machine learning, neural networks, and deep learning [83].

enables more effective product prototypes and efficient predictive analysis of different product life cycles that contribute to product life cycle management and fit circular economy strategies by improving resource utilization and reducing product or service waste production in the use stage [85, 86]. This symbiotic relationship forges sustainability and boosts the decision-making of manufacturing and resource consumption. Biotechnology and biomimicry enable the creation of bio-based materials and coatings that support the culture of sustainability. Applications of castor oil–based polyurethane composites and lignin functionalizing coatings clearly depict how renewable resources can be used to develop high-performance materials [87]. These bio-sourced coatings not only improve adhesion and mechanical characteristics but also are more friendly to the environment. Further, ideas from nature promote sustainable practices through manufacturing, which is evident in bio-inspired materials that are thought to be functional, as they are structurally like their associated natural systems [88]. It also sheds light on the creation of the circular economy concept as a means of effective and economical utilization of resources to reduce waste, which is the requirement of various materials and manufacturing-related applications [89–107].

4.7 CONCLUSION

Sophisticated technologies, material processing, and surface engineering have great significance in tackling environmental challenges within our global village. Many technological and business tools are used in the AM process, including lean manufacturing, additive manufacturing, green product innovation, and life cycle assessments to increase efficiency and reduce waste during the manufacturing process and the lifecycle of the final product. The strategic use of materials and coatings includes lighter composites for enhanced durability, low friction, and low waste generation. The surrounding environment where these advancements are being implemented must be well managed. Some of the processes which are involved in constructing artefacts using 3D printing technology are energy consuming and produce fumes, so they must be well vented/facilitated. CNC machining and laser-based manufacturing are involve certain issues related to waste, energy, and emission.

Life cycle assessment and carbon footprint analysis are critical approaches in the assessment of life cycle environmental impacts of advanced manufacturing. Applying design for circularity, remanufacturing, and closed-loop material systems, along with other circular economy strategies, will therefore greatly improve sustainability. Additional opportunities lie in improving energy efficiency, integrating renewable energy sources, and adopting new technologies including AI, IoT, and biotechnology. These developments underscore the growing importance of technology and environmental considerations in shaping sustainable manufacturing strategies. Ensuring that the social impact is a positive one is therefore of paramount importance to reduce the amount of negative social environmental impact where possible and to work towards achieving sustainable development where this is not already in place through rationalizing national legislation and corporate social responsibility.

REFERENCES

1. Abdulghafor MA, Mahmood MK, Tassery H, Tardivo D, Falguiere A, Lan R. Biomimetic coatings in implant dentistry: A quick update. Journal of Functional Biomaterials. 2023;15:15.
2. Nartita R, Ionita D, Demetrescu I. Sustainable coatings on metallic alloys as a nowadays challenge. Sustainability. 2021;13:10217.
3. Abubakar AA, Arif AFM, Al-Athel KS, Akhtar SS, Mostaghimi J. Modeling residual stress development in thermal spray coatings: Current status and way forward. Journal of Thermal Spray Technology. 2017;26:1115–45.
4. Abu-Warda N, López A, López M, Utrilla M. High temperature corrosion and wear behavior of HVOF-sprayed coating of Al_2O_3-NiAl on AISI 304 stainless steel. Surface Coatings Technology. 2019;359:35–46.
5. Abu-warda N, Tomás L, López A, Utrilla M. High temperature corrosion behavior of Ni and Co base HVOF coatings exposed to NaCl-KCl salt mixture. Surface Coatings Technology. 2021;418:127277.
6. Afrasiabi A, Saremi M, Kobayashi A. A comparative study on hot corrosion resistance of three types of thermal barrier coatings: YSZ, YSZ+ Al_2O_3 and YSZ/Al_2O_3. Materials Science Engineering: A. 2008;478:264–69.
7. Ahledel N, Schulz R, Gariepy M, Hermawan H, Alamdari H. Electrochemical corrosion behavior of Fe3Al/TiC and Fe_3Al-Cr/TiC coatings prepared by HVOF in NaCl solution. Metals. 2019;9:437.
8. Liu L, Wang S, Zhang B, Jiang G, Liu H, Yang J, Wang J, Liu W. Present status and prospects of nanostructured thermal barrier coatings and their performance improvement strategies: A review. Journal of Manufacturing Processes. 2023;97:12–34.
9. Aboutaleb AM, Bian L, Elwany A, Shamsaei N, Thompson SM, Tapia G. Accelerated process optimization for laser-based additive manufacturing by leveraging similar prior studies. IISE Transactions. 2017;49:31–44.

10. Adam L, Lietaer O, Mathieu S, Doghri I. Numerical simulation of additive manufacturing of polymers and polymer-based composites. Structure and Properties of Additive Manufactured Polymer Components. Elsevier; 2020:115–46.
11. Mehta A, Singh G. Consequences of hydroxyapatite doping using plasma spray to implant biomaterials. Journal of Electrochemical Science Engineering Failure Analysis. 2023;13:5–23.
12. Afrasiabi M, Bambach M. Modelling and simulation of metal additive manufacturing processes with particle methods: A review. Virtual Physical Prototyping. 2023;18:e2274494.
13. Ahn D-G. Direct metal additive manufacturing processes and their sustainable applications for green technology: A review. International Journal of Precision Engineering Manufacturing-Green Technology. 2016;3:381–95.
14. Al Rashid A, Ahmed W, Khalid MY, Koc M. Vat photopolymerization of polymers and polymer composites: Processes and applications. Additive Manufacturing. 2021;47:102279.
15. Al Rashid A, Khan SA, Al-Ghamdi SG, Koç M. Additive manufacturing of polymer nanocomposites: Needs and challenges in materials, processes, and applications. Journal of Materials Research Technology. 2021;14:910–41.
16. Al Rashid A, Koç M. Experimental validation of numerical model for thermomechanical performance of material extrusion additive manufacturing process: Effect of process parameters. Polymers. 2022;14:3482.
17. Zhou L, Miller J, Vezza J, Mayster M, Raffay M, Justice Q, Al Tamimi Z, Hansotte G, Sunkara LD, Bernat J. Additive manufacturing: A comprehensive review. Sensors. 2024;24:2668.
18. Alagha AN, Hussain S, Zaki W. Additive manufacturing of shape memory alloys: A review with emphasis on powder bed systems. Materials Design. 2021;204:109654.
19. Alifui-Segbaya F, Ituarte IF, Hasanov S, Gupta A, Fidan I. Opportunities and limitations of additive manufacturing. Springer Handbook of Additive Manufacturing: Springer; 2023:125–43.
20. Mehta A, Vasudev H, Jeyaprakash N. Role of sustainable manufacturing approach: Microwave processing of materials. International Journal on Interactive Design Manufacturing. 2023:1–17.
21. Soori M, Jough FKG, Dastres R, Arezoo B. Sustainable CNC machining operations, a review. Sustainable Operations Computers. 2024;5:73–87.
22. Mehta A, Vasudev H, Singh S. Sustainable manufacturing approach with novel thermal barrier coatings in lowering CO_2 emissions: Performance analysis with probable solutions. International Journal on Interactive Design Manufacturing. 2023;18:5427–5439.
23. Bhattacharyya SS, Atre S. Additive manufacturing technology: Realities and strategic perspectives from India. International Journal of Asian Business Information Management. 2020;11:1–20.
24. Cai Y, Xiong J, Chen H, Zhang G. A review of in-situ monitoring and process control system in metal-based laser additive manufacturing. Journal of Manufacturing Systems. 2023;70:309–26.
25. Calignano F, Manfredi D, Ambrosio EP, Biamino S, Lombardi M, Atzeni E, Salmi A, Minetola P, Iuliano L, Fino P. Overview on additive manufacturing technologies. Proceedings of the IEEE. 2017;105:593–612.
26. Cano-Vicent A, Tambuwala MM, Hassan SS, Barh D, Aljabali AA, Birkett M, Arjunan A, Serrano-Aroca Á. Fused deposition modelling: Current status, methodology, applications and future prospects. Additive Manufacturing. 2021;47:102378.
27. Cao C, Li W, Zhang Z, Yang X, Xu Y. Cold spray additive manufacturing of Ti6Al4V: Special nozzle design using numerical simulation and experimental validation. Coatings. 2022;12:210.

28. Cao Y, Lin X, Kang N, Ma L, Wei L, Zheng M, Yu J, Peng D, Huang W. A novel high-efficient finite element analysis method of powder bed fusion additive manufacturing. Additive Manufacturing. 2021;46:102187.
29. Mehta A, Vasudev H, Singh S, Prakash C, Saxena KK, Linul E, Buddhi D, Xu J. Processing and advancements in the development of thermal barrier coatings: A review. Coatings. 2022;12:1318.
30. Carrillo LEL. Additive Manufacturing of Binary and Ternary Shape Memory Polyester Blends as a Pathway Towards High Entropy Polymer Systems. The University of Texas at El Paso; 2023.
31. Castro e Costa E, Duarte JP, Bartolo P. A review of additive manufacturing for ceramic production. Rapid Prototyping Journal. 2017;23:954–63.
32. Cattenone A, Morganti S, Alaimo G, Auricchio F. Finite element analysis of additive manufacturing based on fused deposition modeling: Distortions prediction and comparison with experimental data. Journal of Manufacturing Science Engineering. 2019;141:011010.
33. Yue W, Zhang Y, Zheng Z, Lai Y. Hybrid laser additive manufacturing of metals: A review. Coatings. 2024;14:315.
34. Chatham CA, Long TE, Williams CB. A review of the process physics and material screening methods for polymer powder bed fusion additive manufacturing. Progress in Polymer Science. 2019;93:68–95.
35. Chen C, Xie Y, Yin S, Li W, Luo X, Xie X, Zhao R, Deng C, Wang J, Liao H. Ductile and high strength Cu fabricated by solid-state cold spray additive manufacturing. Journal of Materials Science Technology. 2023;134:234–43.
36. Alizadeh-Sh M, Marashi S, Ranjbarnodeh E, Shoja-Razavi R. Laser cladding of Inconel 718 powder on a non-weldable substrate: Clad bead geometry-solidification cracking relationship. Journal of Manufacturing Processes. 2020;56:54–62.
37. Al-Ketan O, Rowshan R, Palazotto AN, Abu Al-Rub RK. On mechanical properties of cellular steel solids with shell-like periodic architectures fabricated by selective laser sintering. Journal of Engineering Materials Technology. 2019;141:021009.
38. Allahham N, Fina F, Marcuta C, Kraschew L, Mohr W, Gaisford S, Basit AW, Goyanes A. Selective laser sintering 3D printing of orally disintegrating printlets containing ondansetron. Pharmaceutics. 2020;12:110.
39. Balasubramanian S, Gurumurthy B, Balasubramanian A. Biomedical applications of ceramic nanomaterials: A review. International Journal of Pharmaceutical Science and Research. 2017;8:4950–59.
40. Croitoru C, Bode F, Calotă R, Berville C, Georgescu M. Harnessing nanomaterials for enhanced energy efficiency in transpired solar collectors: A review of their integration in phase-change materials. Energies. 2024;17:1239.
41. Divakaran N, Ajay Kumar PV, Mohapatra A, Alex Y, Mohanty S. Significant role of carbon nanomaterials in material extrusion-based 3D-printed triboelectric nanogenerators. Energy Technology. 2023;11:2201275.
42. Farooq SA, Raina A, Mohan S, Arvind Singh R, Jayalakshmi S, Irfan UL, Haq M. Nanostructured coatings: Review on processing techniques, corrosion behaviour and tribological performance. Nanomaterials. 2022;12:1323.
43. Jampilek J, Placha D. Advances in use of nanomaterials for musculoskeletal regeneration. Pharmaceutics. 2021;13:1994.
44. Kara HT, Anshebo ST, Sabir FK. A novel modified cellulose nanomaterials (CNMs) for remediation of chromium (VI) ions from wastewater. Materials Research Express. 2020;7:115008.
45. Kurapati SK, Mahendar Reddy N, Sujithra R, Kola R, Ramesh GV, Saritha D. Nanomaterials and nanostructures in additive manufacturing: Properties, applications, and technological challenges. Nanotechnology. 2023;1:53–102.

46. Liu H, van Nooten P, Deng L, Cui W. Clinical translation of nanomaterials. Theranostic Bionanomaterials. 2019:75–111.
47. Liu K, Du H, Zheng T, Liu W, Zhang M, Liu H, Zhang X, Si CJGC. Lignin-containing cellulose nanomaterials: Preparation and applications. Green Chemistry. 2021;23:9723–46.
48. Maduraiveeran G, Jin W. Nanomaterials based electrochemical sensor and biosensor platforms for environmental applications. Trends in Environmental Analytical Chemistry. 2017;13:10–23.
49. Manawi YM, Samara A, Al-Ansari T, Atieh MA. A review of carbon nanomaterials' synthesis via the chemical vapor deposition (CVD) method. Materials. 2018;11:822.
50. Nasiri S, Khosravani MR. Machine learning in predicting mechanical behavior of additively manufactured parts. Journal of Materials Research Technology. 2021;14:1137–53.
51. Mishra RK, Ha SK, Verma K, Tiwari SK. Recent progress in selected bio-nanomaterials and their engineering applications: An overview. Journal of Science: Advanced Materials Devices. 2018;3:263–88.
52. Osterberg M, Henn KA, Farooq M, Valle-Delgado JJ. Biobased nanomaterials: The role of interfacial interactions for advanced materials. Chemical Reviews. 2023;123:2200–41.
53. Sandhu K, Singh S, Hussain CM. 3D printing of nanomaterials using inkjet printing. Additive Manufacturing with Functionalized Nanomaterials. Elsevier; 2021:155–92.
54. Soares LMM. Environmental assessment of biomass pretreatment and processing towards bio-nanomaterials and bio-polymers enablers. Additive Manufacturing. 2022;28.
55. Solanki A, Kim JD, Lee K-B. Nanotechnology for regenerative medicine: Nanomaterials for stem cell imaging. Additive Manufacturing. 2008;27.
56. Teodoro KB, Sanfelice RC, Migliorini FL, Pavinatto A, Facure MH, Correa DS. A review on the role and performance of cellulose nanomaterials in sensors. ACS Sensors. 2021;6:2473–96.
57. Velu R, Calais T, Jayakumar A, Raspall F. A comprehensive review on bio-nanomaterials for medical implants and feasibility studies on fabrication of such implants by additive manufacturing technique. Materials. 2019;13:92.
58. Yang G, Kong H, Chen Y, Liu B, Zhu D, Guo L, Wei G. Recent advances in the hybridization of cellulose and carbon nanomaterials: Interactions, structural design, functional tailoring, and applications. Carbohydrate Polymers. 2022;279:118947.
59. Zhao Y, Li J, Yu Q, Li KD, Li Q, Zhou R, Zhou R, Ek M, Ostrikov KK. Fabrication of multidimensional bio-nanomaterials from nanocellulose oxalate. Cellulose. 2023;30:2147–63.
60. Zhu X, Li Y, Gu N. Application of artificial intelligence in the exploration and optimization of biomedical nanomaterials. Nano Biomedicine Engineering. 2023;15.
61. Yilmaz NF, Kurt HI, Oduncuoglu M, Ergul E. Experimental and theoretical analysis of the welding process parameters for UTS with different methods. Materials Research Express. 2018;6:016524.
62. Yilmaz O, Erdogan M, Karakaya I. Combined effects of ALS and SLS on Al_2O_3 reinforced composite nickel coatings. Surface Engineering. 2020;36:477–84.
63. Yin H, He Y, Sun Q. Effect of deformation frequency on temperature and stress oscillations in cyclic phase transition of NiTi shape memory alloy. Journal of the Mechanics Physics of Solids. 2014;67:100–28.
64. Yin S, He P, Liao H, Wang X. Deposition features of Ti coating using irregular powders in cold spray. Journal of Thermal Spray Technology. 2014;23:984–90.
65. Yin X, Bai Y, Zhou S-j, Ma W, Bai X, Chen W-d. Solubility, mechanical and biological properties of fluoridated hydroxyapatite/calcium silicate gradient coatings for orthopedic and dental applications. Journal of Thermal Spray Technology. 2020;29:471–88.

66. Prashar G, Vasudev H, Mehta A. Application of bimodal stuructured thermal sprayed coatings in power plant components. Advanced Ceramic Coatings for Energy Applications. 2024:273–87.
67. Yohe ST, Freedman JD, Falde EJ, Colson YL, Grinstaff MW. A mechanistic study of wetting superhydrophobic porous 3D meshes. Advanced Functional Materials. 2013;23:3628–37.
68. Yoo B-U, Kwon S-C, Cho S-H, Nersisyan HH, Lee J-H. High-temperature stability of CSZ, YSZ, and MSZ ceramic materials in $CaCl_2$-CaF_2-CaO molten salt system. Journal of Alloys and Compounds. 2019;771:924–35.
69. Yoo J, Shoaran M. Neural interface systems with on-device computing: Machine learning and neuromorphic architectures. Current Opinion in Biotechnology. 2021;72:95–101.
70. Yoo M, Na Y, Song H, Kim G, Yun J, Kim S, Moon C, Jo K. Motion estimation and hand gesture recognition-based human–UAV interaction approach in real time. Sensors. 2022;22:2513.
71. You D, Gao X, Katayama S. Review of laser welding monitoring. Science Technology of Welding Joining. 2014;19:181–201.
72. You Y, Zhang Z, Hsieh C-J, Demmel J, Keutzer K. Fast deep neural network training on distributed systems and cloud TPUs. IEEE Transactions on Parallel Distributed Systems. 2019;30:2449–62.
73. You Y, Zhao Y. Discrete element modelling of ellipsoidal particles using super-ellipsoids and multi-spheres: A comparative study. Powder Technology. 2018;331:179–91.
74. Farshidianfar MH, Khodabakhshi F, Khajepour A, Gerlich AP. Closed-loop control of microstructure and mechanical properties in additive manufacturing by directed energy deposition. Materials Science Engineering: A. 2021;803:140483.
75. Li Y, Li X, Zhang G, Horváth I, Han Q. Interlayer closed-loop control of forming geometries for wire and arc additive manufacturing based on fuzzy-logic inference. Journal of Manufacturing Processes. 2021;63:35–47.
76. Fidan I, Naikwadi V, Alkunte S, Mishra R, Tantawi K. Energy efficiency in additive manufacturing: Condensed review. Technologies. 2024;12:21.
77. Djafari Petroudy SR, Chabot B, Loranger E, Naebe M, Shojaeiarani J, Gharehkhani S, Ahvazi B, Hu J, Thomas S. Recent advances in cellulose nanofibers preparation through energy-efficient approaches: A review. Energies. 2021;14:6792.
78. Dong K, Wang ZL. Self-charging power textiles integrating energy harvesting triboelectric nanogenerators with energy storage batteries/supercapacitors. Journal of Semiconductors. 2021;42:101601.
79. Drakos I, Kenny P, Fearn T, Speller R. Multivariate analysis of energy dispersive X-ray diffraction data for the detection of illicit drugs in border control. Crime Science. 2017;6:1–10.
80. Dwivedi SP, Sharma S. Metallic cladding through microwave energy of the mixture of Ni and 15% SiC powder on AISI 304: A green approach in surface engineering. Proceedings of the Institution of Mechanical Engineers, Part E: Journal of Process Mechanical Engineering. 2024;238:2545–56.
81. Gkoutzamanis V, Chatziangelidou A, Efstathiadis T, Kalfas A, Travers A, Chiu JN. Thermal energy storage for gas turbine power augmentation. Journal of the Global Power Propulsion Society. 2019;3:592–608.
82. Han Y, Wang W, Zou J, Li Z, Cao X, Xu S. Self-powered energy conversion and energy storage system based on triboelectric nanogenerator. Nano Energy. 2020;76:105008.
83. Li S, Deng Y-Q, Zhu Z-L, Hua H-L, Tao Z-Z. A comprehensive review on radiomics and deep learning for nasopharyngeal carcinoma imaging. Diagnostics. 2021;11:1523.
84. Fraga-Lamas P, Lopes SI, Fernández-Caramés TM. Green IoT and edge AI as key technological enablers for a sustainable digital transition towards a smart circular economy: An industry 5.0 use case. Sensors. 2021;21:5745.

85. Fraga-Lamas P, Ramos L, Mondéjar-Guerra V, Fernández-Caramés TM. A review on IoT deep learning UAV systems for autonomous obstacle detection and collision avoidance. Remote Sensing. 2019;11:2144.
86. Saxena A, Mehta A, Vasudev H, Prashar G, Jasim NY. A short review on machine learning for the purpose of optimizing and predicting the properties of polymeric nanocomposites. Materials Today: Proceedings. 2023.
87. Singh G, Vasudev H, Arora H. A short note on the processing of materials through microwave route. Advances in Materials Processing: Select Proceedings of ICFMMP 2019. Springer Singapore, 2020: 101–111.
88. Kumar V, Sharma NK, Mittal A, Verma P. The role of IoT and IIoT in supplier and customer continuous improvement interface. Digital Transformation and Industry 40 for Sustainable Supply Chain Performance. Springer; 2023:161–74.
89. Anjaneya G, Sunil S, Kakkeri S, Math MM, Vaibhav MN, Solaimuthu C, Durga Prasad C, Vasudev H. Numerical simulation of microchannel heat exchanger using CFD. International Journal on Interactive Design and Manufacturing (IJIDeM). 2023:1–17.
90. Arora H, Kumar V, Prakash C, Pimenov D, Singh M, Vasudev H, Singh V. Analysis of sensitization in austenitic stainless steel-welded joint. Advances in Metrology and Measurement of Engineering Surfaces: Select Proceedings of ICFMMP 2019. Springer Singapore; 2021:13–23.
91. Bhowmik A, Kumar R, Beemkumar N, Kumar AV, Singh G, Kulshreshta A, Santhosh AJ. Casting of particle reinforced metal matrix composite by liquid state fabrication method: A review. Results in Engineering. 2024;103152.
92. Dutta V, Thakur L, Singh B, Vasudev H. A study of erosion–corrosion behaviour of friction stir-processed chromium-reinforced NiAl bronze composite. Materials. 2022;15(15):5401.
93. Sharanabasava H, Prasad CD, Ramesh MR. Microstructure, mechanical and wear properties of SiC and Mo reinforced NiCr microwave cladding. Advances in Materials and Processing Technologies. 2023;10(4):3620–33. https://doi.org/10.1080/2374068X.2023.2257937.
94. Kapade TT, Shah SK, Shah DS, Kumar NS, Bhor JK, Panwar VS, Singh G. Thermal analysis of v-cut twisted tape fitted heat exchanger: A numerical approach. International Journal on Interactive Design and Manufacturing (IJIDeM). 2024;15:1–7.
95. Kattimani MA, Venkatesh PR, Masum H, Math MM, Bahadurdesai VN, Mustafkhadri S, Prasad CD, Vasudev H. Design and numerical analysis of tensile deformation and fracture properties of induction hardened inconel 718 superalloy for gas turbine applications. International Journal on Interactive Design and Manufacturing (IJIDeM). 2024;18(8):5939–49.
96. Kulkarni SD, Manjunatha CU, Manjunath KV, Prasad CD, Vasudev H. Design and optimization of polyvinyl-nitride rubber for tensile strength analysis. International Journal on Interactive Design and Manufacturing (IJIDeM). 2024;18(8):5893–908.
97. Majji BGR, Vasudev H, Bansal A. A review on the oxidation and wear behavior of the thermally sprayed high-entropy alloys. Materials Today: Proceedings. 2022;50:1447–51.
98. Mahantesh MM, Rao KR, Chandra AP, Vijayakumar MN, Nandini B, Prasad CD, Vasudev H. Design and modeling using finite element analysis for the sitting posture of computer users based on ergonomic perspective. International Journal on Interactive Design and Manufacturing (IJIDeM). 2024;18(8):5875–91.
99. Mehta A, Vasudev H, Jeyaprakash N. Role of sustainable manufacturing approach: Microwave processing of materials. International Journal on Interactive Design and Manufacturing (IJIDeM). 2024;18(8):5283–99.
100. Mehta A, Singh G, Vasudev H. Processing of shape memory alloys research, applications and opportunities: A review. Physica Scripta. 2024. https://iopscience.iop.org/article/10.1088/1402-4896/ad48cb

101. Nagabhushana P, Ramprasad S, Prasad CD, Vasudev H, Prakash C. Numerical investigation on heat transfer of a nano-fluid saturated vertical composite porous channel packed between two fluid layers. International Journal on Interactive Design and Manufacturing (IJIDeM). 2024;18(5):2927–44.
102. Parkash J, Saggu HS, Vasudev H. A short review on the performance of high velocity oxy-fuel coatings in boiler steel applications. Materials Today: Proceedings. 2022;50:1442–46.
103. Poojari M, Hanumanthappa H, Prasad CD, Jathanna HM, Ksheerasagar AR, Shetty P, Shanmugam BK, Vasudev H. Computational modelling for the manufacturing of solar-powered multifunctional agricultural robot. International Journal on Interactive Design and Manufacturing (IJIDeM). 2024;18(8):5725–36.
104. Prashar G, Vasudev H. Surface topology analysis of plasma sprayed Inconel625-Al2O3 composite coating. Materials Today: Proceedings. 2022;50:607–11.
105. Prashar G, Vasudev H. A review on the influence of process parameters and heat treatment on the corrosion performance of Ni-based thermal spray coatings. Surface Review and Letters. 2022;29(1):2230001.
106. Prashar G, Vasudev H, Thakur L, Bansal A. Performance of thermally sprayed hydroxyapatite coatings for biomedical implants: A comprehensive review. Surface Review and Letters. 2023;30(1):2241001.
107. Rao GS, Mukkamala U, Hanumanthappa H, Prasad CD, Vasudev H, Shanmugam B, Kishore Kumar KC. Evaluating and optimizing surface roughness using genetic algorithm and artificial neural networks during turning of AISI 52100 steel. International Journal on Interactive Design and Manufacturing (IJIDeM). 2024;18(8):6151–60.
107. Singh G, Bansal A, Vasudev H. Corrosion and microstructural behaviour of Inconel 625 microwave clad deposited on mild steel. Thermal Claddings for Engineering Applications. CRC Press; 2024:51–62.

5 Durable Microwave Post-Processing of Thermal Spray Coatings

Gaurav Prashar and Hitesh Vasudev

5.1 INTRODUCTION

In the current landscape of engineering, companies are prioritizing environmentally friendly technological manufacturing innovations. Research focused on minimizing power consumption and enhancing efficiency remains a critical need within advanced manufacturing industries. The primary entities in the manufacturing sector are actively pursuing sustainability objectives. These objectives encompass the reduction of carbon emissions and the implementation of environmentally friendly manufacturing technologies. Microwave energy, utilized as microwave heating, has introduced new opportunities in materials manufacturing over the years. Microwaves are utilized in numerous industrial applications beyond food processing, facilitating the treatment of various materials. The microwave heating realm is currently transitioning from applications involving low temperatures to ultra-high-temperature functions.

Engineering components predominantly experience failure as a result of corrosion and wear when functioning in aggressive interacting surroundings. The causes of deterioration in the mechanical properties of the substrate are primarily due to oxidation, hot corrosion, and erosion, all of which contribute to material loss [1–4]. Corrosion is a significant destructive phenomenon that leads to the formation of microcracks, thereby compromising the mechanical strength of metal structures. This deterioration can result in hazardous accidents and substantial losses in terms of resources. Consequently, manufacturing new parts for each failed component is frequently not a practical solution. The primary challenge encountered by numerous engineering components is the significant expenditure required to meet their material demands, often amounting to millions. Surface engineering serves as a method to improve the surface characteristics of the target material, ensuring that its fundamental properties remain intact. The right modification of the surface is necessary through various thermal spray techniques, like plasma spray, detonation guns, high-velocity oxy fuel, and cold spray coatings [5]. However, achieving defect-free coatings using current thermal spray technology for developing homogeneous and dense coating structures is not feasible. For example, cold-sprayed coatings are consistently characterized by weak inter-splat bonding, more brittleness, and limited ductility. These limitations arise from the work hardening that occurs between

DOI: 10.1201/9781003546528-5

particles. To ensure that the as-sprayed coatings fulfill the performance requirements for structural components, it is necessary to conduct additional heat treatments to enhance the microstructure and mechanical properties [6]. Also, when HVOF is carried out in atmospheric conditions, air can easily get stuck within the coating region. This results in potential defects such as pores and voids which are located at the splat boundaries. These defects primarily influence the coatings' susceptibility to wear and corrosive environments. Un-melted particles are observable in the HVOF-deposited process, which additionally induces residual stresses in the deposited coating [7–10]. The thermal spray process must be executed in inert conditions to ensure that air does not infiltrate the coating. This approach is associated with high production costs for coatings [11, 12]. Laser processing and electron beam can both present several disadvantages, including elevated processing costs and significant maintenance expenses. Furthermore, the rapid cooling rate increases the likelihood of crack formation at the time of solidification and results in significant thermal distortion of the underlying base material. In contrast, conventional heat-treating methods are associated with lower costs but require extended processing times [13–15].

Heat treatment is typically performed to achieve a more uniform microstructure by removing splat boundaries and pores [16]. The studies in the literature have employed microwave processing to develop wear-resistant claddings [17]. In addition to its cost-effectiveness, microwave processing facilitated the development of dense claddings exhibiting excellent adherence to the substrate. The processing duration in microwave processing is considerably reduced in comparison to traditional furnace heating, resulting in a notable decrease in energy consumption [18]. Furthermore, microwave processing is advantageous due to its minimal setup cost. Additionally, microwave processing induces volumetric heating, which has the potential to enhance interfacial strength [17, 19–21]. Nevertheless, there exist a limited number of studies that have investigated the application of microwave processing for the post-processing of thermal spray coatings and its effects on the wear resistance [17, 22, 23]. Sharma et al. [24] initially documented the influence of microwave heating on thermal-sprayed ceramic coatings. Coatings processed by microwave exhibit greater densities compared to as-sprayed coatings. This chapter summarizes the different studies related to microwave post-processing of thermal spray coatings.

5.2 MICROWAVE FUSION PRINCIPLES

Microwaves represent a category of electromagnetic waves characterized by a frequency range extending from 0.3 to 300 GHz, with associated wavelengths varying from 1 meter to 1 millimeter. Similar to lasers, they exhibit high coherence and polarization. Microwave heating differs primarily from traditional heating processes. In typical thermal processing, the transfer of energy to the material occurs via three modes of heat transfer: conduction, convection, and radiation. Conversely, microwave energy is transmitted directly to materials via molecular interaction with the electromagnetic field. In the context of heat transfer with the traditional approach, energy is conveyed as a result of thermal gradients. However, microwave heating involves the conversion of electromagnetic energy into thermal energy, categorizing it as energy conversion instead of traditional heat transfer. The molecular structure influences the interaction of microwaves with materials and the transfer of energy. Microwave radiation enters and interacts with molecules, resulting in the transmission of electromagnetic energy all over the material's volume, which leads to volumetric heating. This method, in contrast to traditional processes, guarantees quick and consistent heating of the material. Hence the thermal gradient and heat flow in materials processed by microwave methods are opposite to those observed in materials subjected to conventional heating techniques [24].

5.2.1 Post-Treatment Parameters for Different Materials Using Microwaves

Over-processing led to a distorted substrate, whereas under-processing did not demonstrate adequate enhancement in the microstructure of the developed coatings. Hence time, temperature, power of the oven, and frequency for post-processing of the coatings may vary according to the coating materials. Ceramic composite coatings underwent exposure to microwave radiation at a frequency of 2.45 GHz for a duration of 30 minutes using a 1-KW domestic microwave oven. Due to the transparency of alumina-titania to microwaves at low temperatures, the spray deposits were positioned in an environment with a higher concentration of activated charcoal to enhance hybrid heating efficiency. During irradiation, charcoal initially interacts with microwaves, resulting in an increase in temperature. Upon exceeding the critical temperatures of the ceramics, a coupling with microwaves occurs, resulting in a rapid increase in temperature [24].

Ni and WC coatings underwent post-processing at optimized parameters to achieve microstructural homogenization [25]. The parameters of 2.45 GHz frequency and 900-Watt power were established to initiate the fusing process of the HVOF as-sprayed CoMoCrSi-Cr_3C_2, CoMoCrSi-WC + CrC + Ni, and CoMoCrSi-WC + 12Co coating within a duration of 10 minutes [26, 27]. In another study, flame-sprayed [28] and HVOF-sprayed [29] Ni-based coating underwent beneficial post-processing utilizing microwave hybrid heating. The sprayed coatings underwent microwave post-processing at a frequency of 2.45 GHz using a 1 KW industrial microwave applicator for a duration of 15 minutes. Matthews and Yang [30] carried out microwave post-processing of WC-Co and Cr_3C_2-NiCr carbide coatings at a frequency of 2.45 GHz using a 1.3 KW industrial microwave applicator for durations of 5, 10, 30, and 60 minutes.

Samples of cold-sprayed Cu-Al_2O_3 were subjected to heating using a microwave oven at power output of 900 W. The mechanical properties of developed coatings indicate that an appropriate microwave post-processing technique can significantly enhance both strength and toughness if an appropriate temperature is chosen during post-processing. The preliminary experiment indicates that overheating, defined as temperatures exceeding 500 °C, results in the collapse of the coating. Conversely, temperatures below 400 °C yield only marginal improvements in mechanical properties of cold spray deposits. When subjected to 450 °C, the maximum adhesion strength can be achieved for cold-sprayed Cu-Al_2O_3 coatings [31]. A schematic of the cold-spray process and microwave post-processing is shown in Figure 5.1.

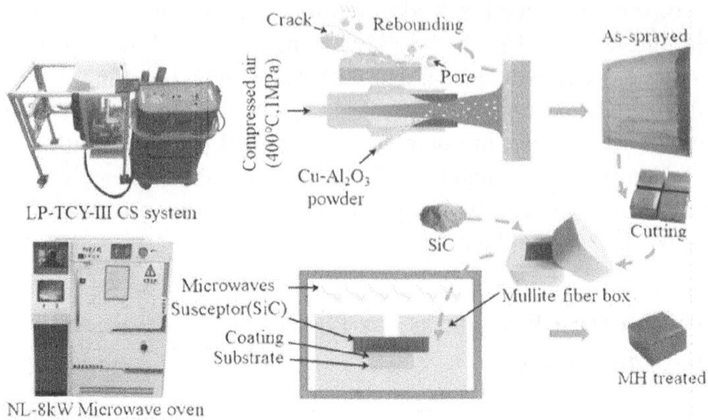

FIGURE 5.1 Schematic of cold-spray process and microwave post-processing [31].

5.2.2 Comparison of Microstructures of As-Sprayed and Microwave Post-Processed Coatings

Figure 5.2 illustrates the cross-sectional microstructure of cold-sprayed coatings following various microwave post-processing treatments. Prior to post-processing, the pores and microcracks are predominantly located in the inter-splat zones, and the coating/substrate interface exhibits weak mechanical bonding characteristics, as illustrated in Figure 5.2a, resulting from the embedding and bounce effects of Al_2O_3 hard particles. The imperfections in the coating and the unbonded interface region are repaired at 400°C, as illustrated in Figure 5.2b. The phenomenon is linked to atomic diffusion, which is caused by heat conduction from the SiC susceptor that absorbs microwave energy. Following the application of microwave 450°C treatment, there is a further reduction in both the quantity and size of pores. Interfacial metallurgical bonding is illustrated in Figure 5.2c. Consequently, the coating exhibits uniform distribution and experiences a rapid increase in temperature comparison to traditional heating methods. This process promotes densification, pore shrinkage, and good metallurgical bonding of cold-sprayed coatings [31]. A significant presence of irregular voids and pore development is observed at 500°C, as illustrated in Figure 5.2. The anti-densification issue is attributed to the Kirkendall effect, which is caused by the non-uniform diffusion rates of elements between copper and Al_2O_3.

Zafar and Sharma [28] present findings from studies into the microstructure of Ni-based coatings. The application of microwave post-processing resulted in the healing of pores and micro-cracks, leading to the homogenization of the microstructure within the coating layer. The microwave post-processing of Ni coatings demonstrates enhanced tribological properties in comparison to the as-sprayed Ni coating layer. Figure 5.3 illustrates the standard cross-sectional microstructure of the deposited and microwave processed samples. The average thickness of 300 μm demonstrates effective adhesion to the substrate. The as-sprayed coating displays a characteristic heterogeneous microstructure, featuring multiple micro-pores, as

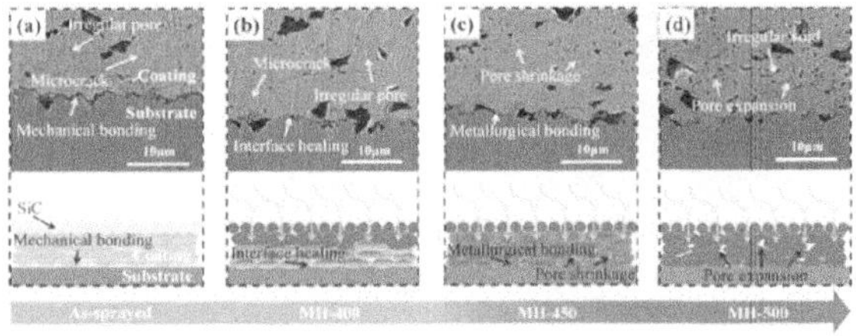

FIGURE 5.2 SEM micrographs and cross-sectional microstructure of cold-spray coatings with various microwave heating temperatures: (a) as-sprayed coating, (b) 400 °C, (c) 450 °C, and (d) 500 °C [31].

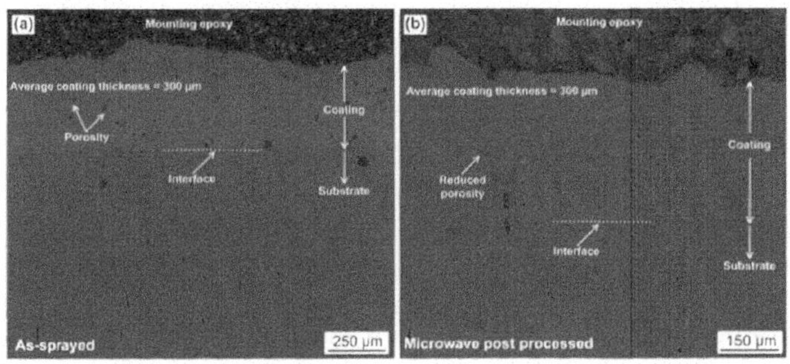

FIGURE 5.3 Cross-sectioned SEM images: (a) as-sprayed, and (b) microwave post-processed [28].

illustrated in Figure 5.3a, whereas the post-processed coating demonstrated a reduction in pore size and a denser microstructure when compared to the as-sprayed samples (Figure 5.3).

Prolonged exposure to microwave radiation resulted in the melting and flow of the splats, potentially leading to the occlusion of the pores. Consequently, the post-processed coupons demonstrated a microstructure that was both dense and homogeneous.

Matthews and Yang [30] investigated the application of microwave post-processing for WC-Co and Cr_3C_2-NiCr thermally sprayed coatings. The study compared the phase development rate resulting from this method to that from traditional furnace treatment. Coatings underwent successful microwave heat treatment, resulting in a phase composition equivalent to that achieved through furnace treatment. Both treatments produced similar outcomes in the Cr_3C_2-NiCr system. The WC-Co system demonstrated a significantly more crystalline structure in a considerably reduced time compared to the sample treated in a typical furnace. Prasad et al. [26] confirmed that microwave-processed HVOF coating demonstrated good metallurgical bonding, a homogeneous structure, reduced porosity, and enhanced hardness when compared to the as-sprayed samples. The microwave treated coating demonstrated superior wear properties compared to the as-sprayed coating. Figures 5.4a and b present the

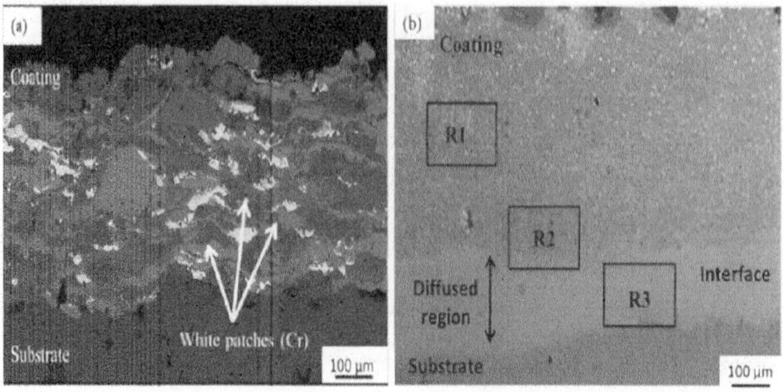

FIGURE 5.4 Cross-sectioned SEM micrographs: (a) as-sprayed, and (b) microwave post-processed [26].

SEM images of the as-coated and microwave-processed CoMoCrSi-Cr_3C_2 coating, respectively, as observed along the cross-section. Figure 5.4a for as-sprayed coating reveals various characteristics, including un-melted particles, imperfections, interfacial cracks, and porosity. The microwave heating approach demonstrated a beneficial impact on the properties of the HVOF-sprayed coating. Figure 5.4b illustrates that the fusing of coated samples results in a homogeneous, refined, and dense microstructure. During a period of 10 minutes, the microwave effectively re-melts the un-melted particles present in the as-sprayed coating. The small cracks, large pores, and lamellar structure have been eliminated. Metallurgical bonding was established between the substrate and the coating, resulting in enhanced adhesion to the substrate. Vasudev et al. [29] also have similar findings in their work related to microwave post-processing of HVOF coatings.

5.3 STUDIES RELATED TO MECHANICAL AND TRIBOLOGICAL PROPERTIES OF MICROWAVE-PROCESSED COATINGS

Sharma and Krishnamurthy [24] conducted an evaluation of the erosive wear behavior of microwave-processed ceramic coatings. The report indicates a reduction in porosity (25.9%) and an enhancement in both wear resistance (33.33%) and micro-hardness (12.31%) values after post-processing. Zafar and Sharma [28] also documented the enhancement of mechanical and tribological properties in microwave-processed coatings (Ni-based) applied to SS-304 grade steel. An enhancement in micro-hardness of 18% and an increase in fracture toughness of 35% were noted in the coatings processed using microwave technology. A decrease in porosity of 8% and a reduction in surface roughness of 44% were observed in the coatings after post-processing. Vasudev et al. [29] documented enhancements in the mechanical and tribological characteristics of microwave-processed HVOF coatings based on Superalloy-718, which were applied to a grey cast iron base material. Grey cast iron is a widely utilized material for the fabrication of machine tool components, such as beds and similar structures. HVOF coatings of Superalloy-718, approximately

250 μm in thickness, were applied to the surface of grey cast iron using a thermal spray gun to enhance their surface properties. The occurrence of micro-porosity is an inherent characteristic linked to HVOF coatings. Consequently, post-processing is conducted via the industrial microwave method. The XRD results confirmed the densification of coatings through the generation of lattice strains within the processed Superalloy-718 coatings. The coatings processed using microwave technology exhibited superior fracture toughness, enhanced micro-hardness, and increased abrasive wear resistance compared to the as-sprayed coatings. Prasad et al. [32] conducted an investigation into the tribological characteristics, specifically the abrasive wear behavior, of microwave-processed HVOF coatings. The coatings processed using microwave technology demonstrated adhesion strength of approximately 72.65 MPa. At temperatures of 200 °C, 400 °C, and 600 °C, wear rate is reduced by factors of 6, 8, and 10, respectively, compared to the as-sprayed coatings.

Microwave post-processing was first proposed by Qian et al. [31] in 2024 to enhance the properties of cold-sprayed Cu-Al_2O_3 composite coatings. The porosity value of as-sprayed coupons drops substantially from 1.32% to 0.75% following an microwave treatment, indicating a transition to metallurgical bonding among particles. The adhesion strength of treated samples was reported to be 43.63 MPa, and bending strengths rose to 1062 MPa following microwave treatment. The findings indicate that microwave treatment possesses significant application prospective for the cold-sprayed coatings. A comparison of different mechanical properties for different coatings before and after microwave treatment is summarized in Table 5.1.

5.4 CHALLENGES AND FUTURE PERSPECTIVES

The microwave post-processing treatment of thermal spray coatings involves various challenges.

Material Compatibility: The response of thermal spray materials to microwave energy varies significantly. Identifying appropriate materials that improve microwave absorption and heating is essential.

Process Optimization: It is essential to develop optimized microwave processing parameters, including power levels and exposure times and temperatures, to prevent substrate damage and attain the desired coating properties. Over-processing led to a distorted substrate, whereas under-processing did not demonstrate adequate enhancement in the microstructure of the developed coatings [31].

Insufficient data: Due to the insufficient availability of experimental data at this time, modeling and simulation of the model present challenges [33].

The discipline of microwave processing is witnessing substantial growth, but the limited dimensions of the microwave applicator restrict the capabilities of microwave systems. Nonetheless, the size constraint remains, although multiple cavities can be utilized to process several small parts simultaneously.

Material Development: Investigating novel materials that demonstrate enhanced microwave absorption and interaction may result in superior coating performance.

TABLE 5.1
Comparison of Different Mechanical Properties for Different Coatings before and after Microwave Treatment

Coatings	Porosity (As-Sprayed) %	Porosity (Microwave Fused) %	Micro-Hardness (As-Sprayed) Hv	Micro-Hardness (Microwave Fused) Hv	Fracture Toughness (As-Sprayed) Mpa $M^{1/2}$	Fracture Toughness (Microwave Fused) Mpa $M^{1/2}$	Adhesion Strength (As-Sprayed) Mpa	Adhesion Strength (Microwave Fused) Mpa
CoMoCrSi-WC + CrC + Ni [27]	1.4	0.4	150	1245	3.77	5.04	60.81	77.6
CoMoCrSi-WC+ 12Co [27]	1.2	0.25	1290	1410	4.23	5.56	64.22	79.45
Ni-based powder [28]	23	15	548	645	3.4	4.6		
Alloy-718 [29]			560	665	3.7	4.4		
Cu-Al_2O_3 [31]	1.32	0.75					11.36	43.63

Hybrid Processes that integrate microwave processing with additional techniques, such as laser treatment, have the potential to improve efficiency and broaden the range of applications.

Application Expansion: Increased utilization across sectors, including aerospace, automotive, and medical devices, where improved coating characteristics can deliver notable enhancements in performance.

Microwave Processing has the potential to decrease energy consumption and emissions when compared to traditional post-processing methods, thereby supporting sustainability objectives in manufacturing.

Microwave Heating in the future presents a viable method for an environmentally friendly powder casting technique aimed at the development of different metal matrix composites, which exhibit improved structure-property correlations.

Volumetric Additive Manufacturing (Vl:Ad.Mfg) technology utilizing visible light has most recently facilitated 3D printing in just one step. There is currently significant interest in broadening the design parameters of Vl:Ad.Mfg. Microwave energy exhibits greater penetration capabilities compared to visible light across a wider range of materials. To effectively utilize microwaves for Vl:Ad.Mfg, it is essential to possess a foundational understanding of the dielectric properties of materials, as well as the propagation of microwave fields [34]. Microwaves have the capability to penetrate materials more effectively than projected light, positioning them as a suitable option for curing various types of resins [35–46]. This capability has the potential to greatly improve the flexibility of 3D printing, enabling the production of more complicated, functional, and bigger components. In recent years, significant research efforts have been directed towards additive manufacturing [47–59]. This emerging method offers significant potential for the manufacturing of composites, eliminating the constraints associated with traditional forming resources and the complexities of multi-step planning processes. The standard composite production method involves intense manual labor, costly production equipment, and significant material waste, all of which can be eliminated. Microwave post-processing has emerged as an innovative and sustainable approach for enhancing the performance of thermal spray coatings by offering superior mechanical and corrosion-resistant properties. Recent studies have demonstrated the potential of microwave energy in modifying the microstructure and improving the adhesion of thermal spray coatings, such as Ni-based and hydroxyapatite coatings, on diverse substrates [60–68]. The role of heat treatment in enhancing the corrosion resistance of Ni-based coatings has been well documented, while the wear and oxidation performance of high-entropy alloys in thermally sprayed coatings have also been explored. Microwave-assisted cladding not only enhances wear resistance but also provides a cost-effective route for large-scale industrial applications. Additionally, advancements in microwave cladding of composite materials, such as Ni-WC and Al_2O_3-reinforced coatings, have demonstrated substantial improvements in tribological and mechanical performance. This chapter underscores the transformative potential of microwave post-processing techniques in achieving durable, efficient, and environmentally sustainable solutions for thermal spray coatings [69–73].

5.5 CONCLUSION

Microwave post-processing is regarded as an innovative method for the post-treatment of surface coatings and is also applicable for enhancing the surface characteristics of soft metals. Post-processing via microwave route yields effective metallurgical bonding with the base material. This technique has demonstrated significant potential and offers viable alternatives for the post-processing of coatings. The popularity of this technique can be attributed to its numerous advantages such as efficient energy utilization and better mechanical and metallurgical properties. The microwave heating procedure demonstrates sustainability; however, numerous future applications in the discipline of microwave material processing equipment remain under development. It is anticipated that scientific groups will introduce innovative ideas and techniques to fully leverage the endless applications of microwave heating in the near future.

REFERENCES

1. Prashar, G. and Vasudev, H., 2024. Erosion performance of direct-aged Al_2O_3-reinforced plasma-sprayed composite coatings at high-temperature conditions. *Surface Topography: Metrology and Properties*, *12*(3), p.035021.
2. Prashar, G. and Vasudev, H., 2024. High-temperature erosion behavior of Al_2O_3 reinforced Inconel625 high velocity oxy-fuel sprayed and direct-aged composite coatings. *Physica Scripta*, *99*(8), p.085939.
3. Prashar, G. and Vasudev, H., 2023. A review on the processing of various coating materials using surface modification techniques for high-temperature solid particle erosion applications. *Results in Surfaces and Interfaces*, p.100167.
4. Prashar, G. and Vasudev, H., 2023. High-temperature oxidation behavior of direct-aged bimodal Al_2O_3-reinforced Inconel625 plasma sprayed composite coatings. *Surface and Coatings Technology*, *475*, p.130156.
5. Vasudev, H., Singh, G., Bansal, A., Vardhan, S. and Thakur, L., 2019. Microwave heating and its applications in surface engineering: A review. *Materials Research Express*, *6*(10), p.102001.
6. Wu, D., Su, Y., Xu, Y., Li, W. and Yang, X., 2023. An induction heat treatment method for improving the mechanical properties and electrical conductivity of cold spray additive manufactured copper deposit. *Materials Letters*, *349*, p.134773.
7. Asl, S.K., Sohi, M.H., Hokamoto, K. and Uemura, M., 2006. Effect of heat treatment on wear behavior of HVOF thermally sprayed WC-Co coatings. *Wear*, *260*(11–12), pp.1203–1208.
8. Picas, J.A., Punset, M., Menargues, S., Martín, E. and Baile, M.T., 2015. Microstructural and tribological studies of as-sprayed and heat-treated HVOF Cr_3C_2–CoNiCrAlY coatings with a CoNiCrAlY bond coat. *Surface and Coatings Technology*, *268*, pp.317–324.
9. Matikainen, V., Bolelli, G., Koivuluoto, H., Sassatelli, P., Lusvarghi, L. and Vuoristo, P., 2017. Sliding wear behaviour of HVOF and HVAF sprayed Cr_3C_2-based coatings. *Wear*, *388*, pp.57–71.
10. Utu, I.D. and Marginean, G., 2017. Effect of electron beam remelting on the characteristics of HVOF sprayed Al_2O_3-TiO_2 coatings deposited on titanium substrate. *Colloids and Surfaces A: Physicochemical and Engineering Aspects*, *526*, pp.70–75.
11. Wang, Y., Li, C.G., Tian, W. and Yang, Y., 2009. Laser surface remelting of plasma sprayed nanostructured Al_2O_3–13wt% TiO_2 coatings on titanium alloy. *Applied Surface Science*, *255*(20), pp.8603–8610.

12. Guo, D.Z., Li, F.L., Wang, J.Y. and Sun, J.S., 1995. Effects of post-coating processing on structure and erosive wear characteristics of flame and plasma spray coatings. *Surface and Coatings Technology*, *73*(1–2), pp.73–78.
13. Prasad, C.D., Joladarashi, S., Ramesh, M.R. and Sarkar, A., 2018, April. High temperature gradient cobalt based clad developed using microwave hybrid heating. In *AIP Conference Proceedings* (Vol. 1943, No. 1). AIP Publishing.
14. Gupta, D. and Sharma, A.K., 2011. Investigation on sliding wear performance of WC10Co_2Ni cladding developed through microwave irradiation. *Wear*, *271*(9–10), pp.1642–1650.
15. Bhattacharya, M. and Basak, T., 2016. A review on the susceptor assisted microwave processing of materials. *Energy*, *97*, pp.306–338.
16. Guo, D.Z., Li, F.L., Wang, J.Y. and Sun, J.S., 1995. Effects of post-coating processing on structure and erosive wear characteristics of flame and plasma spray coatings. *Surface and Coatings Technology*, *73*(1–2), pp.73–78.
17. Babu, A., Arora, H.S., Singh, H. and Grewal, H.S., 2019. Microwave synthesized composite claddings with enhanced cavitation erosion resistance. *Wear*, *422*, pp.242–251.
18. Yuanzheng, Y., Youlan, Z., Zhengyi, L. and Yuzhi, C., 2000. Laser remelting of plasma sprayed Al_2O_3 ceramic coatings and subsequent wear resistance. *Materials Science and Engineering: A*, *291*(1–2), pp.168–172.
19. Nair, R.B., Arora, H.S., Mandal, P., Das, S. and Grewal, H.S., 2018. High-performance microwave-derived multi-principal element alloy coatings for tribological application. *Advanced Engineering Materials*, *20*(9), p.1800163.
20. Nair, R.B., Arora, H.S. and Grewal, H.S., 2019. Microwave synthesized complex concentrated alloy coatings: Plausible solution to cavitation induced erosion-corrosion. *Ultrasonics Sonochemistry*, *50*, pp.114–125.
21. Kaushal, S., Gupta, D. and Bhowmick, H., 2018. On surface modification of austenitic stainless steel using microwave processed Ni/Cr_3C_2 composite cladding. *Surface Engineering*, *34*(11), pp.809–817.
22. Babu, A., Arora, H.S., Behera, S.N., Sharma, M. and Grewal, H.S., 2018. Towards highly durable bimodal composite claddings using microwave processing. *Surface and Coatings Technology*, *349*, pp.655–666.
23. Prasad, C.D., Joladarashi, S., Ramesh, M.R., Srinath, M.S. and Channabasappa, B.H., 2018. Influence of microwave hybrid heating on the sliding wear behaviour of HVOF sprayed CoMoCrSi coating. *Materials Research Express*, *5*(8), p.086519.
24. Sharma, A.K., Aravindhan, S. and Krishnamurthy, R., 2001. Microwave glazing of alumina–titania ceramic composite coatings. *Materials Letters*, *50*(5–6), pp.295–301.
25. Babu, A., Arora, H.S. and Grewal, H.S., 2019. Microwave-assisted post-processing of detonation gun-sprayed coatings for better slurry and cavitation erosion resistance. *Journal of Thermal Spray Technology*, *28*(7), pp.1565–1578.
26. Prasad, C.D., Joladarashi, S., Ramesh, M.R., Srinath, M.S. and Channabasappa, B.H., 2019. Effect of microwave heating on microstructure and elevated temperature adhesive wear behavior of HVOF deposited CoMoCrSi-Cr_3C_2 coating. *Surface and Coatings Technology*, *374*, pp.291–304.
27. Prasad, C.D., Joladarashi, S., Ramesh, M.R., Srinath, M.S. and Channabasappa, B.H., 2020. Comparison of microstructural and sliding wear resistance of HVOF coated and microwave treated CoMoCrSi-WC+ CrC+ Ni and CoMoCrSi-WC+ 12Co composite coatings deposited on titanium substrate. *Silicon*, *12*(12), pp.3027–3045.
28. Zafar, S. and Sharma, A.K., 2017. Microstructure and mechanical properties of microwave post-processed Ni coating. *Journal of Materials Engineering and Performance*, *26*, pp.1382–1390.
29. Vasudev, H., Singh, P., Thakur, L. and Bansal, A., 2020. Mechanical and microstructural characterization of microwave post processed Alloy-718 coating. *Materials Research Express*, *6*(12), p.1265f5.

30. Matthews, S. and Yang, F., 2024. Comparison of microwave versus conventional furnace heat treatments of carbide composite thermal spray coatings. *Journal of Thermal Spray Technology*, pp.1–31.
31. Qian, Z., Zhang, Y., Xu, H., Cheng, H. and Huang, H., 2024. Improving microstructural and mechanical properties of cold-sprayed copper matrix composite coatings by microwave heating treatment. *Materials Letters*, p.136736.
32. Prasad, C.D., Joladarashi, S., Ramesh, M.R., Srinath, M.S. and Channabasappa, B.H., 2018. Influence of microwave hybrid heating on the sliding wear behaviour of HVOF sprayed CoMoCrSi coating. *Materials Research Express*, *5*(8), p.086519.
33. Srivastava, S.C., Murtaza, Q. and Kumar, P., 2020. Microwave cladding on metallic surfaces: A review. *Materials Today: Proceedings*, *21*, pp.1533–1536.
34. Mukherjee, S., Schwartz, J., Baluyot, E., Chang, T., Tringe, J.W., Spadaccini, C.M. and Shusteff, M., 2024. Towards microwave volumetric additive manufacturing: Generation of a computational multi-physics model for localized curing. *Additive Manufacturing Letters*, *10*, p.100209.
35. Prashar, G. and Vasudev, H., 2023. Understanding cold spray technology for hydroxyapatite deposition. *Journal of Electrochemical Science and Engineering*, *13*(1), pp.41–62.
36. Parkash, J., Saggu, H.S. and Vasudev, H., 2022. A short review on the performance of high velocity oxy-fuel coatings in boiler steel applications. *Materials Today: Proceedings*, *50*, pp.1442–1446.
37. Singh, J., Vasudev, H., Kumar, R. and Ubaidullah, M., 2024. Study on wear analysis of Ni-20Al_2O_3 HVOF micron layers using artificial neural network technique. *International Journal on Interactive Design and Manufacturing (IJIDeM)*, *18*(8), pp.5925–5938.
38. Singh, J., Vasudev, H., Szala, M. and Gill, H.S., 2024. Neural computing for erosion assessment in Al-20TiO_2 HVOF thermal spray coating. *International Journal on Interactive Design and Manufacturing (IJIDeM)*, *18*(4), pp.2321–2332.
39. Prakash, C., Gupta, L.R., Mehta, A., Vasudev, H., Tominov, R., Korman, E., Fedotov, A., Smirnov, V. and Kesari, K.K., 2023. Computing of neuromorphic materials: An emerging approach for bioengineering solutions. *Materials Advances*, *4*(23), pp.5882–5919.
40. Singh, M., Vasudev, H. and Kumar, R., 2021. Corrosion and tribological behaviour of BN thin films deposited using magnetron sputtering. *International Journal of Surface Engineering and Interdisciplinary Materials Science (IJSEIMS)*, *9*(2), pp.24–39.
41. Singh, M., Vasudev, H. and Singh, M., 2022. Surface protection of SS-316L with boron nitride based thin films using radio frequency magnetron sputtering technique. *Journal of Electrochemical Science and Engineering*, *12*(5), pp.851–863.
42. Singh, G., Singh, H., Sharma, Y., Vasudev, H. and Prakash, C., 2024. Analysis and optimization of various process parameters and effect on the hardness of SS-304 stainless steel welded joints. *International Journal on Interactive Design and Manufacturing (IJIDeM)*, *18*(5), pp.2889–2896.
43. Singh, G., Mehta, A. and Vasudev, H., 2024. Sustainability of additive manufacturing: A comprehensive review. *Progress in Additive Manufacturing*, *9*, pp.2249–2272.
44. Singh, G., Bansal, A., Vasudev, H. and Mishra, V., 2024. Sliding wear study of the Inconel-625 clad deposits by microwave heating on SS–304. *Physica Scripta*, *99*(6), p.065503.
45. Singh, J., Yadav, A., Gupta, S., Balu, R., Shukla, N., Singh, G. and Singh, S., 2024. Design and computational analysis of continuous wave fiber laser cutting of flat and conformal ITO PET sheets. *International Journal on Interactive Design and Manufacturing (IJIDeM)*, pp.1–9.
46. Singh, G., Singh, R. and Gul, J., 2023. Machinability behavior of human implant materials. *Journal of Electrochemical Science and Engineering*, *13*(1), pp.99–114.
47. Verma, P.K., Vasudev, H., Singh, G. and Singh, S., 2024. A review on the microstructural and biomedical properties of implants manufactured using additive manufacturing. *Journal of the Institution of Engineers (India): Series D*, pp.1–15.

48. Mehta, A., Singh, G. and Vasudev, H., 2024. Processing of shape memory alloys research, applications and opportunities: A review. *Physica Scripta*. https://iopscience.iop.org/article/10.1088/1402-4896/ad48cb
49. Singh, G., Bansal, A. and Vasudev, H., 2024. Corrosion and microstructural behaviour of Inconel 625 microwave clad deposited on mild steel. In *Thermal Claddings for Engineering Applications* (pp. 51–62). CRC Press.
50. Nagabhushana, P., Ramprasad, S., Prasad, C.D., Vasudev, H. and Prakash, C., 2024. Numerical investigation on heat transfer of a nano-fluid saturated vertical composite porous channel packed between two fluid layers. *International Journal on Interactive Design and Manufacturing (IJIDeM)*, *18*(5), pp.2927–2944.
51. Singh, P., Vasudev, H. and Bansal, A., 2024. Effect of post-heat treatment on the microstructural, mechanical, and bioactivity behavior of the microwave-assisted alumina-reinforced hydroxyapatite cladding. *Proceedings of the Institution of Mechanical Engineers, Part E: Journal of Process Mechanical Engineering*, *238*(3), pp.1060–1071.
52. Mahantesh, M.M., Rao, K.R., Chandra, A.P., Vijayakumar, M.N., Nandini, B., Prasad, C.D. and Vasudev, H., 2024. Design and modeling using finite element analysis for the sitting posture of computer users based on ergonomic perspective. *International Journal on Interactive Design and Manufacturing (IJIDeM)*, *18*(8), pp.5875–5891.
53. Singh, J., Singh, R., Singh, S., Vasudev, H. and Kumar, S., 2023. Reducing scrap due to missed operations and machining defects in 90PS pistons. *International Journal on Interactive Design and Manufacturing (IJIDeM)*, *17*(5), pp.2527–2539.
54. Arora, H., Kumar, V., Prakash, C., Pimenov, D., Singh, M., Vasudev, H. and Singh, V., 2021. Analysis of sensitization in austenitic stainless steel-welded joint. In *Advances in Metrology and Measurement of Engineering Surfaces: Select Proceedings of ICFMMP 2019* (pp. 13–23). Springer Singapore.
55. Vasudev, H., 2020. Wear characteristics of Ni-WC powder deposited by using a microwave route on mild steel: Microwave cladding of Ni-WC. *International Journal of Surface Engineering and Interdisciplinary Materials Science (IJSEIMS)*, *8*(1), pp.44–54.
56. Prashar, G., Vasudev, H., Thakur, L. and Bansal, A., 2023. Performance of thermally sprayed hydroxyapatite coatings for biomedical implants: A comprehensive review. *Surface Review and Letters*, *30*(1), p.2241001.
57. Mehta, A., Vasudev, H. and Jeyaprakash, N., 2024. Role of sustainable manufacturing approach: Microwave processing of materials. *International Journal on Interactive Design and Manufacturing (IJIDeM)*, *18*(8), pp.5283–5299.
58. Prashar, G. and Vasudev, H., 2023. Understanding cold spray technology for hydroxyapatite deposition. *Journal of Electrochemical Science and Engineering*, *13*(1), pp.41–62.
59. Kapade, T.T., Shah, S.K., Shah, D.S., Kumar, N.S., Bhor, J.K., Panwar, V.S. and Singh, G., 2024. Thermal analysis of v-cut twisted tape fitted heat exchanger: A numerical approach. *International Journal on Interactive Design and Manufacturing (IJIDeM)*. https://link.springer.com/article/10.1007/s12008-024-01883-2
60. Bhowmik, A., Kumar, R., Beemkumar, N., Kumar, A.V., Singh, G., Kulshreshta, A. and Santhosh, A.J., 2024. Casting of particle reinforced metal matrix composite by liquid state fabrication method: A review. *Results in Engineering*, p.103152.
61. Kulkarni, S.D., Manjunatha, C.U., Manjunath, K.V., Prasad, C.D. and Vasudev, H., 2024. Design and optimization of polyvinyl-nitride rubber for tensile strength analysis. *International Journal on Interactive Design and Manufacturing (IJIDeM)*, *18*(8), pp.5893–5908.
62. Prashar, G. and Vasudev, H., 2022. Surface topology analysis of plasma sprayed Inconel625-Al_2O_3 composite coating. *Materials Today: Proceedings*, *50*, pp.607–611.
63. Yedida, V.S. and Vasudev, H., 2022. A review on the development of thermal barrier coatings by using thermal spray techniques. *Materials Today: Proceedings*, *50*, pp.1458–1464.

64. Sharanabasava, H., Raviprakash, M., Prasad, C.D., Ramesh, M.R., Phanibhushana, M.V., Vasudev, H. and Kumar, S., 2023. Microstructure, mechanical and wear properties of SiC and Mo reinforced NiCr microwave cladding. *Advances in Materials and Processing Technologies*, *10*(4), 3620–3633. https://doi.org/10.1080/2374068X.2023.2257937.
65. Dutta, V., Thakur, L., Singh, B. and Vasudev, H., 2022. A study of erosion–corrosion behaviour of friction stir-processed chromium-reinforced NiAl bronze composite. *Materials*, *15*(15), p.5401.
66. Poojari, M., Hanumanthappa, H., Prasad, C.D., Jathanna, H.M., Ksheerasagar, A.R., Shetty, P., Shanmugam, B.K. and Vasudev, H., 2024. Computational modelling for the manufacturing of solar-powered multifunctional agricultural robot. *International Journal on Interactive Design and Manufacturing (IJIDeM)*, *18*(8), pp.5725–5736.
67. Prashar, G. and Vasudev, H., 2022. A review on the influence of process parameters and heat treatment on the corrosion performance of Ni-based thermal spray coatings. *Surface Review and Letters*, *29*(1), p.2230001.
68. Kattimani, M.A., Venkatesh, P.R., Masum, H., Math, M.M., Bahadurdesai, V.N., Mustafkhadri, S., Prasad, C.D. and Vasudev, H., 2024. Design and numerical analysis of tensile deformation and fracture properties of induction hardened inconel 718 superalloy for gas turbine applications. *International Journal on Interactive Design and Manufacturing (IJIDeM)*, *18*(8), pp.5939–5949.
69. Sunitha, K. and Vasudev, H., 2022. A short note on the various thermal spray coating processes and effect of post-treatment on Ni-based coatings. *Materials Today: Proceedings*, *50*, pp.1452–1457.
70. Majji, B.G.R., Vasudev, H. and Bansal, A., 2022. A review on the oxidation and wear behavior of the thermally sprayed high-entropy alloys. *Materials Today: Proceedings*, *50*, pp.1447–1451.
71. Rao, G.S., Mukkamala, U., Hanumanthappa, H., Prasad, C.D., Vasudev, H., Shanmugam, B. and KishoreKumar, K.C., 2024. Evaluating and optimizing surface roughness using genetic algorithm and artificial neural networks during turning of AISI 52100 steel. *International Journal on Interactive Design and Manufacturing (IJIDeM)*, *18*(8), pp.6151–6160.
72. Singh, J., Gill, H.S. and Vasudev, H., 2023. Computational fluid dynamics analysis on role of particulate shape and size in erosion of pipe bends. *International Journal on Interactive Design and Manufacturing (IJIDeM)*, *17*(5), pp.2631–2646.
73. Anjaneya, G., Sunil, S., Kakkeri, S., Math, M.M., Vaibhav, M.N., Solaimuthu, C., Durga Prasad, C. and Vasudev, H., 2023. Numerical simulation of microchannel heat exchanger using CFD. *International Journal on Interactive Design and Manufacturing (IJIDeM)*, pp.1–17.

6 Evaluation and Insulation Strategies of Heating Systems to Prevent Ice Formation in Cold Environments

Gaurav Prashar and Hitesh Vasudev

6.1 INTRODUCTION

The permafrost zones are seeing active industrialization right now, Arctic marine transport is aiming for new goals, and there's a rise in the need for low-temperature aerodynamically essential surfaces in wind generators and aerospace items. But with each passing year, the issue of ice on engineered structure surfaces becomes increasingly pressing. The aerospace and energy-generating sectors rank icing among their top concerns. The buildup of ice on the exterior of specific equipment or inside pipelines can lead to a number of serious issues that compromise their usefulness, efficiency, and safety. In modern times, the development of advanced structures and smart materials necessitates materials that can fulfill multiple roles, beyond their primary mechanical or functional purpose. These materials must be capable of simultaneously meeting various requirements for supporting a structure during its service life. One of the tasks involved in this project is to provide de/anti-icing for the working structural surfaces. The presence of ice accumulation on ocean-going ships has the potential to alter their balance and subsequently impact their safety. The performance of wind turbines is directly influenced by the condition of the blade surfaces. The purpose of this chapter is to present detailed information regarding coating-based de-icing systems used in pipelines, aircraft, and wind turbine blades. The chapter first provides a general overview of the background of these applications and then presents detailed descriptions of the coating-based heating systems used in these applications to melt ice. The chapter contains diagrams and provides a detailed discussion of the various components of the models that are being developed to showcase the de-icing capabilities of coating-based heating systems. This chapter summarizes the latest studies available on the de-icing of components using heating systems developed via different routes.

DOI: 10.1201/9781003546528-6

6.1.1 Pipelines

The occurrence and buildup of ice within enclosed piping systems are common occurrences in residential and industrial sectors in hilly areas of northern India and in Canada carrying fluids. This typically happens when the pipes are exposed to temperatures that are lower than the freezing point of the water contained within them. The negative effects of freezing, such as the possibility of pipe obstruction or rupture, have prompted numerous researchers from around the world to study the solidification of stagnant water in cylindrical pipes [1]. Gilpin [2] states that dendritic ice, which is thin, plate-like solid crystals, forms when supercooled water is present in pipes. Dendritic ice can build in water pipes and lead to early blockage in some situations, according to research [2]. Rezvani Rad et al. [3] studied the subject of water freezing and the subsequent pressurization of the liquid phase inside pipes, which causes the steel pipes to bulge and burst. The United States alone lost an estimated $450 million per year between 1985 and 1995 as a result of damage caused by frozen pipes [4]. U.S. monetary losses owing to ice, freezing, and other problems have averaged $1.2 billion per year over the previous 20 years [5]. Moreover the breakdown of piping systems caused by internal pressurization is not limited to metallic pipes alone. It can also occur in plastic piping systems [6].

In addition to the financial losses incurred, the bursting of pipes results in the release of a significant amount of stored energy. The occurrence of explosive ruptures, coupled with the potential ejection of steel fragments from compromised sections of pipes and pressure equipment, presents a significant hazard to the safety and welfare of workers. Given the multitude of adverse effects associated with this undesirable phenomenon, it is logical to prioritize the development of an innovative and efficient de-icing system. Such a system would aim to mitigate or eradicate the buildup of ice within pipes and pressure equipment, thereby addressing a significant concern.

6.1.2 Wind Turbine Blades

When considering the aerodynamic characteristics of a wind turbine blade, the leading edge is the most crucial component. Because of the icing process, it is also the area of the blade that gets the most ice accumulated. As a result, preventing or significantly decreasing ice accumulation in that section of the blade is obviously necessary. As a result of ice buildup on wind turbine blades, the machines have completely stopped working, the generator that powers the turbine has been damaged by overloading, and the blades' mechanical integrity has deteriorated [7–9]. Even when ice is just beginning to form, the bulk and aerodynamic imbalance of the turbine blades will be a problem, as demonstrated by Antikainen and Peuranen [10]. An increase in the excitation of edgewise vibrations, as well as increased fatigue and dynamic loads, will result from these imbalances [7, 11, 12]. Ice development on wind turbine blades is the root cause of these major difficulties; hence new heating technologies are desperately needed to reduce or eliminate them. An alternative way to fabricate heating components for fiber reinforced polymer composite (FRPC) site wind turbine blades exposed to cold regions could be through the use of thermal spray techniques.

6.1.3 Airplane Wings

When an airplane's wings are coated with ice, it alters their aerodynamic shape [13], which can harm the flying mechanical steering system [14]. Loss of control of the aircraft could occur in extreme circumstances [15]. Icing has been the root cause of numerous high-profile aviation disasters in the past few years [16]. In 2018, a Russian airplane, An-148, went down when pilots received inaccurate speed readings from an ice-covered airspeed tube. The propeller iced up in 2021, causing an accident in a General Aviation B-10GD Model Precipitation Enhancer, which killed all five crew members. The aerodynamic properties of the plane are significantly impacted by ice buildup issues [17]. A 30% drop in lift and a 40% increase in drag can result from ice buildups (on the leading edge or top surface of the wing) that are no thicker or rougher than coarse sandpaper. We desperately require new methods of de-icing. Heating and vibration are now the two most-used methods for de-icing.

Thermal, mechanical, and chemical active de-icing methods are the three main types [18]. To reduce the likelihood of ice formation, it is usual practice to circulate warm air within the airfoil [19, 20]. This technique works effectively for temperatures near freezing and in generally milder areas. This type of de-icing technology works well at higher temperatures, but it becomes much less effective at lower ones [21]. When compared to alternative ways, thermal de-icing procedures that use electrical heating components are superior [18, 21]. These heaters are either built into the airfoil itself or laminated onto the surface of the blade [18]. Surface engineering is a novel field where coating is applied as a heat tracer to avoid the freezing of water inside pipelines or on turbine blades. This solution serves as an appropriate strategy for protection when the components are exposed to cold environments for short periods of time [1].

6.1.4 Coating/Film-Based Heating Systems

The authors of [22] coated the surface of the FRPC with a nickel-chromium (Ni-Cr) and NiCrAlY layer by flame spray process in order to make it a heating element. Because of the protective sand-epoxy layer, applying the coating using high-temperature thermal technology did not compromise the FRPC laminate's integrity or mechanical qualities. Uniform heating was observed to be provided by the resultant coating. A cold chamber with a volume of 18 m^3 was used for the tests. The experiments were carried out at –25 and –5 °C (±3 °C) when the cold room was running and at 23 °C (±3 °C) when it wasn't. A closed conduit of 2 × 0.66 × 0.48 m was used to hold the samples, as illustrated in Figure 6.1. The results of the tests indicated that the surface temperature of the FRPC remained above 0 °C even after being chilled to –25 °C. Coatings with complicated geometries can be applied to FRPC surfaces using thermal spraying technology, and damaged coatings can be repaired using the same method. These results demonstrate that coatings applied by flame spray can be utilized as heating elements to reduce the buildup of ice on structures. The coatings were heated by applying an electrical current, which raised the surface temperature through joule heating. Joule heating happens when electrons come into contact with the conductive material's body. A voltage differential across

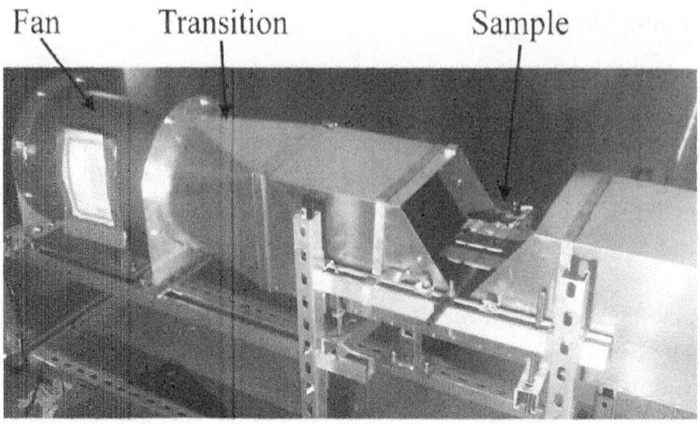

FIGURE 6.1 Experimental setup assembly [22].

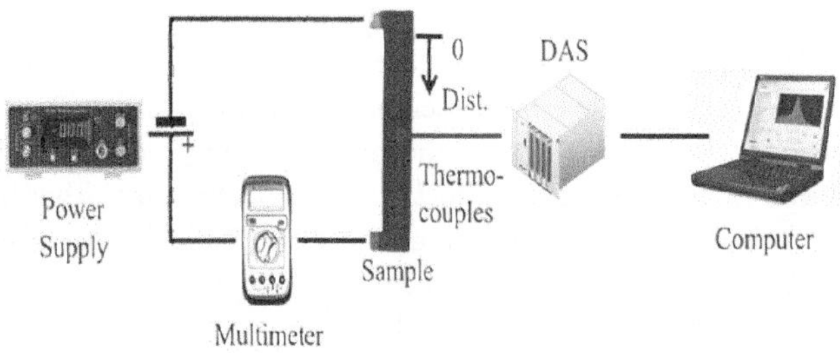

FIGURE 6.2 Experimental setup for conducting joule heating test [22].

two points in a conductor generates an electric field, which in turn imparts kinetic energy to charge carriers and causes them to accelerate in the direction of the field. Joule heating measurements were conducted using the experimental setup depicted in Figure 6.1.

To de-ice carbon steel pipes, heating systems with multiple layers of thermally sprayed coatings were devised by Rad et al. [23]. Researchers examined the viability of applying resistive heating coatings—specifically Ni-20Cr and Ni-50Cr—to carbon steel samples using flame spraying, air plasma spraying, and high-velocity oxygen fuel spraying techniques. To avoid short circuits and the flow of electrons between the layers, alumina was coated between the conductive resistive heating coating and the substrate using flame spraying or suspension plasma spraying. This was necessary due to the high electrical conductivity of carbon steel. In order to assess how well the coating systems that were developed performed, heating experiments were carried out in a cold room under varying climatic conditions. Figure 6.3 shows the finished bi-layered coating system with copper strips attached to the pipe section ends, as well as its overall structure and geometry. The heating element and substrate were protected from short circuits by depositing alumina on a longer stretch than Ni-50Cr.

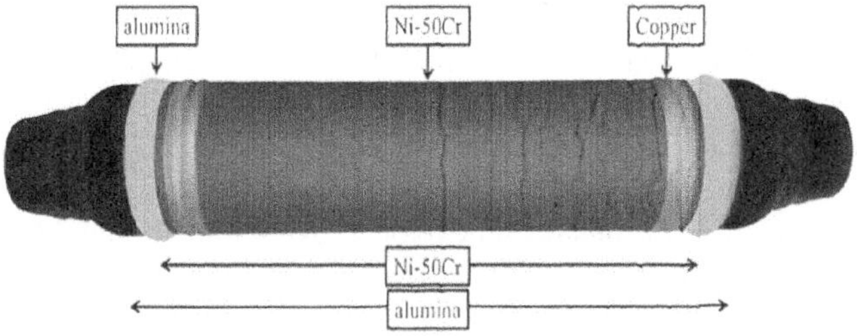

FIGURE 6.3 Coated with a functional bi-layered coating method, the pipe assembly was finished with cold-sprayed copper rings at both ends [23].

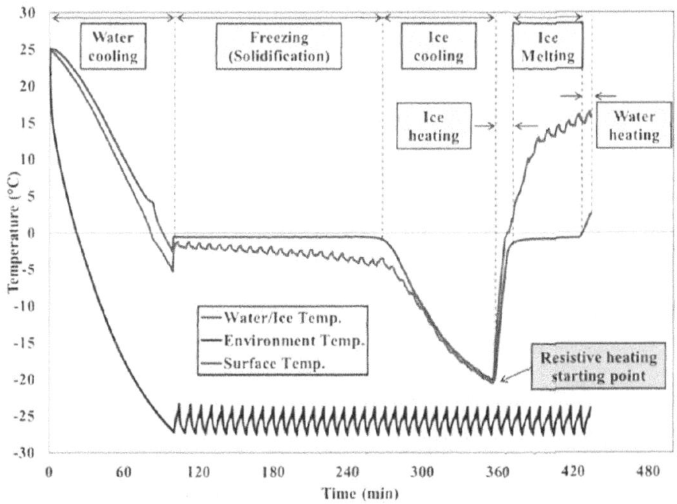

FIGURE 6.4 Results of heating test with power 80 W and voltage 20 V [23].

It was noted that 60 cm^3 of ice could be heated from –20 °C to 0 °C with just 20 watts and even more crucially melted when the pipe was placed directly in cold air at –25 °C. This demonstrates that the manufactured heating system on the pipe is much more efficient than the standard tracers [24]. The close contact of the heating element to the substrate is the main cause of this. On top of that, unlike the commonly utilized heat tracing wires, these heating systems are able to provide a uniform temperature distribution over large surfaces [25]. Figure 6.4 shows the results of the test with 80 W at 20 V, which involved the coated surface and ice/water inside the pipe, as well as their transient temperatures.

6.1.5 De-Icing in the Aviation Industry

Carbon fiber-reinforced polymer (CFRP) technologies have seen a rise in use within the aviation structural industry over the past few years. In their demonstration of an environmentally friendly, cost-effective, and simple method for de-icing aircraft,

Pang et al. highlight the use of induction heating. Electroless nickel plating is employed to generate a nickel-coated carbon fiber-reinforced polymer, which is then utilized as an induction heater for airplane de-icing. The results demonstrate that, with an input voltage of 30 V, the nickel-plated CFRP can be heated to 205 °C and a 30-mm-thick coating of ice can be melted in under 110 seconds [26]. A potential method for de-icing aircraft has been developed using nickel-plated CFRP polymer, which demonstrates a strong de-icing capacity and outstanding electrothermal characteristics. Figure 6.5 displays infrared photographs of the ice melting.

Using flexible graphene film/paper, the researchers for [27] developed a thermal de-icing/anti-icing method for airplane surfaces. Two layers of CFRP laminates held it in place (inset Figure 6.6). In comparison to conventional de-icing methods that use resistance heating pads, the power consumption of this flexible heater paper is as much as nine times lower, and it exhibits superior mechanical and thermal qualities. Experiments proved the de-icing performance was feasible, and the results show that the new electrical heating strategy could be a great way to keep things light while still getting the job done efficiently and reducing pollution from de-icing fluids. In less than 9 minutes, a coating of ice 1 mm thick can be melted with this method. Examining the film's or paper's joule temperature performance to determine its genuine potential as

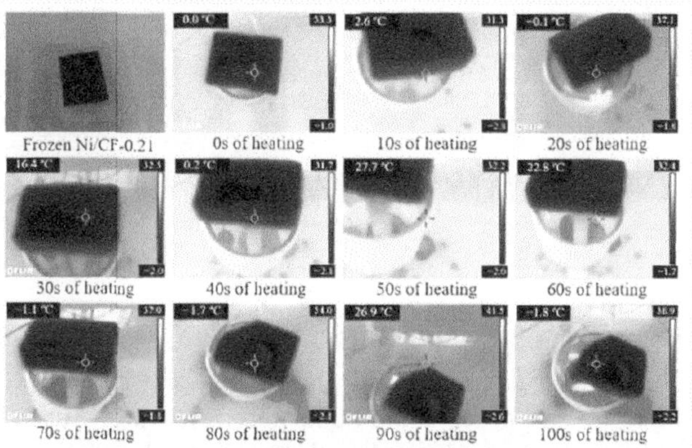

FIGURE 6.5 Infrared photographs of the ice melting [26].

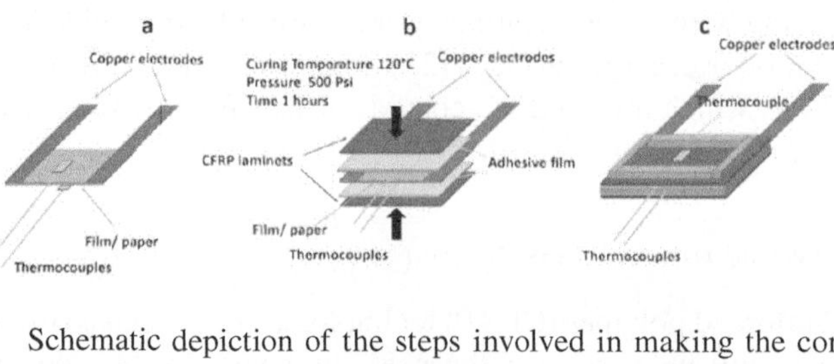

FIGURE 6.6 Schematic depiction of the steps involved in making the composite: (a) the chosen arrangement of electrodes and thermocouples; (b) the steps involved in making the composite; and (c) the final system configuration [27].

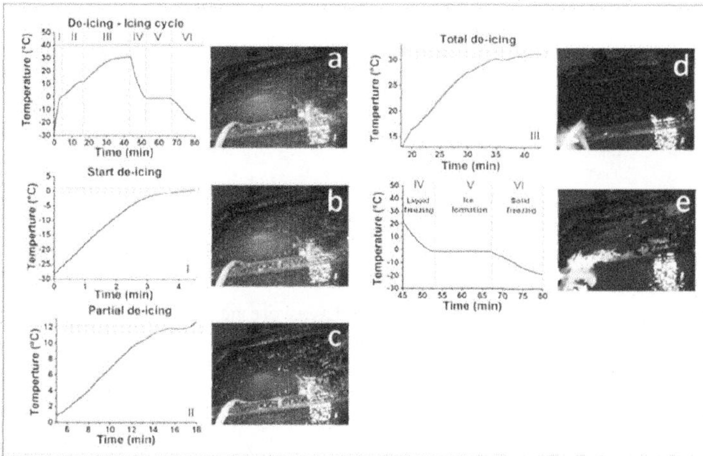

FIGURE 6.7 Measurement of the de-icing process (heat flux density 3706 W/m^2) using an endoscopic camera, with temperature as a function of time during power on-off of the formed composite at an ambient temperature of –32 °C [27].

an efficient de-icing method is a fascinating area of study. The outcomes of the heating and de-icing process are displayed in Figure 6.7. To visually observe the de-icing process, an endoscopic camera is inserted into the freezer. In the first step, known as Zone I, the temperature rose sharply upon power supply and then leveled off at 0 °C, marking the beginning of the de-icing process (inset Figure 6.7b). After holding at 0 °C for 1 minute, the temperature rose once more (Figure 6.7c). This shows that at this stage, the ice that was directly touching the composite melted, but the remaining ice stayed solid. The pace of temperature increase was lower in Zones II and III. All of the ice evaporated towards the end of Zone III. Liquid freezing (Zone IV), ice formation and solid freezing (Zone VI) are the corresponding zones.

Zhao et al. [28] developed an electric heating coating using multi-walled carbon nanotubes (MWCNTs) and a superhydrophobic coating on glass fiber reinforced polymer (GFRP) substrates through a spraying method. Experimental testing of the combined electric heating coating's anti-icing and de-icing capabilities, as well as its electrical conductivity, was conducted using the electric heating performance test and the de-icing performance test. The developed coating system has excellent anti-icing and de-icing performance, as well as high damage resistance, according to the experimental results. In Figure 6.8, we can see the process of making the test samples.

To confirm the same, a fan model consisting of a similar coating system was set. One of the fan blades had an S-EC coating, while the others were GFRP. As demonstrated in Figure 6.9a, the thermal infrared image of the S-EC applied to the surface of the fan blades reveals a highly homogeneous distribution of temperatures. Additionally, the results of the anti-icing test corroborate these claims. At a Pd of 0.35 W/cm^2, the S-EC-coated blade achieved anti-icing; in contrast, the other two blades had thick ice (Figure 6.9b). The versatile anti-icing properties and ease of manufacture of S-EC make it an attractive candidate for a broad range of anti-icing applications.

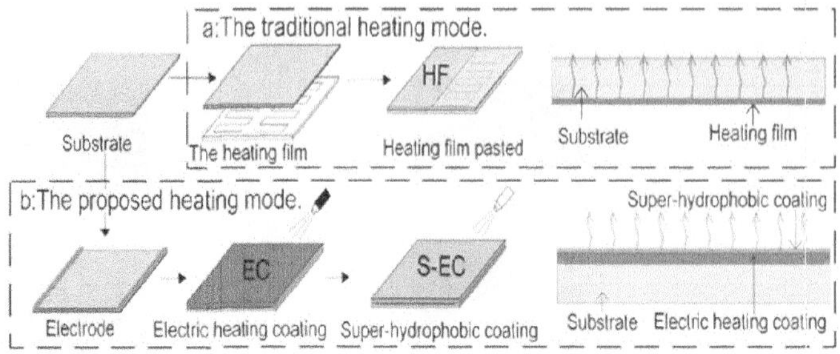

FIGURE 6.8 Two samples, (a) HF and (b) S-EC, were heated in different ways during the fabrication process, as shown in the schematic [28].

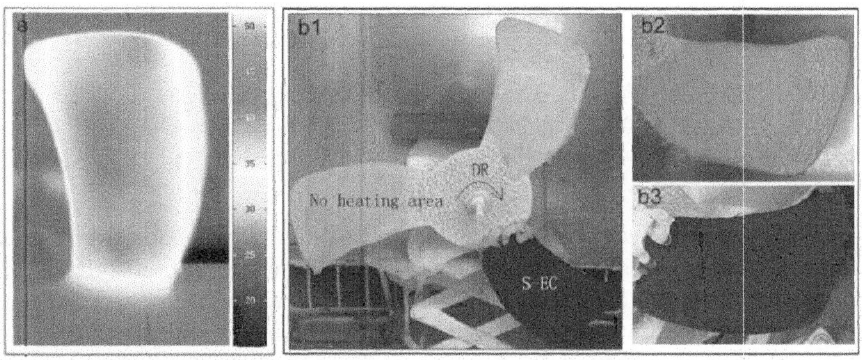

FIGURE 6.9 Coating the surface of a revolving blade. (a) S-EC coating as seen through thermal infrared imaging. Images taken digitally during an anti-icing test in low-icing conditions (b1) and (b2) non-heating, and S-EC–equipped (b3) blades. DR stands for the rotational direction [28].

In contrast to the internal heating approach, which consumed 58% more power, Zhao et al. [28] constructed an electric heating coating to be applied to the outside surfaces of the substrate in order to avoid the buildup of icing, but it is not suitable for complex surfaces. Thanks to recent advances in advanced manufacturing technology, a number of techniques for heating complicated surfaces have been developed [29–33]. Using electro-spinning and electroplating processes, Jo et al. [32] created a flexible heater made of copper fibers. This heater may be applied to surfaces with intricate structures. Nevertheless, these technologies' preparations are complex, which poses a significant obstacle to the actual application of materials. A further method involved coating or spraying nanomaterials directly onto a complex surface in order to heat them [34–38]. Using silver nanowires sprayed via supersonic beam, Lee et al. [34] created a heated coating that adheres to intricate surfaces. However, under realistic icing scenarios, silver nano-wire films can be damaged. Consequently, a practical and trustworthy way to accomplish anti-icing and de-icing duties for aircraft parts with intricate surfaces must be developed.

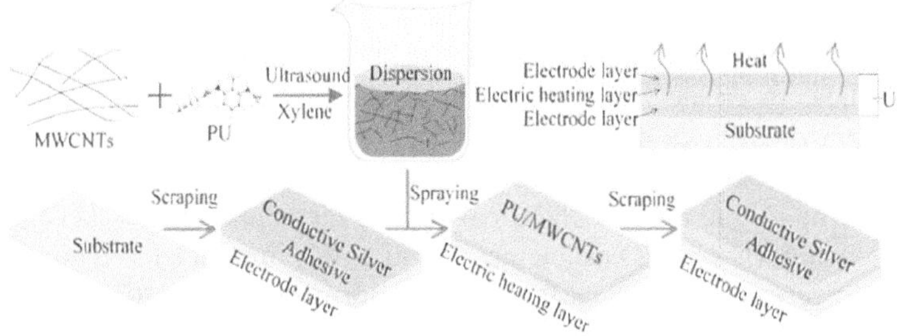

FIGURE 6.10 Steps involved in preparing SEHC [39].

To solve this problem, a spray-on structural electric heating coating (SEHC) was created by [39] for use on intricate surfaces. Electrode layers made with conductive silver adhesives make up the top and bottom layers of the SEHC, while a conductive polymer composite consisting of MWCNTs and a polyurethane resin matrix serves as the middle layer. Figure 6.10 shows the steps involved in preparing the SEHC [39].

Excellent electric heating characteristics, including responsiveness, repeatability, and reliability in the event of local damage, were demonstrated by the SEHC, in addition to its capacity to heat curved and irregular surfaces. These findings will contribute to the practical use of SEHCs in anti-icing and de-icing applications in industry.

6.2 FUTURE SCOPE

The study conducted by [22] revealed that the parameters, flame temperatures, porosity, and resulting oxide content significantly influenced the electrical resistance of manufactured heating elements. It has been suggested by the authors that the electrical resistivity of the heating element is highly sensitive to the spraying process and spray parameters chosen. For instance, according to the S1 set of spraying conditions, the resistivity of flame-sprayed Ni-20Cr was determined to be 4.7 Ω-μm. There was a 68% relative difference between the two sets of data, as 7.9 Ω-μm was determined for the resistivity of the identical material sprayed using the same technique but with different spraying parameters (S2). S1 spraying parameters had higher acetylene and oxygen contents than S2 spraying parameters. This led to a higher combustion flame temperature and a denser, more homogeneous microstructure. Ashrafizadeh et al. [40] found that as coating porosity is reduced, electrical resistance is reduced as well. In addition, the high electrical resistance of oxides, which are formed during flame-spraying, could significantly reduce the electron flow through the coating-based heating element [40]. However, the coating produced by cold spray (CS) has negligible porosity and no oxide content, which are prime requirements for good resistivity. The use of the cold spray process for fabricating heaters in the future will provide numerous benefits compared to thermal spray. These include cost savings in labor and a decrease in environmental impact, as there are no combustion gases produced during the process.

Metal matrix composites (MMCs) have shown great promise in a variety of industrial settings and have found extensive use as a result. Multiple studies have mainly concentrated on using MMCs for thermal management [41, 42]. The use of materials with thermo-mechanical properties within a certain range is essential in the electronics sector due to the excessive heat generated in various device components. Hence, to create heat-sink materials with specific thermal conductivity and coefficient of thermal expansion values, metal matrix composites could be a good choice [43]. Coatings with various ceramic and cermet reinforcing phases were manufactured by [44] to provide electrical heating elements using nickel chromium-aluminum-yttrium (NiCrAlY) as the metal matrix. The manufactured coatings were subjected to a series of joule heating tests to determine their heating effectiveness. This study's findings lend credence to the idea that coating-based de-icing systems can benefit from ceramic reinforcements added to MMC coatings, leading to greater heating efficacy and less energy usage. Further, in order to gauge the heating performance of the metal matrix composite coatings, joule heating tests were carried out. Figures 6.11 and 6.12 depict the higher de-icing performance of MMCs, namely NiCrAlY-alumina and NiCrAlY-titania coatings, as demonstrated by their respective steady-state temperatures of 15.1 °C and 8.4 °C in comparison to other coatings.

The literature has extensively examined the fabrication of heating systems using various thermal spraying methods, specifically involving Ni-Cr and NiCrAlY materials. However, to the best of the authors' knowledge, there has been no previous study on the utilization of CS processes to apply high-entropy alloy (HEA) coatings for the prevention of pipe clogging/bursting. Therefore, in the future, HEAs may be developed and will be tested for the effectiveness of HEA coating systems in heating and melting ice within steel pipes using the CS technique. HEAs are unique materials that consist of 5–13 elements, with 5%–35% molecular weight for each constituent. By decreasing demand for precious metals, these alloys may improve

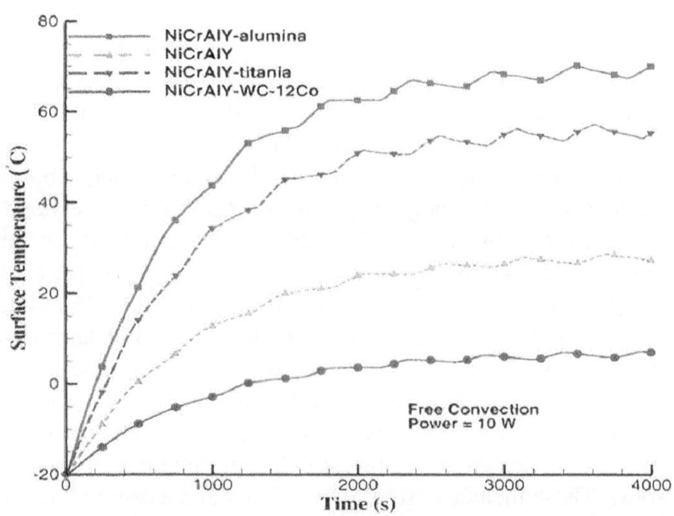

FIGURE 6.11 Middle section surface temperatures of various heating components vs time-free convection [44].

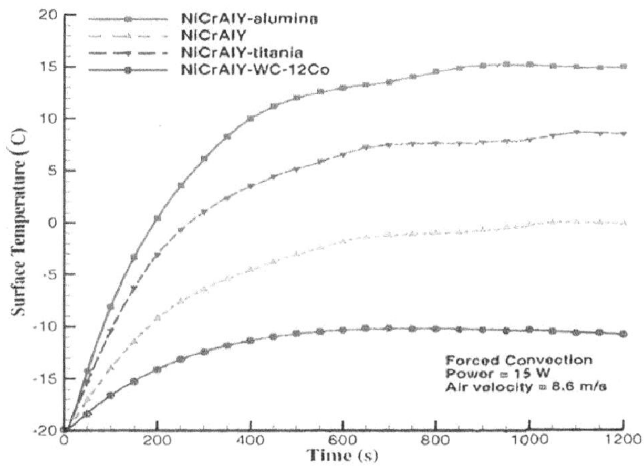

FIGURE 6.12 Middle section surface temperatures of various heating components vs time-forced convection [44].

economic efficiency. Many researchers have concentrated on creating HEA coatings for inexpensive substrate materials because of the remarkable properties of HEAs. Investigating HEAs and related fields has met the need for creating cutting-edge coating materials that work admirably in a variety of settings. Hence, HEAs aim to provide practical and efficient solutions for these applications.

6.2.1 Future Perspectives

Aircraft icing protection system research in the future should center on icing physics, with an emphasis on discovering new principles, cutting energy usage, creating innovative technology, and using novel materials [45–47]. Aircraft icing protection systems will become more efficient and use less energy in the future as more novel principles, technologies, and materials are explored and put into use [48–55]. Consequently, improving aircraft ice prevention systems so they are less harmful to the environment, more dependable, and more efficient will continue to be a significant task [56–61]. Future research on aircraft icing protection systems should prioritize a more comprehensive understanding of icing physics [62]. The focus should be on uncovering new principles, minimizing energy consumption, fostering the development of innovative technologies, and exploring the use of novel materials [63–69]. Extensive research on the thermodynamics and fluid dynamics of ice formation is essential in order to enhance the efficiency of anti-icing and de-icing techniques [70–72]. Research should focus on discovering fundamental principles that can result in significant advancements in passive and active anti-icing technologies [73–79]. Energy efficiency is of the utmost importance. Therefore, it is crucial to investigate low-energy de-icing solutions, such as smart coatings and surface modifications, in order to greatly minimize operational expenses and environmental consequences [80, 81]. Exploring innovative technologies such as electromagnetic, ultrasonic, and superhydrophobic surfaces shows great promise [82–86]. These technologies have the potential to prevent ice adhesion or facilitate its removal with minimal energy input [87–95]. In addition, the use of innovative materials such as nanocomposites and phase change materials can improve the effectiveness and longevity of icing protection systems [96–100]. Interdisciplinary research that combines materials science, aerodynamics, and thermal engineering will be crucial in achieving these objectives [101, 102]. Collaboration among academia, industry, and regulatory bodies will be crucial in driving the innovation required to address the future challenges of aircraft icing [103, 104]. Efficient insulation of heating systems plays a critical role in preventing ice formation in cold environments, where maintaining heat is essential for operational reliability [105–119]. Recent advancements in thermal spray coating techniques have provided new strategies for improving insulation properties. For example, the application of BN thin films on substrates using magnetron sputtering has been shown to enhance the corrosion resistance and tribological behavior of coatings, which can be beneficial in cold-weather applications requiring robust insulation against ice formation [120–131]. In addition, the development of high-entropy alloys and microwave heating for surface cladding has gained attention for its potential to improve the thermal efficiency and wear resistance of materials used in heating systems [132–137]. These techniques optimize heat retention, reduce the risk of freezing, and increase the overall durability of heating systems in challenging environments. Furthermore, computational modeling, including finite-element analysis, has been employed to evaluate the effectiveness of various insulation strategies in preventing ice buildup and improving system performance [138–140]. The combination of advanced coatings, optimized heating materials, and computational simulations offers a promising approach to designing efficient heating systems that can operate reliably in cold conditions, preventing ice formation while maintaining energy efficiency [141–143].

6.3 CONCLUSION

Typical de-icing methods are examined in this chapter, along with their current advancements, concepts, and downsides. The main points are summarized as follows:

- Accumulated ice on surfaces reduces wind turbine efficiency, disrupts flight, and increases the risk of serious injuries and financial losses.
- The parameters, flame temperatures, porosity, and resulting oxide content significantly influence the electrical resistance of manufactured heating elements via thermal spray routes.
- Aircraft icing protection systems will primarily encounter the challenge of standardizing experimental devices and test materials. Based on the available information, it appears that the anti-icing and de-icing systems utilized by aircraft use a variety of test materials, voltages, and frequency settings that are not standardized or unified. As a result, the effectiveness of these systems is compromised.
- HEA may be developed and will be tested for the effectiveness of HEA coating systems in heating and melting ice within steel pipes using the CS technique.

REFERENCES

1. Gordon, J.R., 1996. An investigation into freezing and bursting water pipes in residential construction. *Research Report 96–1.*
2. Gilpin, R.R., 1977. The effects of dendritic ice formation in water pipes. *International Journal of Heat and Mass Transfer*, *20*(6), pp.693–699.
3. Rezvani Rad, M., Marsden, R. and McDonald, A., 2018. Testing analysis of freezing phenomenon in conventional carbon steel pipes.
4. Vine, E., Mills, E. and Chen, A., 2000. Energy-efficiency and renewable energy options for risk management and insurance loss reduction. *Energy*, *25*(2), pp.131–147.
5. Carrns, A., 2014. Avoiding costly home damage from Winter's cold. *The New York Times*, December 2.
6. Edwards, D.B., Smith, K.M., Jr., Duvall, D.E. and Grzetic, J.F., 2009, January. Analysis and testing of freezing phenomena in plastic piping systems. *ASME International Mechanical Engineering Congress and Exposition*, *43864*, pp.201–206.
7. Parent, O. and Ilinca, A., 2011. Anti-icing and de-icing techniques for wind turbines: Critical review. *Cold Regions Science and Technology*, *65*(1), pp.88–96.
8. Dalili, N., Edrisy, A. and Carriveau, R., 2009. A review of surface engineering issues critical to wind turbine performance. *Renewable and Sustainable Energy Reviews*, *13*(2), pp.428–438.
9. Jasinski, W.J., Noe, S.C., Selig, M.S. and Bragg, M.B., 1998. Wind turbine performance under icing conditions.
10. Singh, G. and Vasudev, H., 2025. A Review on smart welding systems: AI integration and sensor-based process optimization. *Journal of Advanced Manufacturing Systems.*
11. Tammelin, B. and Seifert, H., 2014, January. Wind energy production in cold climate. In *1999 European Wind Energy Conference* (pp. 1005–1008). Routledge.
12. Laakso, T. and Peltola, E., 2005. *Wind Energy in Cold Climates-IEA Wind Energy Annual Report, International Energy Agency-Wind Energy.* Helsinki, Finland.
13. Dai, H., Zhu, C., Zhao, N., Zhu, C. and Cai, Y., 2021. An unsteady model for aircraft icing based on tightly-coupled method and phase-field method. *Aerospace*, *8*(12), p.373.
14. Cao, Y., Tan, W. and Wu, Z., 2018. Aircraft icing: An ongoing threat to aviation safety. *Aerospace Science and Technology*, *75*, pp.353–385.
15. Alam, M.F., Thompson, D.S. and Walters, D.K., 2015. Hybrid Reynolds-averaged Navier–Stokes/large-Eddy simulation models for flow around an iced wing. *Journal of Aircraft*, *52*(1), pp.244–256.
16. Li, Z., Xu, H., Xue, Y., Zhang, Z. and Duan, X., 2019. On flight risk quantitative evaluation under icing conditions. *Mathematical Problems in Engineering*, *2019.*
17. Raj, L.P. and Myong, R.S., 2016. Computational analysis of an electro-thermal ice protection system in atmospheric icing conditions. *Journal of Computational Fluids Engineering*, *21*(1), pp.1–9.
18. Petrenko, V.F., Sullivan, C.R., Kozlyuk, V., Petrenko, F.V. and Veerasamy, V., 2011. Pulse electro-thermal de-icer (PETD). *Cold Regions Science and Technology*, *65*(1), pp.70–78.
19. Seifert, H., 2005. Technical requirements for rotor blades operating in cold climate. *DEWI-Magazin.*
20. Horbaty, R., 2005. Wind energy in cold climates: The swiss experience. In *Proceedings of BOREAS VII*, seventh ed. Levi, Finland, March 7–8, pp.1–10.
21. Baring-Gould, I., Tallhaug, L., Ronsten, G., Horbaty, R., Cattin, R., Laakso, T., Durstewitz, M., Lacroix, A., Peltola, E. and Wallenius, T., 2010. Recommendations for wind energy projects in cold climates.
22. Lopera-Valle, A. and McDonald, A., 2015. Application of flame-sprayed coatings as heating elements for polymer-based composite structures. *Journal of Thermal Spray Technology*, *24*, pp.1289–1301.

23. Rad, M.R., Bajgiran, M.M., Moreau, C. and McDonald, A., 2020. Fabrication of thermally sprayed coating systems for mitigation of ice accumulation in carbon steel pipes and prevention of pipe bursting. *Surface and Coatings Technology*, *397*, p.126013.
24. Rad, M.R., Ngaokere, K., Kumar, A., McDonald, A. and Hayden, D., 2019, May. Techno-economic assessment of coating-based resistive heating systems versus conventional heat tracers. In *ITSC2019* (pp. 520–526). ASM International.
25. Prudenziati, M., 2008. Development and the implementation of high-temperature reliable heaters in plasma spray technology. *Journal of Thermal Spray Technology*, *17*, pp.234–243.
26. Pang, J., Qin, W., You, W., Zhong, M. and Meng, Q.A., 2023. Induction heating characteristics of electroless Ni-coated CFRP for deicing and anti-icing. *Coatings*, *13*(9), p.1531.
27. Vertuccio, L., De Santis, F., Pantani, R., Lafdi, K. and Guadagno, L., 2019. Effective de-icing skin using graphene-based flexible heater. *Composites Part B: Engineering*, *162*, pp.600–610.
28. Zhao, Z., Chen, H., Liu, X., Liu, H. and Zhang, D., 2018. Development of high-efficient synthetic electric heating coating for anti-icing/de-icing. *Surface and Coatings Technology*, *349*, pp.340–346.
29. Liu, J., Xiao, L., Rao, Z., Dong, B., Yin, Z. and Huang, Y., 2018. High-performance, micrometer thick/conformal, transparent metal-network electrodes for flexible and curved electronic devices. *Advanced Materials Technologies*, *3*(8), p.1800155.
30. Seong, B., Yoo, H., Nguyen, V.D., Jang, Y., Ryu, C. and Byun, D., 2014. Metal-mesh based transparent electrode on a 3-D curved surface by electrohydrodynamic jet printing. *Journal of Micromechanics and Microengineering*, *24*(9), p.097002.
31. Huang, Y., Wu, H., Xiao, L., Duan, Y., Zhu, H., Bian, J., Ye, D. and Yin, Z., 2019. Assembly and applications of 3D conformal electronics on curvilinear surfaces. *Materials Horizons*, *6*(4), pp.642–683.
32. Jo, H.S., An, S., Lee, J.G., Park, H.G., Al-Deyab, S.S., Yarin, A.L. and Yoon, S.S., 2017. Highly flexible, stretchable, patternable, transparent copper fiber heater on a complex 3D surface. *NPG Asia Materials*, *9*(2), p.e347.
33. Wang, Y., Yu, Z., Mao, G., Liu, Y., Liu, G., Shang, J., Qu, S., Chen, Q. and Li, R.W., 2019. Printable liquid-metal@ PDMS stretchable heater with high stretchability and dynamic stability for wearable thermotherapy. *Advanced Materials Technologies*, *4*(2), p.1800435.
34. Lee, J.G., Lee, J.H., An, S., Kim, D.Y., Kim, T.G., Al-Deyab, S.S., Yarin, A.L. and Yoon, S.S., 2017. Highly flexible, stretchable, wearable, patternable and transparent heaters on complex 3D surfaces formed from supersonically sprayed silver nanowires. *Journal of Materials Chemistry A*, *5*(14), pp.6677–6685.
35. Lee, J.G., An, S., Kim, T.G., Kim, M.W., Jo, H.S., Swihart, M.T., Yarin, A.L. and Yoon, S.S., 2017. Self-cleaning anticondensing glass via supersonic spraying of silver nanowires, silica, and polystyrene nanoparticles. *ACS Applied Materials & Interfaces*, *9*(40), pp.35325–35332.
36. Kim, T.G., Park, C.W., Woo, D.Y., Choi, J. and Yoon, S.S., 2020. Efficient heat spreader using supersonically sprayed graphene and silver nanowire. *Applied Thermal Engineering*, *165*, p.114572.
37. Sohn, H., Park, C., Oh, J.M., Kang, S.W. and Kim, M.J., 2019. Silver nanowire networks: Mechano-electric properties and applications. *Materials*, *12*(16), p.2526.
38. Ma, Z., Kang, S., Ma, J., Shao, L., Wei, A., Liang, C., Gu, J., Yang, B., Dong, D., Wei, L. and Ji, Z., 2019. High-performance and rapid-response electrical heaters based on ultraflexible, heat-resistant, and mechanically strong aramid nanofiber/Ag nanowire nanocomposite papers. *ACS Nano*, *13*(7), pp.7578–7590.
39. Zhao, Z., Chen, H., Liu, X., Wang, Z., Zhu, Y. and Zhou, Y., 2020. Novel sandwich structural electric heating coating for anti-icing/de-icing on complex surfaces. *Surface and Coatings Technology*, *404*, p.126489.

40. Ashrafizadeh, H., McDonald, A. and Mertiny, P., 2016. Deposition of electrically conductive coatings on castable polyurethane elastomers by the flame spraying process. *Journal of Thermal Spray Technology*, *25*, pp.419–430.
41. Clyne, T.W., 2018. 4.7 Thermal and electrical conduction in metal matrix composites. *Materials Science and Engineering A*, *4*, pp.188–212.
42. Qu, X.H., Zhang, L., Mao, W.U. and Ren, S.B., 2011. Review of metal matrix composites with high thermal conductivity for thermal management applications. *Progress in Natural Science: Materials International*, *21*(3), pp.189–197.
43. Jord´a, J.M.M., 2014. Metal matrix composites for thermal management: In *Metal Matrix Composites: Materials, Manufacturing and Engineering* (Vol. 3). Walter de Gruyter GmbH & Co KG.
44. Dehaghani, S.T., Dolatabadi, A. and McDonald, A., 2021. Thermally sprayed metal matrix composite coatings as heating systems. *Applied Thermal Engineering*, *196*, p.117321.
45. Gent, R.W., Dart, N.P. and Cansdale, J.T., 2000. Aircraft icing. *Philosophical Transactions of the Royal Society of London. Series A: Mathematical, Physical and Engineering Sciences*, *358*, pp.2873–2911.
46. Bragg, M.B., Broeren, A.P. and Blumenthal, L.A., 2005. Iced-airfoil aerodynamics. *Progress in Aerospace Sciences*, *41*(5), pp.323–362.
47. Genc, A.T. and Cebeci, T., 2004. Heat transfer and ice accretion on wings. *International Journal of Heat and Fluid Flow*, *25*(6), pp.1026–1035.
48. Cao, Y., Tan, C. and Xu, W., 2014. Numerical simulations of unsteady heat conduction in icing process on airfoil. *International Journal of Thermal Sciences*, *78*, pp.158–170.
49. Tsao, J. and Shen, C., 2003. Anti-icing performance of aircraft deicing fluids under simulated icing conditions. *Journal of Aircraft*, *40*(6), pp.1252–1260.
50. Virk, H.S. and Reddy, S., 2012. Novel passive anti-icing surface coatings for wind turbines. *Renewable Energy*, *38*(1), pp.122–128.
51. Shin, S. and Kang, S., 2012. Energy-efficient de-icing system using electrothermal heating elements. *Journal of Electromagnetic Waves and Applications*, *26*(14–15), pp.2000–2010.
52. Zhao, D. and Xie, Y., 2014. Electromagnetic anti-icing systems for wind turbines: A review. *Renewable and Sustainable Energy Reviews*, *38*, pp.946–953.
53. Hauer, P. and Halverson, S., 2011. Ultrasound anti-icing method for aircraft. *Journal of Aircraft*, *48*(3), pp.896–905.
54. Zhang, J. and Xue, Y., 2014. Superhydrophobic surfaces for anti-icing: Recent progress and future outlooks. *Journal of Materials Chemistry A*, *2*(37), pp.14700–14715.
55. Kulinich, S.A. and Farzaneh, M., 2011. Ice adhesion on superhydrophobic surfaces. *Applied Surface Science*, *255*(18), pp.8153–8157.
56. Collier, C. and Frost, D.L., 2014. Advances in phase change materials for aircraft anti-icing systems. *International Journal of Heat and Mass Transfer*, *75*, pp.663–673.
57. Dou, Z. and Ma, H., 2012. Application of nanocomposites in aircraft icing protection. *Materials Science and Engineering: A*, *556*, pp.364–370.
58. Kim, H.J. and Lee, J.H., 2011. Aerodynamic heating analysis of aircraft wing anti-icing systems. *Journal of Aircraft*, *48*(4), pp.1321–1329.
59. Tang, G.H. and Wang, X., 2010. Ice accretion and melting simulation on aircraft surfaces. *Computers & Fluids*, *39*(5), pp.906–915.
60. Lawson, D.K. and Rausch, R., 2001. Interdisciplinary approach to the design of ice protection systems. *Journal of Aircraft*, *38*(3), pp.501–510.
61. Kim, M.H. and Lee, S.J., 2012. Thermo-fluidic characteristics of a composite anti-icing system for aircraft wings. *Composite Structures*, *94*(8), pp.2359–2366.
62. Silveira, F.N. and Branco, J.R., 2012. Computational fluid dynamics analysis of aircraft wing icing. *International Journal of Heat and Fluid Flow*, *34*, pp.74–83.
63. Boban, M. and Hill, S., 2011. Development of a robust anti-icing system for UAVs. *Unmanned Systems*, *1*(2), pp.117–127.

64. Vasanth, S. and Subramanian, V., 2012. Application of phase change materials for energy-efficient aircraft de-icing systems. *Applied Thermal Engineering*, *39*, pp.108–116.
65. Pandya, D.R. and Meshram, S., 2013. Numerical simulation of icing on wind turbine blades. *Renewable Energy*, *56*, pp.51–59.
66. Korb, R. and Granik, Y., 2001. Impact of surface roughness on ice accretion and drag increase. *Journal of Aircraft*, *38*(3), pp.496–500.
67. Guo, Z. and Wen, X., 2009. Multiscale simulation of ice formation and growth on aircraft surfaces. *Journal of Computational Physics*, *228*(19), pp.6961–6976.
68. Ishak, M.F. and Amir, N., 2013. Heat transfer enhancement using nanofluids in an aircraft anti-icing system. *International Journal of Heat and Mass Transfer*, *61*, pp.132–140.
69. Fukusako, S. and Yamada, M., 2000. Dynamic behavior of water droplets on cold surfaces: A review. *International Journal of Heat and Mass Transfer*, *43*(1), pp.51–60.
70. Etemad, S.G. and Ghasemi, M., 2012. Experimental study on ice adhesion strength on superhydrophobic surfaces. *Colloids and Surfaces A: Physicochemical and Engineering Aspects*, *410*, pp.45–50.
71. Xie, X. and Yao, Z., 2012. Development of a hybrid electro-thermal aircraft de-icing system. *Journal of Electrostatics*, *70*(1), pp.27–32.
72. Khalid, M.N. and Fawzy, E., 2011. Investigation of icing on aircraft wings using particle image velocimetry. *Journal of Aircraft*, *48*(1), pp.55–63.
73. Bian, L. and Zhang, X., 2013. Simulation of ice growth on wind turbine blades and the influence on aerodynamic performance. *Renewable Energy*, *59*, pp.69–76.
74. Shin, J. and Yoon, S., 2012. Development of a smart anti-icing coating for aircraft wings. *Smart Materials and Structures*, *21*(10), p.105007.
75. Wang, X. and Fan, J., 2009. Electromagnetic de-icing system for power lines: Design and testing. *International Journal of Electrical Power & Energy Systems*, *31*(10), pp.551–558.
76. Shen, H. and Lin, Z., 2014. Heat transfer analysis of aircraft wing leading edge for anti-icing applications. *International Journal of Thermal Sciences*, *85*, pp.93–100.
77. Zhu, Y. and Tang, L., 2011. Study of the anti-icing performance of a composite material surface. *Materials Letters*, *65*(23–24), pp.3460–3463.
78. Duan, C. and Li, Y., 2013. Application of graphene-based materials for anti-icing coatings. *Surface and Coatings Technology*, *235*, pp.658–662.
79. Long, Z. and Sun, J., 2012. Electrothermal de-icing system for aircraft windshields: Design and testing. *Journal of Aircraft*, *49*(6), pp.1997–2005.
80. Qian, G. and Xie, L., 2012. Numerical investigation of mixed ice accretion on aircraft surfaces. *International Journal of Heat and Mass Transfer*, *55*(13–14), pp.3792–3800.
81. Lin, L. and Zhang, X., 2010. Ice accretion simulation on airfoils with different roughness. *Journal of Wind Engineering and Industrial Aerodynamics*, *98*(6–7), pp.321–329.
82. Ahmed, N.A. and Zhang, Z., 2012. Investigation of the aerodynamics of ice-contaminated airfoils. *Aerospace Science and Technology*, *22*(1), pp.8–19.
83. Kim, J. and Kim, M., 2014. Thermal performance of a composite electrothermal anti-icing system for helicopter rotor blades. *Composites Part B: Engineering*, *66*, pp.314–323.
84. Song, Y. and Zhang, J., 2013. Experimental study of the anti-icing performance of a flexible electrothermal film. *Journal of Aircraft*, *50*(3), pp.762–769.
85. Lian, Y. and Wang, X., 2012. Ice formation and shedding on rotating surfaces. *Journal of Wind Engineering and Industrial Aerodynamics*, *104*, pp.113–120.
86. Gao, L. and Zhang, J., 2011. Investigation of anti-icing methods for offshore structures. *Marine Structures*, *24*(3), pp.241–259.
87. Sun, G. and Wang, Y., 2014. Nanostructured materials for anti-icing coatings: A review. *Progress in Organic Coatings*, *77*(2), pp.251–261.
88. Zhu, L. and Shi, Z., 2012. Thermal analysis of aircraft wing icing protection systems. *Applied Thermal Engineering*, *48*, pp.14–24.

89. Ma, H. and Gao, W., 2013. Design and testing of a piezoelectric de-icing system for aircraft. *Journal of Intelligent Material Systems and Structures*, *24*(11), pp.1313–1321.
90. Kang, S. and Kim, J., 2014. A study on the efficiency of a thermoelectric de-icing system for aircraft. *Journal of Electromagnetic Waves and Applications*, *28*(1), pp.123–133.
91. Zhu, H. and Liu, Z., 2012. Numerical analysis of the icing and de-icing processes on a wind turbine blade. *Renewable Energy*, *41*, pp.261–270.
92. Wang, S. and Dong, W., 2012. Application of micro/nano-textured surfaces for icephobicity and anti-icing. *Surface and Coatings Technology*, *206*(12–13), pp.3115–3122.
93. Xie, J. and Yu, H., 2011. Numerical simulation of rime ice accretion on airfoil. *International Journal of Heat and Mass Transfer*, *54*(11–12), pp.2454–2461.
94. Zhou, Q. and Wang, L., 2013. Electromagnetic field analysis of an anti-icing system for transmission lines. *Journal of Applied Physics*, *113*(22), p.224908.
95. Zhao, X. and Wang, X., 2011. Ice adhesion reduction using low surface energy materials. *Journal of Adhesion Science and Technology*, *25*(3), pp.277–290.
96. Lu, Y. and Zhang, H., 2012. Application of phase change materials in aircraft anti-icing systems. *Energy Conversion and Management*, *56*, pp.122–129.
97. Zuo, X. and Ma, H., 2014. Performance analysis of a smart anti-icing system for aircraft wings. *Smart Materials and Structures*, *23*(5), p.055011.
98. He, Y. and Li, Y., 2012. Study on the thermal properties of anti-icing nanocomposite materials. *Composites Part B: Engineering*, *43*(2), pp.613–618.
99. Li, C. and Zhu, X., 2013. Application of self-heating materials for anti-icing of high-voltage insulators. *Applied Surface Science*, *284*, pp.601–607.
100. Liu, J. and Jiang, X., 2011. Thermoelectric de-icing system for railway contact wires: Design and testing. *Energy Conversion and Management*, *52*(4), pp.1950–1956.
101. Zhang, Y. and Chen, H., 2014. Investigation of superhydrophobic coatings for anti-icing applications. *Colloids and Surfaces A: Physicochemical and Engineering Aspects*, *451*, pp.109–114.
102. Qi, J. and Xue, M., 2012. Electrothermal anti-icing system for aircraft leading edges. *Journal of Aircraft*, *49*(4), pp.1254–1262.
103. Liu, Z. and Hu, J., 2013. Development of a multifunctional composite material for anti-icing and structural health monitoring. *Composite Structures*, *99*, pp.207–213.
104. Chen, Y. and Yang, L., 2014. Numerical simulation of ice accretion on helicopter rotor blades. *Journal of Aircraft*, *51*(5), pp.1447–1456.
105. Singh, M., Vasudev, H. and Kumar, R., 2021. Corrosion and tribological behaviour of BN thin films deposited using magnetron sputtering. *International Journal of Surface Engineering and Interdisciplinary Materials Science (IJSEIMS)*, *9*(2), pp.24–39.
106. Singh, G., Bansal, A. and Vasudev, H., 2024. Corrosion and microstructural behaviour of Inconel 625 microwave clad deposited on mild steel. In *Thermal Claddings for Engineering Applications* (pp. 51–62). CRC Press.
107. Kapade, T.T., Shah, S.K., Shah, D.S., Kumar, N.S., Bhor, J.K., Panwar, V.S. and Singh, G., 2024. Thermal analysis of v-cut twisted tape fitted heat exchanger: A numerical approach. *International Journal on Interactive Design and Manufacturing (IJIDeM)*. https://link.springer.com/article/10.1007/s12008-024-01883-2
108. Bhowmik, A., Kumar, R., Beemkumar, N., Kumar, A.V., Singh, G., Kulshreshta, A. and Santhosh, A.J., 2024. Casting of particle reinforced metal matrix composite by liquid state fabrication method: A review. *Results in Engineering*, p.103152.
109. Kulkarni, S.D., Manjunatha, C.U., Manjunath, K.V., Prasad, C.D. and Vasudev, H., 2024. Design and optimization of polyvinyl-nitride rubber for tensile strength analysis. *International Journal on Interactive Design and Manufacturing (IJIDeM)*, *18*(8), pp.5893–5908.
110. Prashar, G. and Vasudev, H., 2022. Surface topology analysis of plasma sprayed Inconel625-Al_2O_3 composite coating. *Materials Today: Proceedings*, *50*, pp.607–611.

111. Yedida, V.S. and Vasudev, H., 2022. A review on the development of thermal barrier coatings by using thermal spray techniques. *Materials Today: Proceedings*, *50*, pp.1458–1464.
112. Sharanabasava, H., Prasad, C.D. and Ramesh, M.R., 2023. Microstructure, mechanical and wear properties of SiC and Mo reinforced NiCr microwave cladding. *Advances in Materials and Processing Technologies*, *10*(4), pp.3620–3633. https://doi.org/10.1080/2374068X.2023.2257937.
113. Prashar, G. and Vasudev, H., 2023. Understanding cold spray technology for hydroxyapatite deposition. *Journal of Electrochemical Science and Engineering*, *13*(1), pp.41–62.
114. Prashar, G. and Vasudev, H., 2023. Understanding cold spray technology for hydroxyapatite deposition. *Journal of Electrochemical Science and Engineering*, *13*(1), pp.41–62.
115. Parkash, J., Saggu, H.S. and Vasudev, H., 2022. A short review on the performance of high velocity oxy-fuel coatings in boiler steel applications. *Materials Today: Proceedings*, *50*, pp.1442–1446.
116. Singh, J., Vasudev, H., Kumar, R. and Ubaidullah, M., 2024. Study on wear analysis of Ni-20Al_2O_3 HVOF micron layers using artificial neural network technique. *International Journal on Interactive Design and Manufacturing (IJIDeM)*, *18*(8), pp.5925–5938.
117. Singh, J., Vasudev, H., Szala, M. and Gill, H.S., 2024. Neural computing for erosion assessment in Al-20TiO_2 HVOF thermal spray coating. *International Journal on Interactive Design and Manufacturing (IJIDeM)*, *18*(4), pp.2321–2332.
118. Prakash, C., Gupta, L.R., Mehta, A., Vasudev, H., Tominov, R., Korman, E., Fedotov, A., Smirnov, V. and Kesari, K.K., 2023. Computing of neuromorphic materials: An emerging approach for bioengineering solutions. *Materials Advances*, *4*(23), pp.5882–5919.
119. Dutta, V., Thakur, L., Singh, B. and Vasudev, H., 2022. A study of erosion–corrosion behaviour of friction stir-processed chromium-reinforced NiAl bronze composite. *Materials*, *15*(15), p.5401.
120. Poojari, M., Hanumanthappa, H., Prasad, C.D., Jathanna, H.M., Ksheerasagar, A.R., Shetty, P., Shanmugam, B.K. and Vasudev, H., 2024. Computational modelling for the manufacturing of solar-powered multifunctional agricultural robot. *International Journal on Interactive Design and Manufacturing (IJIDeM)*, *18*(8), pp.5725–5736.
121. Prashar, G. and Vasudev, H., 2022. A review on the influence of process parameters and heat treatment on the corrosion performance of Ni-based thermal spray coatings. *Surface review and letters*, *29*(1), p.2230001.
122. Kattimani, M.A., Venkatesh, P.R., Masum, H., Math, M.M., Bahadurdesai, V.N., Mustafkhadri, S., Prasad, C.D. and Vasudev, H., 2024. Design and numerical analysis of tensile deformation and fracture properties of induction hardened inconel 718 superalloy for gas turbine applications. *International Journal on Interactive Design and Manufacturing (IJIDeM)*, *18*(8), pp.5939–5949.
123. Sunitha, K. and Vasudev, H., 2022. A short note on the various thermal spray coating processes and effect of post-treatment on Ni-based coatings. *Materials Today: Proceedings*, *50*, pp.1452–1457.
124. Majji, B.G.R., Vasudev, H. and Bansal, A., 2022. A review on the oxidation and wear behavior of the thermally sprayed high-entropy alloys. *Materials Today: Proceedings*, *50*, pp.1447–1451.
125. Rao, G.S., Mukkamala, U., Hanumanthappa, H., Prasad, C.D., Vasudev, H., Shanmugam, B. and KishoreKumar, K.C., 2024. Evaluating and optimizing surface roughness using genetic algorithm and artificial neural networks during turning of AISI 52100 steel. *International Journal on Interactive Design and Manufacturing (IJIDeM)*, *18*(8), pp.6151–6160.
126. Singh, J., Gill, H.S. and Vasudev, H., 2023. Computational fluid dynamics analysis on role of particulate shape and size in erosion of pipe bends. *International Journal on Interactive Design and Manufacturing (IJIDeM)*, *17*(5), pp.2631–2646.
127. Anjaneya, G., Sunil, S., Kakkeri, S., Math, M.M., Vaibhav, M.N., Solaimuthu, C., Durga Prasad, C. and Vasudev, H., 2023. Numerical simulation of microchannel heat exchanger using CFD. *International Journal on Interactive Design and Manufacturing (IJIDeM)*, pp.1–17.

128. Nagabhushana, P., Ramprasad, S., Prasad, C.D., Vasudev, H. and Prakash, C., 2024. Numerical investigation on heat transfer of a nano-fluid saturated vertical composite porous channel packed between two fluid layers. *International Journal on Interactive Design and Manufacturing (IJIDeM)*, *18*(5), pp.2927–2944.
129. Singh, P., Vasudev, H. and Bansal, A., 2024. Effect of post-heat treatment on the microstructural, mechanical, and bioactivity behavior of the microwave-assisted alumina-reinforced hydroxyapatite cladding. *Proceedings of the Institution of Mechanical Engineers, Part E: Journal of Process Mechanical Engineering*, *238*(3), pp.1060–1071.
130. Mahantesh, M.M., Rao, K.R., Chandra, A.P., Vijayakumar, M.N., Nandini, B., Prasad, C.D. and Vasudev, H., 2024. Design and modeling using finite element analysis for the sitting posture of computer users based on ergonomic perspective. *International Journal on Interactive Design and Manufacturing (IJIDeM)*, *18*(8), pp.5875–5891.
131. Singh, J., Singh, R., Singh, S., Vasudev, H. and Kumar, S., 2023. Reducing scrap due to missed operations and machining defects in 90PS pistons. *International Journal on Interactive Design and Manufacturing (IJIDeM)*, *17*(5), pp.2527–2539.
132. Arora, H., Kumar, V., Prakash, C., Pimenov, D., Singh, M., Vasudev, H. and Singh, V., 2021. Analysis of sensitization in austenitic stainless steel-welded joint. In *Advances in Metrology and Measurement of Engineering Surfaces: Select Proceedings of ICFMMP 2019* (pp. 13–23). Springer Singapore.
133. Vasudev, H., 2020. Wear characteristics of Ni-WC powder deposited by using a microwave route on mild steel: Microwave cladding of Ni-WC. *International Journal of Surface Engineering and Interdisciplinary Materials Science (IJSEIMS)*, *8*(1), pp.44–54.
134. Prashar, G., Vasudev, H., Thakur, L. and Bansal, A., 2023. Performance of thermally sprayed hydroxyapatite coatings for biomedical implants: A comprehensive review. *Surface Review and Letters*, *30*(1), p.2241001.
135. Mehta, A., Vasudev, H. and Jeyaprakash, N., 2024. Role of sustainable manufacturing approach: Microwave processing of materials. *International Journal on Interactive Design and Manufacturing (IJIDeM)*, *18*(8), pp.5283–5299.
136. Singh, M., Vasudev, H. and Singh, M., 2022. Surface protection of SS-316L with boron nitride based thin films using radio frequency magnetron sputtering technique. *Journal of Electrochemical Science and Engineering*, *12*(5), pp.851–863.
137. Singh, G., Singh, H., Sharma, Y., Vasudev, H. and Prakash, C., 2024. Analysis and optimization of various process parameters and effect on the hardness of SS-304 stainless steel welded joints. *International Journal on Interactive Design and Manufacturing (IJIDeM)*, *18*(5), pp.2889–2896.
138. Singh, G., Mehta, A. and Vasudev, H., 2024. Sustainability of additive manufacturing: A comprehensive review. *Progress in Additive Manufacturing*, pp.1–24.
139. Singh, G., Bansal, A., Vasudev, H. and Mishra, V., 2024. Sliding wear study of the Inconel-625 clad deposits by microwave heating on SS–304. *Physica Scripta*, *99*(6), p.065503.
140. Singh, J., Yadav, A., Gupta, S., Balu, R., Shukla, N., Singh, G. and Singh, S., 2024. Design and computational analysis of continuous wave fiber laser cutting of flat and conformal ITO PET sheets. *International Journal on Interactive Design and Manufacturing (IJIDeM)*, pp.1–9.
141. Singh, G., Singh, R. and Gul, J., 2023. Machinability behavior of human implant materials. *Journal of Electrochemical Science and Engineering*, *13*(1), pp.99–114.
142. Verma, P.K., Vasudev, H., Singh, G. and Singh, S., 2024. A review on the microstructural and biomedical properties of implants manufactured using additive manufacturing. *Journal of the Institution of Engineers (India): Series D*, pp.1–15.
143. Mehta, A., Singh, G. and Vasudev, H. 2024. Processing of shape memory alloys research, applications and opportunities: A review. *Physica Scripta*. https://iopscience.iop.org/article/10.1088/1402-4896/ad48cb

7 Fabrication and Characterization of Wear-Resistant Surfaces

A Study on Powder-Mixed EDM

Satish Kumar, Rahul Mehra, Omar Mbrouk, Pankaj Sharma, and Gurbhej Singh

7.1 INTRODUCTION

Recently, a lot of research has been done on nickel-based coatings as a protective layer against wear and corrosion of metal substrates [1–5]. For instance, precision valves created and utilized in the aerospace, drill pipe, and oil sector casing industries are often coated with nickel-based materials as protective layers [5, 6]. Among the many options for nickel-based coatings, the recently developed Ni–W coating has attracted attention due to its higher hardness, better wear, and corrosion resistance and is a potential replacement for existing carcinogenic hard chromium coating systems [7–9]. Among them, pulse electrodeposition technology is the sought-after technology for producing Ni–W coating on oil casings, drill pipes, and bearing surfaces [10, 11]. The capacity of a material to tolerate volume loss from its surface as a result of mechanical operations such as sliding, rubbing, or scraping is known as wear resistance. It's a crucial mechanical characteristic that influences a material's performance and longevity. Basically, the capacity of plastic materials to tolerate surface defects with regular use is known as wear resistance. Examples of wear include abrasion on our clothes, degradation of plastic bearings over time, plastic clutches, medical devices (hip joints), electrical wires, and plastic insulation. The working surfaces of stressed components deteriorate over time due to abrasion, which eventually reduces the equipment's and machine parts' performance characteristics. Hard, mostly mineral-based particles contact with the part's working surface to generate this abrasive deterioration process. Material abrasion is a major problem. It modifies the component surfaces and dimensions and may result in failure that poses risks. According to some estimates, the expenses of abrasive wear alone account for several percent of the national gross product in the industrialized world, making wear a big economic concern as well. Furthermore, wear increases the number of components that need to be replaced, which has indirect ecological effects. In this

 DOI: 10.1201/9781003546528-7

regard, materials with greater durability in challenging environments, such as steel and Inconel-800, provide a chance to reduce the quantity of material used. Abrasion-resistant steels not only reduce the need for replacement parts but also allow for thinner material, which results in lighter equipment. Experience has shown that the presence of hard particles between two working surfaces that are moving relative to one another is another instance of abrasive wear. This kind of material crushing and grinding is common. On the other hand, sliding pairs—where hard contaminants seep into inadequately sealed work surfaces—can also contain it. Adhesive wear may also result in the formation of hard particles that abrasively impact one or both surfaces. Here, we provide the findings from our analysis of Inconel-800HT's surface composition, surface morphology, wear, and corrosion behavior when boron carbide was used as a powder addition in the EDM.

Moreover, the aerospace industry was the driving force behind the creation of Inconel-800, a super alloy, in the past. At that time, there was a critical demand for novel materials with high strength to weight ratios that lay between those of iron and aluminum. Owing to its exceptional qualities, Inconel-800 has broad applications in several emerging scientific and technological domains, such as nuclear power plants, biomedical, chemical, automotive, and aerospace industries, but precise design and tight tolerances are necessary for these applications. Due to their increased mechanical characteristics, these alloys cannot be machined in acceptable tolerance limits using standard production processes [12]. The great toughness, high hardness, and poor thermal characteristics of these alloys make them extremely challenging to process using traditional machining techniques. Modern machining techniques like electric discharge machining (EDM) are used to machine Inconel 718 in order to get over these restrictions [13–15]. EDM is a technical procedure that has been widely used in industry. Although the process's technological limitations have restricted its use, the conditions required to provide good surface quality over a large area are established by adding a concentration of powder to the dielectric [16–19]. Researchers examined the efficacy of enhancing a number of performance parameters in aluminum matrix composite abrasive PMEDM and found that the depth of material removal rose with powder [20]. Others found that adding graphite powder to EDM fluid increases machining rate [21]. Investigators machined three titanium alloys using three different electrode materials, a mixed dielectric fluid that contained powdered tungsten and manganese, and different machine input parameters. The electrode was shown to have the largest influence on the SR [22–25]. Researchers studied the effects of adding kerosene on specimen surface quality and found that the additives enhanced the specimen's surface quality while also increasing material removing rate (MRR) and lowering TWR [26]. Other studies have demonstrated the use of electro discharge machining (EDM) techniques with various dielectric fluids. The findings showed that adding Al or SiC powder to kerosene in EDM improved the SR. When mixed with kerosene, silicon carbide can achieve a greater material elimination depth than aluminum [27–30]. When it comes to producing excellent surfaces, the effectiveness of powder mixed dielectric and traditional EDM was compared. It was shown that using PMEDM encourages the decrease of SR [31]. It was suggested to use EDM with metal in powder form. Additionally, a thin electrode and a rotating disc electrode with a column were built, according to the experimental findings, which demonstrated two suitable geometries [32]. The use of tungsten carbide powder was recommended as an additive to modify the surface characteristics. The findings showed that on carbon steel, a 150-μm-thick, 1600-Hv hard tungsten carbide layer develops [33–36]. Examined the workings of powder-mixed EDM. According to researchers, the addition of conductive powder decreased the insulating power of the EDM fluid, resulting in a more stable machining process and improved reactions [37]. investigated the effects of process factors on machining performance and surface properties using multi-objective algorithms. The NSGA-II multi-objective optimization program in MATLAB and Pareto frontiers were used to determine the best solution sets. Based on the given goal functions, the LINMAP approach was used to choose the optimum solution set outcomes [38–41].

The literature makes it evident that not much research has been done on the PMEDM of Inconel–800. Research on the machinability of Inconel-800 material in combination with boron carbide as a powder addition has been published sparingly. Previous studies have reported on the machinability of different elements, along with different powders. This work aims to compare the response characteristics resulting from machining in the presence and absence of powder in EDM oil.

7.1.1 Methods and Materials

The entire experimental project was completed using a commercial CNC electric discharge machine, which is accessible from the Central Institute of Hand Tools (a government organization) in Jalandhar, Punjab, India. As a dielectric, Ferrolac 3M EDM oil, which is manufactured and provided by the Oscarmax EDM Company, was utilized. To guarantee flatness, the pieces of metal and tools were smoothed with various types of emery paper prior to the trials. The study employed Inconel-800, a super alloy, as the workpiece material and boron carbide (B_4C) and a 12 mm diameter graphite (Gr) rod as the electrode material. To find out how three process variables, Ip, T_{on}, and T_{off}, affect the chosen responses, MRR and TWR, we chose and adjusted them. The one factor at a time (OFAT) technique was employed to conduct the tests during the investigation. The OFAT technique keeps other parameters constant by varying one factor at a time at various levels rather than all at once.

TABLE 7.1
Operating Range and Levels of the Process Parameters

S. No	1	2	3	4	5
Symbols	A	B	C	D	E
Input Factors	Current (Ip)	Pulse on-time (T_{on})	Pulse off-time (T_{off})	Tool material	Powder particles
Levels	4	60	30	Graphite	Boron carbide
	8	90	45		
	12	120	60		
Units	Ampere	μs	μs		

7.2 RESULTS AND DISCUSSION

Utilizing MINITAB software's Taguchi technique, ANOVA was performed. Table 7.2 displays MRR values for an orthogonal L9 array based on experimental findings from three repetitions, both with and without powder mixing, as well as the S/N ratio. The total mean of MRR is 7.950 mm^3/min when powder is used and 6.435 mm^3/min when powder is not used. With MRR powder, the mean signal to noise ratio is 14.42, but without powder, it is 13.15. The MRR values of R^2 and R^2 adj. are 99.2% and 96.9% when powder is mixed and 75.76% and 61.21% when powder is not mixed. Equations (1) and (2) provide the regression equation with and without powder mixing, and Table 7.4 and 7.5 display the values of R^2 and R^2 (adj.) for both scenarios. The average TWR values are displayed in Table 7.3 together with the S/N ratio, both with the traditional and hybrid method.

Regression Equation with Powder Mixing

$$\text{Mean Value of MRR} = -8.75 + 0.0679\ T_{on} - 0.075\ T_{off} + 1.743\ I_p \quad (1)$$

Regression Equation without Powder Mixing

$$\text{Mean Value of MRR} = -6.75 + 0.0545\ T_{on} - 0.075\ T_{off} + 1.743\ I_p \quad (2)$$

Acetone was used to clean the electrode and work parts before cutting. Table 7.2 displays Inconel-800's chemical composition. The total TWR mean is 1.269 without powder and 0.548 with powder. The S/N ratio's mean when TWR powder is used is 6.437; however, it is −0.407 when powder is not used. In terms of TWR, R^2 and R^2 adj. are 96.0% and 84.0% when powder is mixed and 82.9% and 31.7% when powder is not mixed. It can be seen from Tables 7.3 and 7.6 that adding B_4C as a powder enhances the average signal to noise ratio. This indicates that adding B_4C powder to the EDM oil enhanced MRR. The addition of B_4C in powder form results in a decrease not only in the overall TWR means but also in the S/N ratio means. The values of R^2 and R^2 adj. are also increased by adding B_4C powder to the EDM oil. Tables 7.7 and 7.8 display the MRR and TWR findings both with the traditional and hybrid method. In comparison to other components like T_{on} and T_{off}, current is the most important factor for both responses, according to ANOVA and F-test results. As seen in Tables 7.4 and 7.5, the residual error for MRR decreased with the introduction of powder additive, from 24.24% to 5.91%, but it slightly rose for TWR. It is evident from the ANOVA table that current has the greatest influence on the work specimen.

TABLE 7.2
Inconel-800's Chemical Composition

Element	Ni	Fe	Cr	C	MO	S	Si	Cu
	% Base material	39.5	20.5	<0.1	1.11	<0.015	<1	<0.75

TABLE 7.3
S/N Ratio and Average MRR Findings with Powder Mixing

Exp. No	Levels of Parameters			Material Removal Rate			Mean Value	S/N Ratio
	A	B	C	R-1	R-2	R-3		
MRR (mm^3/min) Using EDM Oil and Powder Mixing								
1	1	1	1	1.887	1.901	1.887	1.892	5.537
2	1	2	2	7.819	7.777	7.819	7.805	17.847
3	1	3	3	7.304	7.304	7.304	7.304	17.271
4	2	1	2	2.926	2.963	2.963	2.951	9.398
5	2	2	3	20.647	20.609	20.532	20.596	26.276
6	2	3	1	1.777	1.772	1.777	1.775	4.986
7	3	1	3	19.244	19.286	19.244	19.258	25.692
8	3	2	1	1.666	1.67	1.666	1.667	4.440
9	3	3	2	8.318	8.283	8.318	8.306	18.388
Overall mean of MRR = 7.950					mean of S/N ratio m = 14.42			
MRR (mm^3/min) without Powder Mixing with EDM Oil								
1	1	1	1	1.254	1.268	1.268	1.263	2.030
2	1	2	2	6.516	6.539	6.539	6.531	16.300
3	1	3	3	6.967	6.918	6.918	6.934	16.820
4	2	1	2	8.379	8.354	8.354	8.362	18.447
5	2	2	3	11.735	11.735	11.735	11.735	21.390
6	2	3	1	1.264	1.279	1.279	1.274	2.103
7	3	1	3	13.587	13.587	13.587	13.587	22.662
8	3	2	1	1.233	1.218	1.223	1.225	1.760
9	3	3	2	6.991	7.039	6.991	7.007	16.911
Overall mean of MRR = 6.435					Mean of S/N ratio m = 13.15			

TABLE 7.4
Regression with Powder Mixing

S	R^2	R^2(adj.)	R^2(pred.)
4.55343	99.2%	96.90%	98.61%

TABLE 7.5
Regression without Powder Mixing

S	R^2	R^2(adj.)	R^2(pred.)
4.55343	75.76%	61.21%	24.69%

TABLE 7.6
TWR Averages with and without Powder Mixing, as Well as the S/N Ratio

TWR (mm^3/min) Using EDM Oil and Powder Mixing									
Exp. No	Levels of Parameters			Tool Wear Rate			Mean Value	S/N Ratio
	A	B	C	R-1	R-2	R-3		
1	1	1	1	0.228	0.254	0.203	0.228	12.829
2	1	2	2	0.73	0.73	0.657	0.706	3.028
3	1	3	3	0.764	0.764	0.849	0.792	2.022
4	2	1	2	0.345	0.259	0.345	0.316	9.997
5	2	2	3	1.069	1.069	1.069	1.069	−0.580
6	2	3	1	0.191	0.191	0.191	0.191	14.379
7	3	1	3	0.582	0.582	0.727	0.630	4.009
8	3	2	1	0.466	0.417	0.417	0.433	7.264
9	3	3	2	0.563	0.563	0.563	0.563	4.990
Overall mean of TWR = 0.548							Mean of S/N ratio m = 6.437	
TWR (mm^3/min) without Powder Mixing with EDM Oil								
1	1	1	1	0.973	0.997	0.95	0.973	0.235
2	1	2	2	2.347	2.428	2.347	2.374	−7.510
3	1	3	3	1.877	1.791	1.877	1.848	−5.336
4	2	1	2	2.083	2.17	2.083	2.112	−6.494
5	2	2	3	1.253	1.127	1.253	1.211	−1.663
6	2	3	1	0.65	0.632	0.65	0.644	3.822
7	3	1	3	1.366	1.366	1.366	1.366	−2.709
8	3	2	1	0.246	0.246	0.246	0.246	12.181
9	3	3	2	0.922	0.09	0.922	0.645	3.813
Overall mean of TWR = 1.269							Mean of S/N ratio m = −0.407	

TABLE 7.7
MRR ANOVA Analysis with and without Powder

ANOVA for MRR without Powder Mixing in EDM Oil							
Control Factors	DOF	Adj. SS	Adj. MS	F	P	% PC	Remarks
Regression	3	323.984	107.995	5.21	0.054	75.75	
Pulse on time	1	24.933	24.933	1.2	0.323	5.83	√√
Pulse off time	1	7.514	7.514	0.36	0.573	1.75	√
Current	1	291.537	291.537	14.06	0.013	68.17	√√√
Error	5	103.669	20.734			24.24	
Total	8	427.653					
ANOVA for MRR in EDM Oil Using Powder Mixing							
Regression	3	154.358	51.453	26.49	0.002	94.08	
Pulse-on-time	1	8.377	8.377	4.31	0.092	5.1	√
Pulse-off-time	1	10.66	10.66	5.49	0.066	6.49	√√
Current	1	135.321	135.321	69.66	0.001	82.47	√√√
Error	5	9.713	1.943			5.91	
Total	8	164.071					

Notes: x: not significant, √: less significant, √√: significant, √√√: most significant.

TABLE 7.7 (CONTINUED)
MRR ANOVA Analysis with and without Powder

Control Factors	DOF	Adj. SS	Adj. MS	F	P	% PC	Remarks
ANOVA for TWR with Powder Mixing in EDM Oil							
Regression	3	0.47	0.157	4.21	0.078	71.63%	
Pulse on time	1	0.001	0.001	0.04	0.842	0.25%	√
Pulse off time	1	0.022	0.022	0.61	0.469	3.48%	√√
Current	1	0.447	0.447	11.96	0.018	67.89%	√√√
Error	5	0.187	0.037			28.37%	
Total	8	0.659				100.00%	
TWR ANOVA Analysis without Powder							
Regression	3	2.821	0.94	3.43	0.109	67.26	
Pulse-on-time	1	1.439	1.439	5.24	0.071	34.31	√√
Pulse-off-time	1	0.287	0.287	1.05	0.353	6.84	√
Current	1	1.094	1.094	3.98	0.102	26.08	√√√
Error	5	1.372	0.274			32.71	
Total	8	4.194					

For MRR, the F-value for current is significantly higher with the addition of powder particles (69.66) compared to without powder (14.06), indicating that material removal rate is enhanced when powder particles are used. When it comes to MRR, the P-value of the current without powder particles is 0.013, but when powder particles are present, it is 0.001 among the three variables. Both values are less than the significance level, which indicates both have a significant effect, particularly when powder particles are used, which produces more significant findings.

Without powder mixing, increasing T_{on} from 90 μs to 120 μs increases the machining time, whereas when powder mixing is used, this change has no effect on MRR. When considering T_{on} and peak current, the effects of T_{off} on MRR are negligible in both scenarios. As powder mixing machining's peak current increases, more material is removed from the workpiece because the conducting powder weakens the EDM oil's insulating properties and intensifies the spark between the electrode and the material being machined. MRR rises in tandem with an increase in discharge energy. As a result, high discharge energy leads to high melting temperatures, which in turn generates evaporation and more impulsive forces acting on the machine surface, which raises MRR [30, 31]. A3B1C3 are the ideal parameters found for MRR using powder additives.

Additionally, a normal probability plot shows that the response is dependent on the input parameters and that the data are normally distributed.

Table 7.8 lists the response values for the S/N ratio for each level of identifying factors. It also displays the factor level values and ranking of each factor. In the case of MRR, the delta value of the current is higher when the powder is mixed with EDM oil. Conversely, the TWR instance shows the opposite impact.

Compared to machining without powder, tool wear rate (TWR) is reduced at a current of 4 amps during the initial and middle phases; however, this trend reverses at the final phase, where TWR increases at the same current level. Compared to

TABLE 7.8
Factor Level Values of Each Factor and Ranking

Delta/Rank for Powder-Free MRR				Delta/Rank for Powder-Free TWR			
Level	**T_{on}**	**T_{off}**	**Current**	**Level**	**T_{on}**	**T_{off}**	**Current**
1	13.55	13.54	4.98	1	5.95	8.94	11.49
2	13.55	16.18	15.21	2	7.93	3.23	6.00
3	16.17	13.54	23.08	3	5.42	7.13	1.81
Delta	2.62	2.64	18.09	Delta	2.51	5.70	9.67
Rank	3	2	1	Rank	3	2	1
Delta/Rank for Powdered MRR				**Delta/Rank for Powdered TWR**			
1	11.717	14.38	1.96	1	−4.20	−2.98	5.41
2	13.98	13.15	17.21	2	−1.44	1.00	−3.39
3	13.77	11.94	20.29	3	4.42	0.76	−3.23
Delta	2.26	2.43	18.32	Delta	8.63	3.99	8.8
Rank	3	2	1	Rank	2	3	1

machining without powder, very low TWR occurs in the early and intermediate phases at current of 8 amps. The final current value, however, demonstrates very little effect. There is a discernible change in TWR at 12 amp current, especially in the early and late phases. But in the intermediate phases, there is little effect.

7.2.1 Microstructure Analysis

Using a scanning electron microscope, the machined specimens were inspected. A 500× magnification was used. The input parameters used were as follows: for Experiment No. 1—Ton = 60 µs, Toff = 30 µs, and Ip = 4 amps; for subsequent experiments—Ton = 90 µs, Toff = 30 µs, and Ip = 8 amps; Ton = 120 µs, Toff = 30 µs, and Ip = 12 amps; and Ton = 120 µs, Toff = 60 µs, and Ip = 8 amps. A graphite electrode was used for both specimens in all experiments.

7.2.2 EDX Analysis

The machined specimens' elemental composition is identified using this procedure. Not all of the material particles are removed from the specimen by the continuous flow of dielectric fluid; some are still present on the surface of the machined specimen.

7.3 FABRICATION OF WEAR-RESISTANT SURFACES USING PMEDM

1. **Function of Powder Additives in Surface Modification**: The inclusion of powders like boron carbide (B_4C), silicon carbide (SiC), and aluminum oxide (Al_2O_3) in the dielectric fluid improves surface finish and decreases micro-cracks in EDM-machined material. The powders enhance the material removal rate while minimizing the tool wear rate to provide greater efficiency in machining superalloys like Inconel-800.

2. **Surface Hardening and Microstructure Enhancement**: PMEDM allows powder particles to be deposited on the machined surface, forming a hardened top layer that offers better wear resistance. The outcome is the development of recast layers, less residual stress, and better tribological characteristics. The use of B4C in PMEDM has manifested considerable decreases in crater formation and roughness.
3. **Process Parameters' Effect on Surface Properties**: PMEDM efficiency is a function of different process parameters such as voltage, pulse duration, current intensity, powder concentration, and dielectric fluid type. Optimal selection of these parameters is important in order to attain desirable surface alterations. For example:
   - Increased voltage and pulse duration enable deeper powder particle penetration, improving surface hardness but possibly at the cost of increased surface roughness.
   - Higher powder concentration enhances the homogeneity of the recast layer but can result in excessive buildup of debris, influencing machining stability.
   - Dielectric fluid content (e.g., kerosene, deionized water, or synthetic oils) affects the breakdown voltage and energy loss, impacting overall surface characteristics.
4. **Industrial Applications of PMEDM for Wear-Resistant Coatings**: PMEDM has seen extensive use in those industries that demand high wear resistance and longer component life. Some major applications are:
   - Aerospace: improved turbine blades, landing gear parts, and extreme-condition wear-resistant coatings.
   - Biomedical: Production of orthopedic implants with enhanced wear resistance and biocompatibility.
   - Automotive: Machining high-performance engine parts exposed to severe stress and friction.
   - Tooling Industry: Design of cutting tools and dies of high hardness with increased durability.
5. **Future Prospects and Advancements:** Current studies in PMEDM center on hybrid methods combining ultrasonic vibration, laser machining, and nanofluid-based dielectric to further improve surface integrity and machining efficiency. Moreover, the application of machine learning and artificial intelligence for real-time parameter optimization is anticipated to transform PMEDM by enhancing process stability and minimizing energy intake. Current studies in PMEDM center on hybrid methods combining ultrasonic vibration, laser machining, and nanofluid-based dielectric to further improve surface integrity and machining efficiency. Moreover, the application of machine learning and artificial intelligence for real-time parameter optimization is anticipated to transform PMEDM by enhancing process stability and minimizing energy intake.

7.4 CONCLUSION

The primary goal of this research is to improve machinability of Inconel-800, a hard-to-cut alloy, in EDM by adding powder blended with a dielectric fluid. To achieve the goals, a super alloy called Inconel-800 was machined using graphite as the tool material and a mixed dielectric fluid made of B_4C powder, varying the initial parameters. Per the experimental strategy, every experiment was conducted twice: first, traditional machining was done, and second, the same process was repeated with powder-mixed EDM oil. Using the Taguchi method, the machining parameters were examined in terms of MRR and TWR. The two main factors that increase the MRR are the peak current and powder concentration. As the Ip and powder concentration rise, so does the MRR. The result shows that with powder mixed 6 g/l in EDM oil, the overall mean of the MRR is increased from 6.435 to 7.950 mm^3/min. Comparing the improvement to the absence of powder concentration (Cp = 0 g/l), it is approximately 19.05%. The outcome demonstrates that mixing dielectric fluid with powder will improve the efficiency of the machining process. Additionally, it is discovered that, within the given conditions, a powder concentration of 6 g/l yields the highest material removal rate. In the case of TWR, erosion of electrodes increased when we machined with the traditional method. Compared to machining with 6 g/L powder concentration, the material removal rate decreased from 1.269 to 0.548 mm^3/min, representing a reduction of 56.81%. Also, it is observed that most vital factor that affects specimen integrity and TWR is current, followed by T_{on} and T_{off}.

REFERENCES

[1] N. Jiang. Corrosion resistance of nickel-phosphorus/nano-ZnO composite multi-layer coating electrodeposited on carbon steel in acidic chloride environments. Int. J. Electrochem. Sci. (2020).

[2] F. Xia et al. Effect of pulse current density on microstructure and wear property of Ni-TiNnanocoatings deposited via pulse electrodeposition. Appl. Surf. Sci. (2021).

[3] Y.-D. Xu et al. Improving cycling performance and rate capability of Ni-rich LiNi0.8Co0.1Mn0.1O2 cathode materials by $Li_4Ti_5O_{12}$ coating. Electrochim. Acta (2018).

[4] B. Liu et al. Preparation of Ni-W-eGO composite coatings and investigation of its mechanical properties and corrosion resistance. Colloids Surf. A (2021).

[5] A.O. Ijaola et al. Superhydrophobic coatings for steel pipeline protection in oil and gas industries: A comprehensive review. J. Nat. Gas Sci. Eng. (2020).

[6] S. Shajahan et al. Corrosion, oxidation and wear study of electro-co-deposited ZrO_2-TiO_2 reinforced Ni-W coatings. Surf. Coat. Technol. (2020).

[7] R. Rao, V. Yadava. Multi-objective optimization of Nd: YAG laser cutting of thin superalloy sheet using grey relational analysis with entropy measurement. Optic. Laser Technol. 41:922–930 (2009).

[8] P.S. Bains et al. On the machinability and properties of Ti–6Al–4V biomaterial with n-Hap powder–mixed ED machining. Proc. Inst. Mech. Eng. Part H J. Eng. Med. 234:232–242 (2019).

[9] V.D. Bui et al. Antibacterial coating of Ti-6Al-4V surfaces using silver nano-powder mixed electrical discharge machining. Surf. Coat. Technol. 383:125254 (2019).

[10] M.H. Allahyarzadeh et al. Ni-W electrodeposited coatings: Characterization, properties and applications. Surf. Coat. Technol. (2016).

[11] B. Li et al. Electrodeposition of Ni W/ZrO_2 nano crystalline film reinforced by CeO_2 nanoparticles: Structure, surface properties and corrosion resistance Mater. Chem. Phys. (2019).
[12] M.H. Allahyarzadeh et al. Electrodeposition mechanism and corrosion behavior of multilayer nanocrystalline nickel-tungsten alloy. Electrochim. Acta (2017).
[13] M.H. Allahyarzadeh et al. Mechanical properties and load bearing capability of nano-crystalline nickel-tungsten multilayered coatings. Surf. Coat. Technol. (2020).
[14] C. Prakash et al. Experimental investigations in powder mixed electric discharge machining of Ti–35Nb–7Ta–5Zr _-titanium alloy. Mater. Manuf. Process. 32:274–285 (2016).
[15] A. Amin et al. Multiple-objective optimization of hydroxyapatite-added EDM technique for processing of 316L-steel. Mater. Manuf. Process. 1–12 (2021).
[16] A. Amin et al. Multi-objective optimization of process variables for MWCNT-added electro-discharge machining of 316L steel. Int. J. Adv. Manuf. Technol. 115:179–198 (2021).
[17] R. Wandra. Finite element analysis of powder mixed electric discharge machining on Ti-alloy. Mater. Today Proc. 50:593–599 (2021).
[18] V.N.R. Jampana et al. Experimental and thermal investigation on powder mixed EDM using FEM and artificial neural networks. Adv. Mater. Sci. Eng. (2021).
[19] B. Surekha et al. Response surface modelling and application of fuzzy grey relational analysis to optimize the multi-response characteristics of EN- 19 machined using powder mixed EDM. Austr. J. Mech. Eng. 19(1):19–29 (2019).
[20] T. Tahsin et al. Particle migration and surface modification on Ti6Al4V in SiC powder mixed electrical discharge machining. J. Manuf. Process. 31:744–758 (2018).
[21] S. Kumar, A.K. Dhingra. Multiresponse optimization of process variables of powder mixed electrical discharge machining on Inconel-600 using taguchi methodology. Int. J. Mech. Prod. Eng. Res. Dev. J. Scopus Index. J. ICV 60:6 (2015).
[22] S. Kumar, A.K. Dhingra. Effect of machining parameters on performance characteristics of powder mixed EDM of Inconel-800. Int. J. Automot. Mech. Eng. Univ. Malaysia Pahang. Malay. 15(2):5221–5237 (2018). Scopus Indexed Journal; https://doi.org/10.15282/ijame.15.2.2018.6.0403, ISSN: 2229–8649 (Print); ISSN: 2180–1606 (Online).
[23] S. Kumar, A.K. Dhingra. Parametric optimization of powder mixed electrical discharge machining for nickel-based superalloy Inconel-800 using response surface methodology. Mech. Adv. Mater. Modern Process. 3:7 (2017). https://doi.org/10.1186/s40759-017-0022-4. SpringerPublications, USA.
[24] S. Kumar, S. Kumar. Experimental investigation on surface characteristics of nickel-based super alloy Inconel-600 in powder mixed electric discharge machining by using response surface methodology. *Advances in Engineering Materials*, Springer:281–285 (2021).
[25] S. Kumar et al. To evaluate the effect of boron carbide (B4C) powder mixed EDM on the machining characteristics of INCONEL-600. *Materials Today: Proceedings*, Elsevier. https://doi.org/10.1016/j.matpr.2021.10.096.
[26] S. Kumar et al. Parametric evaluation of PMEDM for the machining of inconel-800 using response surface methodology. Mater. Today Proc. https://doi.org/10.1016/j.matpr.2023.02.048.
[27] A.K. Singh et al. Effect of the addition of conductive powder in dielectric on the surface properties of superalloy Super Co 605 by EDM process. Int. J. Adv. Manuf. Technol. 77(1–4):99–106 (2015).
[28] A.K. Singh et al. Document details: Optimization of parameters using conductive powder in dielectric for EDM of super Co 605 with multiple quality characteristics. Mater. Manuf. Process. 29(3):267–273 (4 March 2014).

[29] S. Rajesha et al. On electro discharge machining of Inconel718 with hollow tool. J. Mater. Eng. Perform. 21(6):882 (2012).

[30] P. Pecas, E. Henriques. Influence of silicon powder-mixeddielectric on conventionalelectrical discharge machining. Int. J. Mach. Tools Manuf. 43:1465–1471 (2003).

[31] S. Singh, M.F. Yeh. Optimization of abrasive powder mixed EDM of aluminum matrix composites with multiple responses using gray relational analysis. J. Mater. Eng. Perform. 21(4):481–491 (2012).

[32] A. Kumar et al. Realizing potential of graphite powder in enhancing machining rate in AEDM of nickel based super alloy 718. Int. J. Prod. Ind. Eng. 1(1):48–51 (2010).

[33] S. Kumar et al. Study the surface characteristics of cryogenically treated tool-electrodes in powder mixed electric discharge machining process. Mater. Sci. Forum. 808:19–33 (2015).

[34] H. Singh et al. Microstructure and cavitation erosion performance of cold-sprayed WC-12Co and WC-17Co coatings on hydraulic turbine steels. J. Mater. Eng. Perform. (2024). https://doi.org/10.1007/s11665-024-10097-2.

[35] H. Singh et al. Deposition behaviour of WC-17Co cold-sprayed composite coating on hydro turbine steels. Surf. Rev. Lett. (2024). https://doi.org/10.1142/S0218625X25500209.

[36] H. Singh et al. Cavitation erosion behaviour of WC-Co coatings on turbine steel deposited by cold spray. Surf. Rev. Lett. (2024). https://doi.org/10.1142/S0218625X25500052.

[37] H. Singh et al. Slurry erosion performance of cold-sprayed WC-12Co composite coatings. Surf. Rev. Lett. (2024). https://doi.org/10.1142/S0218625X2450050.

[38] K. Furutania et al. Accretion of titanium carbide by electrical discharge machining with powder suspended in working fluid. Precis. Eng. 25(2):138–144 (2001).

[39] H.K. Kansal et al. Technology and research developments inpowder mixed electric discharge machining (PMEDM). J. Mater. Process. Technol. 184:32–41 (2007).

[40] S. Kumar, S. Kumar. Experimental investigation on surface characteristics of nickel-based super alloy Inconel-600 in powder mixed electric discharge machining by using response surface methodology. *Advances in Engineering Materials*, Springer:281–285 (2021).

[41] S. Kumar et al. To evaluate the effect of boron carbide (B_4C) powder mixed EDM on the machining characteristics of INCONEL-600. *Materials Today: Proceedings*, Elsevier. https://doi.org/10.1016/j.matpr.2021.10.096.

8 Characterization and Testing of Surface-Processed Materials Using Friction Stir Processing

Harmandeep Singh, Gurbhej Singh, Manpreet Singh, and Pankaj Sharma

8.1 INTRODUCTION

Friction stir processing (FSP) is an application of medium carbon steel which is used in the automobile industry. Medium carbon steel is mostly used for high strength, high hardness, and durability. In this method, different heat treatment techniques are used like normalizing, quenching, and annealing that enhance the mechanical properties and microstructure of the steel. FSP uses parameters such as speed of rotation, feed rate, and longituidinal speed. Other methods of modifying steel properties include heat treatment, surface plastic deformation, shot peening, and surface mechanical attrition treatment. FSP enhances the steel properties through significant microstructural refinement of the processed zone; the resulting steel exhibits improved hardness, fatigue limit, wear, strength, and stability.

Figure 8.1 shows four regions in the process. The nugget zone is where the grain size is thermomechanically disfigured to refine it. The heat affected zone retains the same grain structure as the base material (BM) because it is unaffected by the procedure. Research on FSPed steels remains limited because the processing is challenging and the phase changes are complex. However, FSP can improve the mechanical properies of medium carbon steel by producing ultrafine-grained phase structures. FSP is typically performed in the strengthened condition. The process generates high surface heat, compressing the grains in that region. FSP alters the physical properties of the steel without changing its phase, which enables outcomes such as high strain-rate superplasticity. The grain refinement occurs in the BM, enhancing the material's properties [1–3].

The FSP apparatus uses EN31 steel. The tool is fixed to the VMC machine bed, and the work piece is mounted on the tool; then, the machine rotates at high speed to process the steel surface. This procedure produces a microstructure with fine

 DOI: 10.1201/9781003546528-8

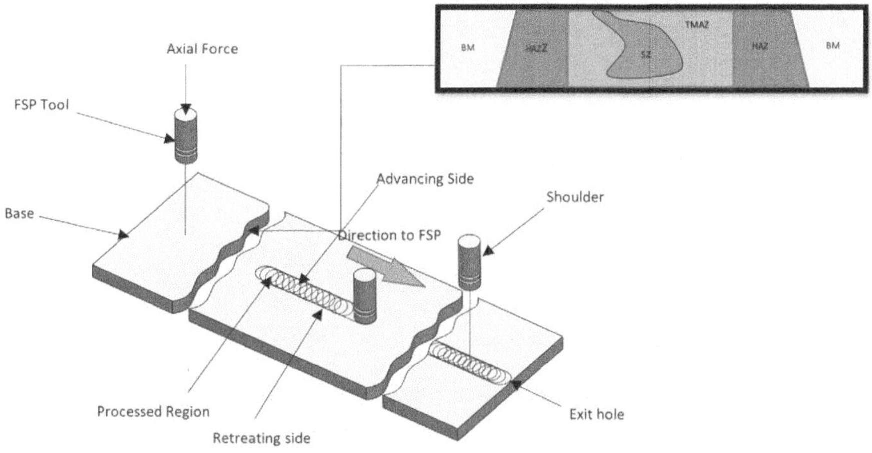

FIGURE 8.1 The friction stirring process.

equiaxed grains and material blending without changing the phase. FSP allows for measuring properties such as surface and mechanical characteristics. It combines the benefits of solid-state processing and mechanical stirring to modify the microstructure and surface characteristics of materials, making it suitable for applications requiring improved wear and corrosion resistance. The characterization of surface-processed materials using FSP involves analyzing their microstructure, mechanical properties, and tribological behavior. Researchers have explored the effects of FSP on materials like stainless steels, superalloys, and composites, showing improvements in surface performance under different operating conditions [4–11].

8.2 EXPERIMENTAL PROCEDURE AND TECHNIQUES

8.2.1 The Experimental Materials

AISI 4337 steel is a high-strength, low-alloy steel belonging to the SAE-AISI 4300 series and is classified as a medium carbon steel with 0.34% carbon content (Table 8.1). It exhibits mechanical properties such as tensile strength, bending resistance, and hardness and is primarily used in engineering and industrial applications requiring structural integrity and durability under demanding conditions. The AISI 4337 steel in the current work was obtained from Guru Nanak Dev Thermal Power Plant in Bathinda. The uniquely designed EN31 fixture was obtained from Chandigarh University, Gharuan (Figure 8.2).

8.2.2 The Experimental Procedure

The material used in this experiment was 3 mm thick AISI 4337 steel plates with dimensions of 76 × 50 × 3 mm. The FSP was performed on a vertical milling machine using a tungsten carbide tool 10 mm in diameter; the workpieces were clamped using

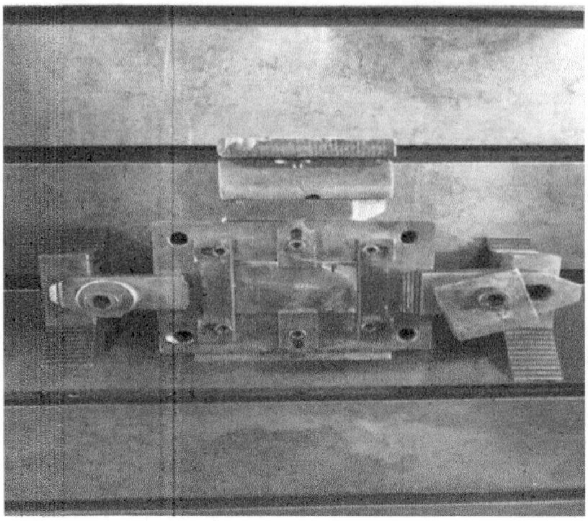

FIGURE 8.2 The experimental fixture.

TABLE 8.1
Chemical Composition of AISI 4337 Steel

Elements	Composition
C%	Carbon –0.40
Mn%	Manganese –0.80
Si%	Silicon –0.35
P%	Phosphorus –0.035
S%	Sulfur –0.04
Cr%	Chromium –0.90
Mo%	Molybdenum –0.30
Ni%	Nickel –2.00
Fe%	Iron –95.76

the specially designed EN31 steel fixture. During FSP, the rotating tool generated high surface heat, improving the mechanical properties compared with the BM. After the FSP, rapid cooling was applied to the processed surface with dry ice, inducing brittleness. This quenching process was repeated for all FSPed samples. The material properties were influenced by changing the rpm while holding the feed rate constant at 40 mm/min. The experimental parameters are shown in Table 8.2.

8.2.3 The Workpiece and Fixture Setup

Figure 8.3 shows the experimental tool, and Figure 8.4 shows the fixture with the workpiece clamped in place. The steel is cut to the size of the fixture, 76 × 50 mm. The EN31 fixture is set on the bed of the vertical milling machine, and the workpiece is fixed so that it does not move. Figure 8.5 shows the fixture with the rotating tool.

TABLE 8.2
Parameters of the Experiment

S no.	Rotational Speed (rpm)	Feed Rate (mm/min)	Plunge Depth (mm)	Tool Diameter (mm)	Passes
S1	100	40	1	10	2
S2	200	50	1	10	3
S3	100	40	1	10	4
S4	200	50	1	10	5
S5	100	40	1	10	6
S6	200	50	1	10	9

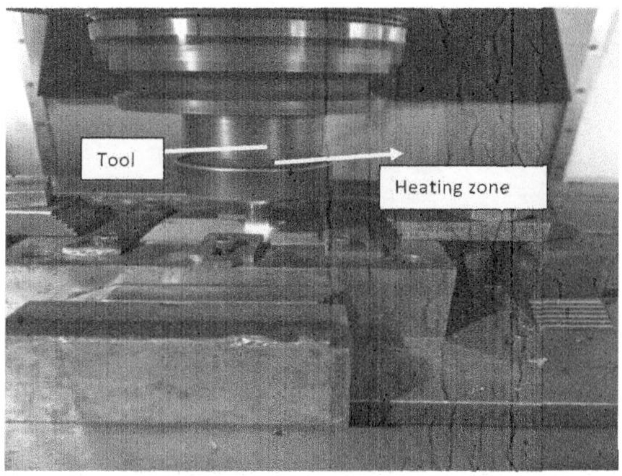

FIGURE 8.3 The FSP rotating tool.

FIGURE 8.4 The workpiece mounted on the fixture.

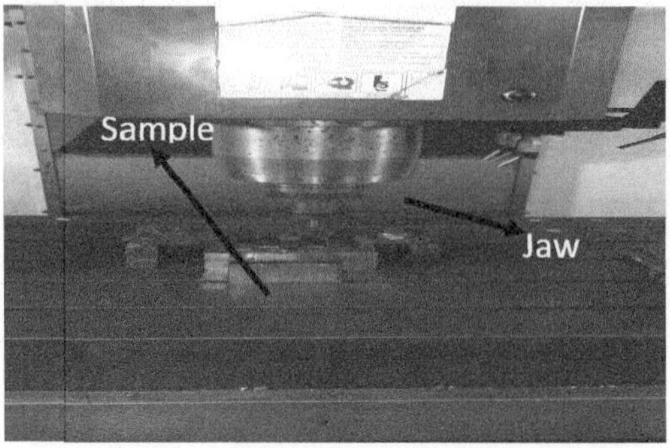

FIGURE 8.5 The fixture setup with the tool.

FIGURE 8.6 Rapid cooling in the friction stirring process.

Studies on the corrosion and tribological behavior of thin films, such as boron nitride deposited using magnetron sputtering, highlight the impact of surface treatments on material performance. Such surface protection techniques for stainless steel give significant improvements in mechanical properties and corrosion resistance [12–19]. These findings are crucial for considering FSP as a method for modifying the surface integrity of materials [20–25].

8.2.4 Rapid Cooling

FSP changes the grain dimensions and structure through high temperatures, which increases the risk of phase changes. To preserve the desired phase, rapid cooling is applied; in this experiment, dry ice was used for the cooling (Figure 8.6). In materials science, this rapid cooling is called quenching and typically uses water, air, or oil to obtain different material properties. This process prevents low-temperature transformations. Immediate quenching reduces crystal grain size in both plastics and metals, enhancing their hardness.

8.3 RESULTS AND DISCUSSION

8.3.1 Microstructure

Optical microscopy is a tool for visualizing the microstructures of materials. Figure 8.7 compares the experimental steel sample before and after FSP; note that the grain boundaries increase with increased rotational speed. The average measured grain size is 15 µm. The figure images highlight the interfaces between the unprocessed and processed zones. The FSP resulted in significant changes that changed the structure into combined ferrite and bainite phases. The polished and etched sections reveal clear variations caused by the friction stir processing. After rapid cooling, martensite precipitated within the processed zone, which led to a dual-phase microstructure composed of martensite and ferrite. This transformation improved strength and ductility but also led to decreased strain hardening and ductility because of increased dislocation density. The tool and shoulder maintained proper contact with the surface zone, generating uniform fine grains without phase changes.

Figure 8.8 shows optical microscope images of the sectional samples, highlighting the microstructures of both the natural and processed zones. The transition zone clearly denotes the boundary between the prepared and untreated regions, and the images demonstrate the interface between the FSPed and unprocessed steel. High strain rates led to plastic deformation, accelerating the breakdown of internal structures, which multiple researchers have observed. It is well established that during

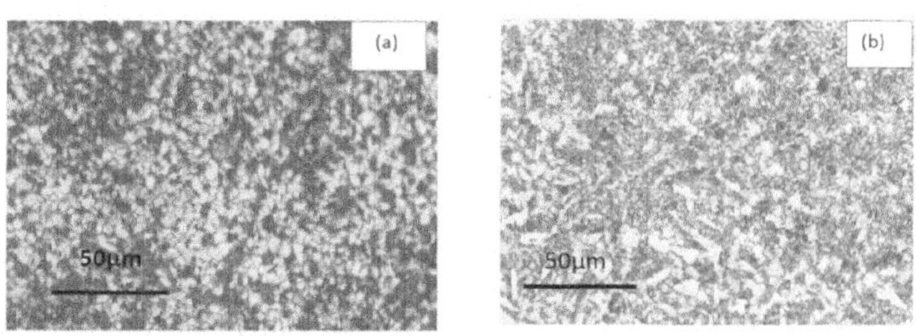

FIGURE 8.7 Optical microstructures of (a) unprocessed and (b) processed AISI 4337 steel.

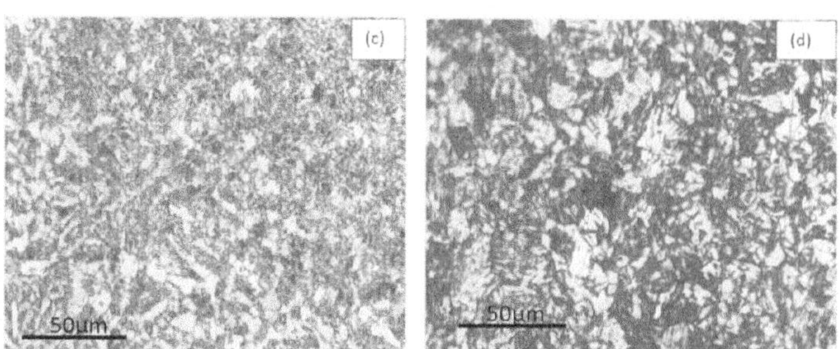

FIGURE 8.8 The images display the improved features of the steel after the FSP.

FSP, the martensite phase forms from high-temperature austenite and contains more carbon atoms than the ferrite phase. After tempering, precipitation occurs along grain boundaries within the martensite phase. The microstructure of the FSPed samples consists of four distinct regions: the base steel sample, the thermomechanically affected zone, the heat-affected zone, and the stir zone. These zones show the surface characteristics on the advancing and retreating sides of the tool when the friction stirring is applied unevenly.

During the FSP, the localized frictional heating of the steel raised the austenizing point, enabling the formation of martensite following the rapid cooling. This transformation varied across the stir zone, as demonstrated by the fine grain size of the martensite; austenite was also detected in some areas. Grain refinement occurred through plastic deformation and transformation of austenite to ultra-fine martensite. Some researchers have reported the formation of sigma phase bands on the advancing side of the nugget zone in FSPed austenite steels; in this study, the bands were observed but not clearly formed. Delta ferrite formation at high temperatures followed by the rapid cooling can influence sigma phase precipitation under high strain conditions; recrystallization was also introduced during processing.

8.3.2 Microhardness

Figures 8.9 to 8.12 graphically show the hardnesses of the BM compared with each of the treated samples, S1–S12; Table 8.3 lists all 13 hardness values. The BM hardness is 344 Hv, and in Figure 8.9, hardness for S1 to S3 ranged from 400 to 550 Hv. Microhardness was uniformly distributed within the processed zones (PZs) of S1 and S3, averaging 481 Hv. The figure also shows that S2 and S3 did not differ significantly from the BM. Cross-sectional hardness distributions along the longitudinal and transverse directions are also shown.

The increase in microhardness is attributed to the Hall–Petch relationship [26]. Indentation tests confirmed higher resistance in FSPed steel than in untreated BM. The hardness improved when the austenitic phase changed into martensite after quenching. The yield strength increased notably following FSP: The AISI 4337 steel yield was 712 Mpa after processing, nearly double the BM yield of 550 Mpa (data not

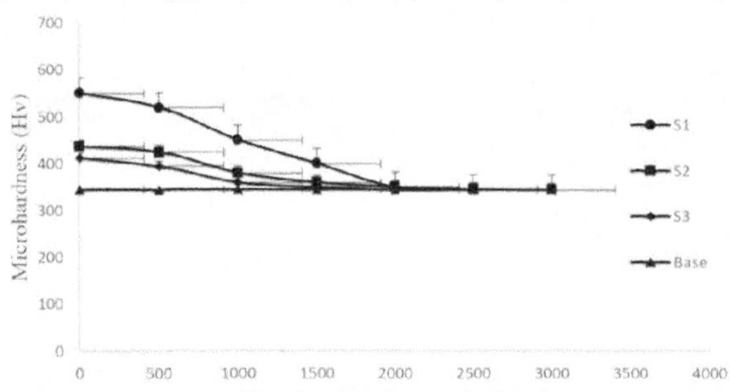

FIGURE 8.9 Microhardness of FSPed AISI 4337 samples S1, S2, and S3 compared with the base material (BM).

TABLE 8.3
Hardnesses of All Samples

Sample	Hardness
BM	334
S1	550
S2	437
S3	413
S4	560
S5	491
S6	433
S7	530
S8	432
S9	405
S10	522
S11	460
S12	422

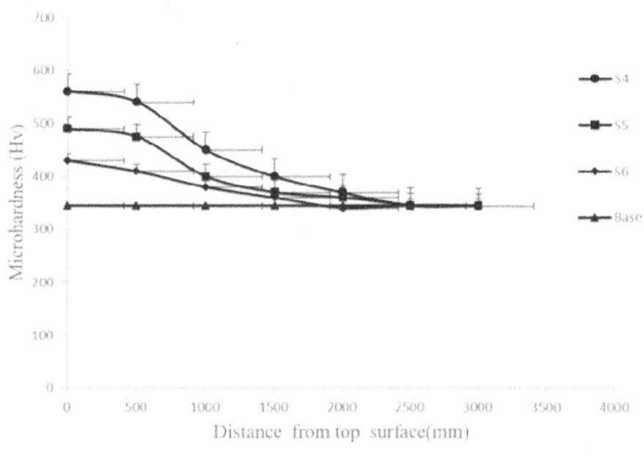

FIGURE 8.10 Microhardness of S4, S5, and S6 compared with BM.

shown), representing a 7% increase. However, work capacity decreased, potentially because of the smaller grains, which other researchers observed as well. Grain size was also inversely correlated with strain rate, consistent with the Zener–Holloman relationship.

Figure 8.10 compares BM (344 Hv) with S4, S5, and S6; hardness again ranged from 400 to 550 Hv. Microhardness was again uniformly distributed within the PZs of S4 and S6, averaging 496 Hv. S5 and S6 did not differ significantly from BM, as shown in the figure. Microhardness was measured across the cross-sections of both samples, which had undergone multipass FSP. Each pass increased the processing depth by approximately 1 mm; this experiment involved three passes in total. Microhardness profiles were recorded along both the transverse and longitudinal directions. Increased hardness was observed on the advancing side, likely due to smaller grain size; the advancing side experiences higher deformation and temperatures during FSP, which promotes severe recrystallization and grain refinement.

Figure 8.11 compares BM (344 Hv) with S7 to S9; hardness again ranged from 400 to 550 Hv. Microhardness was again uniformly distributed within the PZs of S7 and S9, averaging 467 Hv. The figure shows that S8 and S9 also did not differ significantly from BM, and microhardness profiles were again recorded along both the transverse and longitudinal directions. The FSP zone formed a hard martensite phase because of the high temperature reached during the processing. The rapid cooling applied to the material's surface increased the steel's hardness by transforming austenite into martensite. FSP was demonstrated to be an effective technique for repairing material and improving strength, potentially extending service life.

Figure 8.12 compares BM (344 Hv) with S10 to S12; hardness again ranged from 400 to 550 Hv. Microhardness distribution was uniform within the PZs of S10 and S12, averaging 472 Hv, and S11 and S12 again did not differ significantly from BM; microhardness profiles were again recorded across the transverse and longitudinal lines. In this work, higher dislocation density in the stir zone contributed to lower strain hardening in the processed samples. Compared with BM, the FSPed steel exhibited coarser grains. Instability in grain deformation also contributed to reduced ductility in the processed steel plates. Limited accommodation of fine grains in the

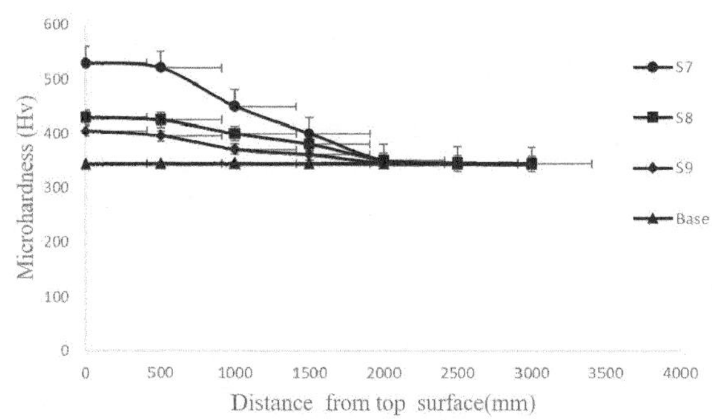

FIGURE 8.11 Microhardness of S7, S8, and S9 compared with BM.

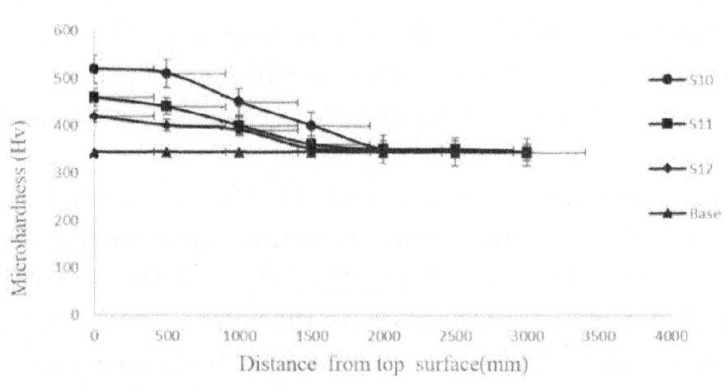

FIGURE 8.12 Microhardness of S10, S11, and S12 compared with BM.

nugget zone further results in lowwer strain hardening and reduced ductility. In the FSPed samples, the steel's hardness depended on the tool's rotating speed, which influenced both ultimate and yield strength.

The results show that friction stir processing offers an alternate method for producing a hard martensitic structure in AISI 4337 steel. This technique allows localized surface treatment, making it suitable for applications like machine knives, where only the cutting edge requires high wear resistance while the body remains tough.

The process surface transforms into a brittle martensite phase without the need to heat the entire part, resulting in increased hardness. Surface hardening through FSP occurs via three mechanisms: grain refinement, phase transformation, and improved carbon solubility. The tool temperature in this work reached 940 °C, with recorded peak temperatures of 1000 °C during processing. Rapid cooling reduced the temperature to 780 °C, promoting the transformation of austenite into the much harder martensite. As a result, the FSPed steel exhibited finer recrystallized grains with the highest hardness.

Further research on the sliding wear behavior of Inconel-625 clad deposits indicates that combining microwave heating with FSP can improve wear resistance and microstructural characteristics of materials [27–32]. Additionally, studies on the influence of process parameters on the hardness of welded joints have provided valuable insights into optimizing FSP conditions to achieve desired mechanical properties [33–36]. These advancements emphasize the importance of FSP in tailoring material properties for specific industrial applications, particularly with respect to improving surface hardness, wear resistance, and overall durability. Through a combination of experimental testing and computational modeling, FSP continues to demonstrate promise for surface engineering across industries such as aerospace, automotive, and biomedical [26, 37–39].

8.4 CONCLUSION

The study in this chapter was an investigation of the effects of friction stir processing (FSP) on the microstructure and mechanical properties of AISI 4337 steel. FSP resulted in significant grain refinement and the formation of a dual-phase martensite–ferrite structure in the processed zone. Microhardness increased from 344 HV in the base material to 548 HV in the FSPed steel. Yield strength also improved from 550 MPa to 720 MPa, although a slight reduction in elongation was observed. The enhanced properties were attributed to grain refinement, phase transformation, and increased dislocation density. Optical microscopy and electron backscatter diffraction were used to analyze the microstructural changes. The study highlights the potential of FSP for improving the performance of AISI 4337 steel in automotive and engineering applications.

REFERENCES

1. Singh, M., Vasudev, H. and Kumar, R., 2021. Corrosion and tribological behaviour of BN thin films deposited using magnetron sputtering. *International Journal of Surface Engineering and Interdisciplinary Materials Science (IJSEIMS)*, *9*(2), 24–39.
2. Singh, M., Vasudev, H. and Singh, M., 2022. Surface protection of SS-316L with boron nitride based thin films using radio frequency magnetron sputtering technique. *Journal of Electrochemical Science and Engineering*, *12*(5), 851–863.

3. Singh, G., Singh, H., Sharma, Y., Vasudev, H. and Prakash, C., 2024. Analysis and optimization of various process parameters and effect on the hardness of SS-304 stainless steel welded joints. *International Journal on Interactive Design and Manufacturing (IJIDeM), 18*(5), 2889–2896.
4. Singh, G., Mehta, A. and Vasudev, H., 2024. Sustainability of additive manufacturing: A comprehensive review. *Progress in Additive Manufacturing*, 1–24.
5. Singh, G., Bansal, A., Vasudev, H. and Mishra, V., 2024. Sliding wear study of the Inconel-625 clad deposits by microwave heating on SS–304. *Physica Scripta*, *99*(6), 065503.
6. Singh, J., Yadav, A., Gupta, S., Balu, R., Shukla, N., Singh, G. and Singh, S., 2024. Design and computational analysis of continuous wave fiber laser cutting of flat and conformal ITO PET sheets. *International Journal on Interactive Design and Manufacturing (IJIDeM)*, 1–9.
7. Singh, G., Singh, R. and Gul, J., 2023. Machinability behavior of human implant materials. *Journal of Electrochemical Science and Engineering*, *13*(1), 99–114.
8. Verma, P.K., Vasudev, H., Singh, G. and Singh, S., 2024. A review on the microstructural and biomedical properties of implants manufactured using additive manufacturing. *Journal of the Institution of Engineers (India): Series D*, 1–15.
9. Mehta, A., Singh, G. and Vasudev, H., 2024. Processing of shape memory alloys research, applications and opportunities: A review. *Physica Scripta*. https://iopscience.iop.org/article/10.1088/1402-4896/ad48cb
10. Singh, G., Bansal, A. and Vasudev, H., 2024. Corrosion and microstructural behaviour of Inconel 625 microwave clad deposited on mild steel. In *Thermal Claddings for Engineering Applications* (pp. 51–62). CRC Press.
11. Kapade, T.T., Shah, S.K., Shah, D.S., Kumar, N.S., Bhor, J.K., Panwar, V.S. and Singh, G., 2024. Thermal analysis of v-cut twisted tape fitted heat exchanger: A numerical approach. *International Journal on Interactive Design and Manufacturing (IJIDeM)*. https://link.springer.com/article/10.1007/s12008-024-01883-2
12. Bhowmik, A., Kumar, R., Beemkumar, N., Kumar, A.V., Singh, G., Kulshreshta, A. and Santhosh, A.J., 2024. Casting of particle reinforced metal matrix composite by liquid state fabrication method: A review. *Results in Engineering*, 103152.
13. Kulkarni, S.D., Manjunatha, C.U., Manjunath, K.V., Prasad, C.D. and Vasudev, H., 2024. Design and optimization of polyvinyl-nitride rubber for tensile strength analysis. *International Journal on Interactive Design and Manufacturing (IJIDeM), 18*(8), 5893–5908.
14. Prashar, G. and Vasudev, H., 2022. Surface topology analysis of plasma sprayed Inconel625-Al_2O_3 composite coating. *Materials Today: Proceedings*, *50*, 607–611.
15. Yedida, V.S. and Vasudev, H., 2022. A review on the development of thermal barrier coatings by using thermal spray techniques. *Materials Today: Proceedings*, *50*, 1458–1464.
16. Sharanabasava, H., Prasad, C.D. and Ramesh, M.R., 2023. Microstructure, mechanical and wear properties of SiC and Mo reinforced NiCr microwave cladding. *Advances in Materials and Processing Technologies*, *10*(4), 3620–3633. https://doi.org/10.1080/2374068X.2023.2257937.
17. Dutta, V., Thakur, L., Singh, B. and Vasudev, H., 2022. A study of erosion–corrosion behaviour of friction stir-processed chromium-reinforced NiAl bronze composite. *Materials*, *15*(15), 5401.
18. Poojari, M., Hanumanthappa, H., Prasad, C.D., Jathanna, H.M., Ksheerasagar, A.R., Shetty, P., Shanmugam, B.K. and Vasudev, H., 2024. Computational modelling for the manufacturing of solar-powered multifunctional agricultural robot. *International Journal on Interactive Design and Manufacturing (IJIDeM)*, *18*(8), 5725–5736.

19. Prashar, G. and Vasudev, H., 2022. A review on the influence of process parameters and heat treatment on the corrosion performance of Ni-based thermal spray coatings. *Surface Review and Letters*, *29*(1), 2230001.
20. Kattimani, M.A., Venkatesh, P.R., Masum, H., Math, M.M., Bahadurdesai, V.N., Mustafkhadri, S., Prasad, C.D. and Vasudev, H., 2024. Design and numerical analysis of tensile deformation and fracture properties of induction hardened inconel 718 superalloy for gas turbine applications. *International Journal on Interactive Design and Manufacturing (IJIDeM)*, *18*(8), 5939–5949.
21. Sunitha, K. and Vasudev, H., 2022. A short note on the various thermal spray coating processes and effect of post-treatment on Ni-based coatings. *Materials Today: Proceedings*, *50*, 1452–1457.
22. Majji, B.G.R., Vasudev, H. and Bansal, A., 2022. A review on the oxidation and wear behavior of the thermally sprayed high-entropy alloys. *Materials Today: Proceedings*, *50*, 1447–1451.
23. Rao, G.S., Mukkamala, U., Hanumanthappa, H., Prasad, C.D., Vasudev, H., Shanmugam, B. and Kishore Kumar, K.C., 2024. Evaluating and optimizing surface roughness using genetic algorithm and artificial neural networks during turning of AISI 52100 steel. *International Journal on Interactive Design and Manufacturing (IJIDeM)*, *18*(8), 6151–6160.
24. Singh, J., Gill, H.S. and Vasudev, H., 2023. Computational fluid dynamics analysis on role of particulate shape and size in erosion of pipe bends. *International Journal on Interactive Design and Manufacturing (IJIDeM)*, *17*(5), 2631–2646.
25. Anjaneya, G., Sunil, S., Kakkeri, S., Math, M.M., Vaibhav, M.N., Solaimuthu, C., Durga Prasad, C. and Vasudev, H., 2023. Numerical simulation of microchannel heat exchanger using CFD. *International Journal on Interactive Design and Manufacturing (IJIDeM)*, 1–17.
26. Singh, J., Vasudev, H., Kumar, R. and Ubaidullah, M., 2024. Study on wear analysis of Ni-20Al_2O_3 HVOF micron layers using artificial neural network technique. *International Journal on Interactive Design and Manufacturing (IJIDeM)*, *18*(8), 5925–5938. https://link.springer.com/article/10.1007/s12008-023-01433-2
27. Nagabhushana, P., Ramprasad, S., Prasad, C.D., Vasudev, H. and Prakash, C., 2024. Numerical investigation on heat transfer of a nano-fluid saturated vertical composite porous channel packed between two fluid layers. *International Journal on Interactive Design and Manufacturing (IJIDeM)*, *18*(5), 2927–2944.
28. Singh, P., Vasudev, H. and Bansal, A., 2024. Effect of post-heat treatment on the microstructural, mechanical, and bioactivity behavior of the microwave-assisted alumina-reinforced hydroxyapatite cladding. *Proceedings of the Institution of Mechanical Engineers, Part E: Journal of Process Mechanical Engineering*, *238*(3), 1060–1071.
29. Mahantesh, M.M., Rao, K.R., Chandra, A.P., Vijayakumar, M.N., Nandini, B., Prasad, C.D. and Vasudev, H., 2024. Design and modeling using finite element analysis for the sitting posture of computer users based on ergonomic perspective. *International Journal on Interactive Design and Manufacturing (IJIDeM)*, *18*(8), 5875–5891.s
30. Singh, J., Singh, R., Singh, S., Vasudev, H. and Kumar, S., 2023. Reducing scrap due to missed operations and machining defects in 90PS pistons. *International Journal on Interactive Design and Manufacturing (IJIDeM)*, *17*(5), 2527–2539.
31. Arora, H., Kumar, V., Prakash, C., Pimenov, D., Singh, M., Vasudev, H. and Singh, V., 2021. Analysis of sensitization in austenitic stainless steel-welded joint. In *Advances in Metrology and Measurement of Engineering Surfaces: Select Proceedings of ICFMMP 2019* (pp. 13–23). Springer Singapore.
32. Vasudev, H., 2020. Wear characteristics of Ni-WC powder deposited by using a microwave route on mild steel: Microwave cladding of Ni-WC. *International Journal of Surface Engineering and Interdisciplinary Materials Science (IJSEIMS)*, *8*(1), 44–54.

33. Prashar, G., Vasudev, H., Thakur, L. and Bansal, A., 2023. Performance of thermally sprayed hydroxyapatite coatings for biomedical implants: A comprehensive review. *Surface Review and Letters*, *30*(1), 2241001.
34. Mehta, A., Vasudev, H. and Jeyaprakash, N., 2024. Role of sustainable manufacturing approach: Microwave processing of materials. *International Journal on Interactive Design and Manufacturing (IJIDeM)*, *18*(8), 5283–5299.
35. Prashar, G. and Vasudev, H., 2023. Understanding cold spray technology for hydroxyapatite deposition. *Journal of Electrochemical Science and Engineering*, *13*(1), 41–62.
36. Prashar, G. and Vasudev, H., 2023. Understanding cold spray technology for hydroxyapatite deposition. *Journal of Electrochemical Science and Engineering*, *13*(1), 41–62.
37. Parkash, J., Saggu, H.S. and Vasudev, H., 2022. A short review on the performance of high velocity oxy-fuel coatings in boiler steel applications. *Materials Today: Proceedings*, *50*, 1442–1446.
38. Singh, J., Vasudev, H., Szala, M. and Gill, H.S., 2024. Neural computing for erosion assessment in Al-20TiO_2 HVOF thermal spray coating. *International Journal on Interactive Design and Manufacturing (IJIDeM)*, *18*(4), 2321–2332.
39. Prakash, C., Gupta, L.R., Mehta, A., Vasudev, H., Tominov, R., Korman, E., Fedotov, A., Smirnov, V. and Kesari, K.K., 2023. Computing of neuromorphic materials: An emerging approach for bioengineering solutions. *Materials Advances*, *4*(23), 5882–5919.

9 Environmentally Friendly Machining of Aluminum Alloys Using a Coated Tool Under Dry and MQL Conditions

Jasjeevan Singh, Atul Agnihotri, Suresh Gudala, Pankaj Sharma, and Gurbhej Singh

Abbreviations

Al	Aluminum
BUL	Built up layer
BUE	Built up edge
MQL	Minimum quantity lubrication
TBL	Triple bottom line
MF	Metalworking fluid
DOC	Depth of cut
PVD	Physical vapor deposition
CVD	Chemical vapor deposition
GRA	Grey relational analysis
TiAlN	Titanium aluminum nitride
AlCrN	Aluminum chromium nitride
SEM	Scanning electron microscopy
DSS	Duplex stainless steel

9.1 INTRODUCTION

Aluminum alloys are used instead of steel in lightweight and non-maintenance applications, particularly in the aviation sector [1]. AA2024-T3 has found a wide range of application due to its high reliability in such crucial structural applications [2]. It is often used in aircraft fuselages and wings and in auto parts, sporting apparatus, and military equipment [2, 3].

The primary elements in AA2024-T3 include copper, magnesium, and manganese among others [3]. It has a very good strength-to-weight ratio and superior

DOI: 10.1201/9781003546528-9

fatigue resistance, making it highly suited for demanding applications [4]. However, dry machining aluminum alloys is very challenging [5]. Aluminum tends to stick on dry cutting tools, building up in layers and on edges [6]. These accumulated deposits aggravate tool wear and limit the performance of machining [6]. Sanchez et al. reported on tool wear in dry machining of AA2024 and AA7050 [7]. Dry machining avoids the use of coolant, but it does not perform well with aluminum due to its high sticking [8].

To address this, there is potential in minimum quantity lubrication (MQL) [9]. In MQL, a limited amount of oil is sent to the cutting area with the use of compressed air [9, 10]. The rates of oil flows normally vary between 10 and 100 mL/hour [10]. MQL enhances the recycling of the chips and alleviates the problem of coolant contamination [11]. It also improves the surface finish, decreases power consumption, and minimizes tool wear and hardness in turning aluminum [12].

9.1.1 Sustainable Machining

Standard machining processes are used in different metal-cutting machines to form parts by cutting off the redundant material. New technology in manufacturing costs less than conventional machining, which is still prevalent in automotive and aviation industries. It enables producing complex geometries with superior surface finish and high dimensional.

The commonest traditional machining techniques include turning, grinding, milling, and drilling, and similar to other manufacturing processes, conventional machining produces a great deal of waste material. This wastage affects the well-being of operators and is disastrous to the environment [13]. Sustainable machining embraces the idea of environmental friendliness in machining, recycling, and utilization of cleaner sources of energy [14]. The concept was developed based on policies, available tools, and eco-friendly practices. Saving energy reduces greenhouse gases and lowers operation costs as well. Conventional machining also uses a great deal of electricity, which heightens its impact on the environment [15].

The manufacturers are under pressure to increase productivity, quality and reduce cost. However, they must also address the health and environmental impacts associated with machining [16]. Environmental regulations are also on the increase, which has prompted the waste minimization and sustainable manufacturing, which also focuses on low environmental impacts and energy consumption and worker safety [16].

It also incorporates the triple bottom line (TBL), that is, environmental, economic, and social responsibility. TBL is presented as a Venn diagram and the input-output of machining in Figure 9.1 [17]. The machining inputs comprise a workpiece, a machine, the cutting tool, coolant, and energy. Products are final products together with waste, which consist of chips, coolant mist, dust, and noise. This waste can harm worker health and cause soil, water, and air pollution.

The figure shows that under the TBL, decreasing the use of metal working fluid (MWF) during cutting increases the sustainability of the manufacturing process. Reducing fluid use affects all three components of the TBL—environmental, economic, and social.

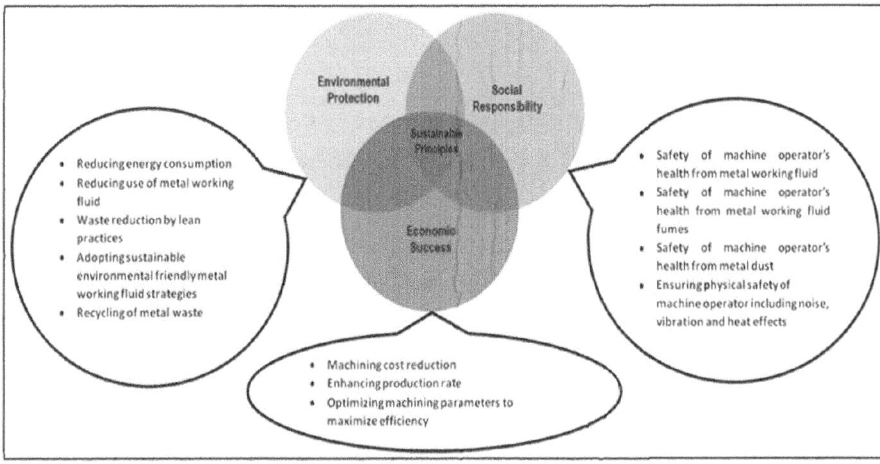

FIGURE 9.1 The triple bottom line applied in machining processes.

9.1.2 Metal Working Fluid Strategies

MWFs are important in machining for cooling and lubricating the cutting tools and materials, eliminating chips, and guarding against corrosion. Effective MWF use can significantly improve machining efficiency, tool longevity, and workpiece quality. Turning experiments were carried out employing two approaches to determine which if either would eliminate or greatly reduce the amount of fluid used. The two approaches were dry machining and MQL; dry machining used no fluid, and edible soya bean oil was used for MQL. MQL creates an aerosol by mixing a small amount of oil with compressed air that is then misted onto the cutting region for lubrication and cooling. The cutting coolant was supplied at the chip–tool interface at 90 mL/h with compressed air at 6×105 N/m^2.

9.1.3 Coated Tools

Coating is a surface modification approach to fulfill market demands. It was developed in the 1960s and has become widely used in metal-cutting tool manufacturing. The introduction of high-speed cutting methods resulted in increased use of coatings, which became an essential component of high-speed cutting tool manufacturing. Numerous parameters can influence a tool's cutting performance and workpiece quality, including cutting speed, feed, depth of cut (DOC), lubrication regimen, and tool malfunctions. Researchers have investigated cutting tool behavior with respect to appropriate coatings (Figure 9.2).

9.1.3.1 Coating accumulation techniques for coated tools

Physical vapor deposition (PVD) is a process that deposits material under vacuum conditions through either evaporation or sputtering. Chemical vapor deposition (CVD) is the heat-driven chemical reaction that occurs on the substrate that has been

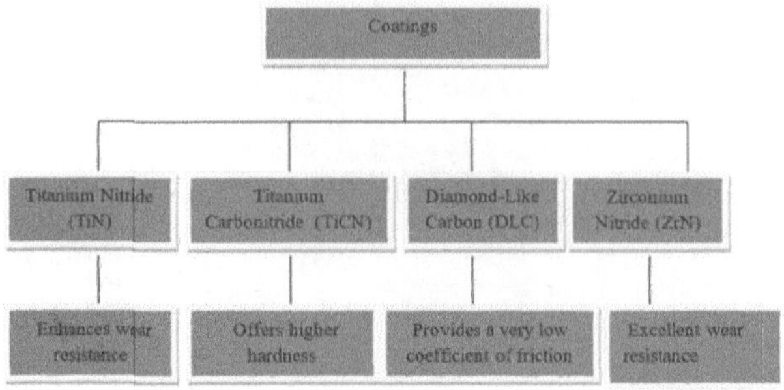

FIGURE 9.2 Types of coatings.

preheated. PVD and CVD are based on type of material, film characteristics, and compatibility with the substrate.

Ramanujam et al. research worked on PVD coated and uncoated carbide tools during turning Inconel 825 [18]. Wear and surface finish of TiAlN-AlCrN bilayered coatings were better than single coatings. The authors calculated the best machining parameters under multi-response criteria using grey relational analysis.

Koseki et al. discussed TiN-coated tools prepared employing different deposition techniques [19]. The results indicated that CVD coatings performed better when applied with continuous machining Ni-based superalloys. Thakur and Gangopadhyay reported dry turning TiN/TiAlN PVD-coated tool on Incoloy 825 [20]. They measured force, temperature, wear, and surface quality to compare dry, flood, and MQL machining. High temperature and PVD tools produced a better finish through dry machining than with flood and MQL machinine. Under dry conditions, the lowest cutting forces were obtained when the PVD coating was used, which promoted environmentally friendly operations.

Martinho et al. also compared CVD- and PVD-coated inserts in duplex stainless steel (DSS) [21]. Tool performance was measured based on surface roughness, tool wear, and vibration. CVD (TiN/TiCN/A12O3) coatings were demonstrated to be suitable for rough DSS milling. During dry machining of Ti alloys, PVD coatings provided the least wear and the most prolonged tool life [22]. The coating type had a stronger influence on tool wear than cutting speed and feed rate. The CVD-coated tools were superior to the PVD-coated tools because of relatively thick Al_2O_3–TiCN layers shielding the substrates [23]. The PVD TiN-coated AISI H13 offered more heat, with greater force and surface roughness than with CVD [24]. SEM analysis showed more adhesion and abrasion on CVD Al_2O_3 tools than on PVD TiN tools.

Coatings provide enhanced benefits to both the cutting tool and the machining process, increasing tool life and improving cutting performance. They reduce cutting heat and aluminum adhesion and enhance the surface finish of the machined part. The choice of a coating that will best suit the requirements depends on particular machining parameters and conditions.

Gupta et al. machined AA2024-T351 using MQL, LN_2, and CO_2 cooling [25]. All three cooling methods improved the alloy machinability compared with dry cutting.

Surface roughness improved by 39.28% (CO_2), 30.95% (LN_2), and 11.90% (MQL) over dry machining. In particular, Paresi et al. investigated orthogonal machining of AA2024-T351 with uncoated carbide tools [26] by varying cutting and feed rates and holding DOC constant. The machinability was determined based on cutting forces, chip morphology, and coefficient of friction (COF); cutting and feed forces rose and COF decreased at higher feed rates. Uncut chip thickness and tool rake angle had an effect on the chip segmentation [27]. Blocked-up edges formed based on variations in friction at the tool–work interface along with the ductility of AA2024-T351.

9.2 EXPERIMENTAL SETUP AND METHOD

9.2.1 Workpiece, Cutting Tool, and Machine

For the experiment for this chapter, AA2024-T3 aluminum used as the workpiece material was provided by Bharat Aerospace Metal, Mumbai, with the dimensions of 35 mm diameter and 230 mm length. Table 9.1 provides the mechanical features. The turning machine was an Okuma with spindle speeds of 30–4100 rpm and a power of 16 hp. Cutting insert CNMG120408MS is shown in Figure 9.3, and the experimental setup is shown in Figure 9.4. The cutting insert specifications were cutting edge length, 16.5 mm; thickness, 3.97 mm; corner radius, 0.8 mm; and hole size, 4.4 mm, and Table 9.2 summarizes the experimental parameters.

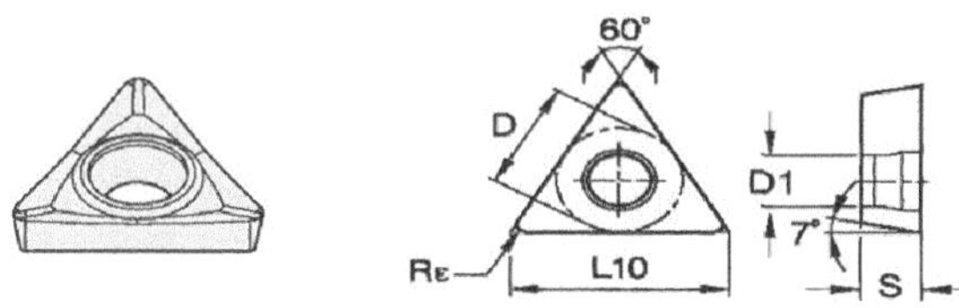

FIGURE 9.3 Diagram of the cutting insert used in the study.

TABLE 9.1
Testing Parameters

Inputs	Details
Machine	Okuma Lathe Machine
Cutting speed (m/min)	350, 450
Feed rate (mm/rev)	0.1
Depth of cut (mm)	1
Cooling/lubrication strategy	Dry, MQL
MQL parameters	Air pressure 8 bar, flow rate 100mL/h

FIGURE 9.4 Experimental setup.

TABLE 9.2
Chip Morphology with Different MWF Approaches

Dry	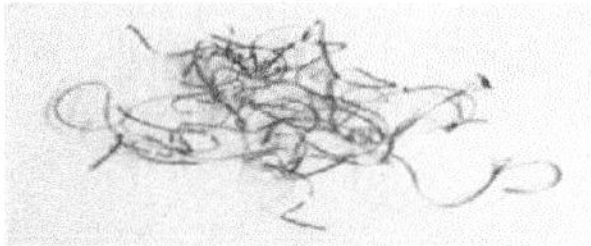
MQL	

9.3 RESULTS AND DISCUSSION

9.3.1 Maximum Temperature

Cutting temperatures significantly impact key machining indicators like surface roughness, chip formation, and tool wear, all of which are linked to machining performance [28]. Figure 9.5 shows that dry machining delivered the highest cutting temperature at cutting speeds of 350 and 450 m/min. However, excessive cutting temperatures in the region being cut caused substantial damage to both the cutting tool and the workpiece, and cooling techniques are favored for reducing the temperature at the cutting area. Dry machining uses no coolant or lubricant, which results in higher temperatures; in contrast, MQL reduced cutting temperature by 15% in comparison. Figure 9.5 indicates that regardless of the cooling and lubrication conditions, tool temperature increases as cutting speed increases because of the increased strain rate, which Resis et al. observed as well [29].

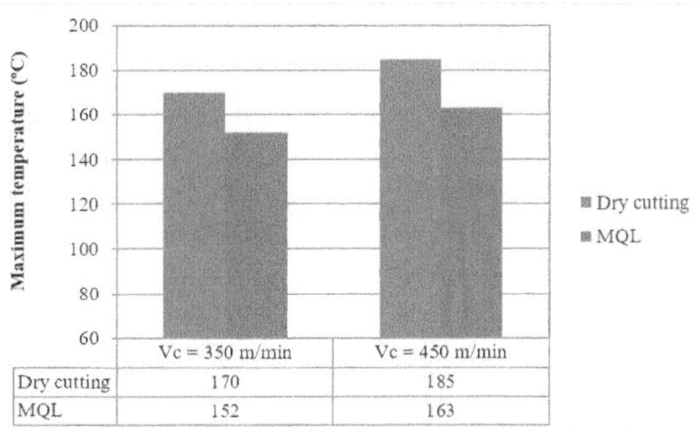

FIGURE 9.5 Maximum temperatures under dry cutting and MQL at cutting speeds of 350 and 450 mm/min.

9.3.2 Tool Wear

Tool wear is the most essential determinant of tool life, and decreasing tool wear extends productivity; the condition of the tools affects the quality of made parts and has cost implications as well. Figure 9.6 shows microscopic views of flank wear under dry machining and MQL at 350 mm/min, and Figure 9.7 depicts the effects of the MWF conditions on flank wear VBmax at both speeds (350 and 450 mm/min). Flank wear increases because of elevated temperatures, material deformation, and tool rubbing on the workpiece. MQL uses cooling fluids to disperse heat from the cutting region and lubrication to reduce tool–workpiece friction, and as a result, there is less flank wear and less friction during machining, particularly at the cutting area. The figures show that tool wear increased rapidly at first and then rose gradually until it reached its limiting value.

9.3.3 Chip Morphology

Chip morphology refers to the size, form, and kind of chips created through machining operations like cutting, grinding, and drilling. Analyzing chip morphology is essential in manufacturing because it offers information about machining process efficiency, flank wear, and material behavior. Table 9.3 shows chip morphology under MQL and dry machining. Dry conditions produce continuous-type chips, potentially because of low feed rates and ductile workpieces. MQL chips are more discontinuous: The process isolates the chips, minimizes friction, and prevents the material from sticking.

Both dry and MQL machining (coated tools) of aluminum alloys have become popular. Changes accomplished using these methods lessen environmental effects and enhance operation in industrial machining [30–39]. The application of coatings such as boron nitride (BN) on the tool surface increases surface tribology features and corrosion resistance. BN coating reduced wear in problem machining conditions and prolonged tool life [40–48]. The magnetron sputtering-deposited BN thin films enhanced the protection of stainless steel tools from wear and corrosion. BN

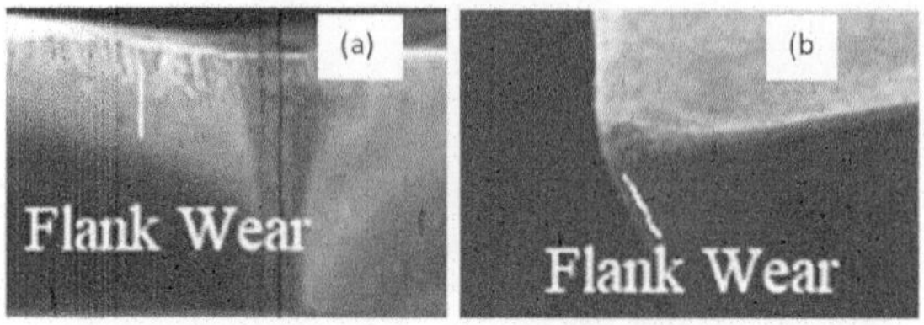

FIGURE 9.6 Tool flank wear at 350 mm/min: (a) dry machining; (b) MQL.

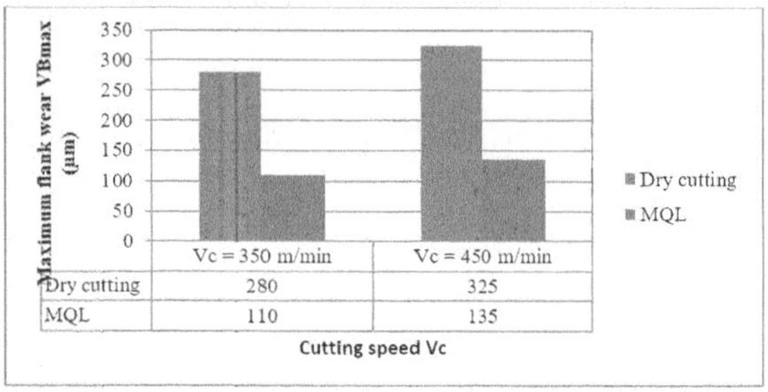

FIGURE 9.7 The effects of metal working fluid on flank wear at 350 and 450 mm/min.

coating also reduces tool wear and enhances the surface finish in machining aluminum alloys [49–53].

MQL is a green alternative to conventional flood lubrication. It lowers the consumption of cutting fluids and supports environmentally friendly manufacturing. MQL used with BN-coated tools reduces friction and heat to improve machining, tool performance, and surface quality of aluminum alloy components [54–63]. Dry and MQL machining using coated tools are less harmful to the environment than flood cooling. The operations result in lower emissions, fluid discharge, and use of resources. Therefore, the integration of dry or MQL machining with coated facilitates green manufacturing [64–68].

9.4 CONCLUSION

MQL results in less tool wear than dry machining by reducing cutting temperatures at the tool–workpiece interface by up to 15% over dry turning. VBmax is lower with MQL than with dry cutting, and tool flank wear increases with cutting speed under both cutting conditions. However, although MQL can minimize tool replacement costs and improve surface quality, it does incur certain expenditures for the lubricant and the MQL system.

REFERENCES

1. Singh, J., Gill, S.S. and Mahajan, A., 2023. Experimental investigation and optimizing of turning parameters for machining of al7075-t6 aerospace alloy for reducing the tool wear and surface roughness. *Journal of Materials Engineering and Performance*, pp.1–12.

2. Yücel, A., Yıldırım, Ç.V., Sarıkaya, M., Şirin, Ş., Kıvak, T., Gupta, M.K. and Tomaz, Í.V., 2021. Influence of MoS2 based nanofluid-MQL on tribological and machining characteristics in turning of AA 2024 T3 aluminum alloy. *Journal of Materials Research and Technology*, 15, pp.1688–1704.
3. Singh, J., Gill, S.S., Dogra, M., Sharma, S., Singh, M., Dwivedi, S.P., Li, C., Singh, S., Muhammad, S., Salah, B. and Shamseldin, M.A., 2022. Effect of Ranque-Hilsch vortex tube cooling to enhance the surface-topography and tool-wear in sustainable turning of Al–5.6 Zn–2.5 Mg–1.6 Cu–0.23 Cr-T6 aerospace alloy. Materials, 15(16), p.5681.
4. Reis, D.A., Couto, A.A., Domingues, N.I., Jr., Hirschmann, A.C., Zepka, S. and de Moura Neto, C., 2012 May. Effect of artificial aging on the mechanical properties of an aerospace aluminum alloy 2024. In *Defect and Diffusion Forum* (Vol. 326, pp. 193–198). Trans Tech Publications Ltd.
5. Li, K.M. and Liang, S.Y., 2007. Performance profiling of minimum quantity lubrication in machining. *The International Journal of Advanced Manufacturing Technology*, 35, pp.226–233.
6. Singh, J., Gill, S.S., Dogra, M. and Singh, R., 2021. A brief study on machinability of aluminium alloys. In *Advances in Engineering Materials: Select Proceedings of FLAME 2020* (pp. 579–591). Springer Singapore.
7. Carrilero, M.S., Bienvenido, R., Sánchez, J.M., Alvarez, M., González, A. and Marcos, M., 2002. A SEM and EDS insight into the BUL and BUE differences in the turning processes of AA2024 Al–Cu alloy. *International Journal of Machine Tools and Manufacture*, 42(2), pp.215–220.
8. Zare Chavoshi, S., 2011. Tool flank wear prediction in CNC turning of 7075 AL alloy SiC composite. *Production Engineering*, 5, pp.37–47.
9. Yıldırım, Ç.V., Kıvak, T., Sarıkaya, M. and Erzincanlı, F., 2017. Determination of MQL parameters contributing to sustainable machining in the milling of nickel-base superalloy waspaloy. *Arabian Journal for Science and Engineering*, 42, pp.4667–4681.
10. Singh, J., Gill, S.S., Dogra, M. and Singh, R., 2021. A review on cutting fluids used in machining processes. *Engineering Research Express*, 3(1), p.012002.
11. Boubekri, N. and Shaikh, V., 2015. Minimum quantity lubrication (MQL) in machining: Benefits and drawbacks. *The Journal of Industrial Information Integration*, 3(3).
12. Singh, J. and Singh Gill, S., 2023. Turning of Al 7075-T6 aerospace alloy under different sustainable metalworking fluid strategies by coated carbide tools. *Surface Engineering*, 39(3), pp.275–294.
13. Inasaki, I., 2012. Towards symbiotic machining processes. *International Journal of Precision Engineering and Manufacturing*, 13, pp.1053–1057.
14. Haleem, A., Javaid, M., Singh, R.P., Suman, R. and Qadri, M.A., 2023. A pervasive study on green manufacturing towards attaining sustainability. *Green Technologies and Sustainability*, 1(2), p.100018.
15. Elemure, I., Dhakal, H.N., Leseure, M. and Radulovic, J., 2023. Integration of lean green and sustainability in manufacturing: A review on current state and future perspectives. *Sustainability*, 15(13), p.10261.
16. Byrne, G. and Scholta, E., 1993. Environmentally clean machining processes—A strategic approach. *CIRP Annals*, 42(1), pp.471–474.
17. Haider, J. and Hashmi, M.S.J., 2014. 8.02 Health and environmental impacts in metal machining processes. *Comprehensive Materials Processing*, 8, pp.7–33.
18. Ramanujam, R., Vignesh, M., Tamiloli, N., Sharma, N., Srivastava, S. and Patel, A., 2018. Comparative evaluation of performances of TiAlN, AlCrN, TiAlN/AlCrN coated carbide cutting tools and uncoated carbide cutting tools on turning Inconel 825 alloy using grey relational analysis. *Sensors and Actuators A: Physical*, 279, pp.331–342.

19. Koseki, S., Inoue, K., Morito, S., Ohba, T. and Usuki, H., 2015. Comparison of TiN-coated tools using CVD and PVD processes during continuous cutting of Ni-based superalloys. *Surface and Coatings Technology*, 283, pp.353–363.
20. Thakur, A. and Gangopadhyay, S., 2016. Dry machining of nickel-based super alloy as a sustainable alternative using TiN/TiAlN coated tool. *Journal of Cleaner Production*, 129, pp.256–268.
21. Martinho, R.P., Silva, F.J.G., Martins, C. and Lopes, H., 2019. Comparative study of PVD and CVD cutting tools performance in milling of duplex stainless steel. *The International Journal of Advanced Manufacturing Technology*, 102, pp.2423–2439.
22. Raja Abdullah, R.I., Redzuwan, B.I., Abdul Aziz, M.S. and Kasim, M.S., 2017. Comparative study of tool wear in milling titanium alloy (Ti-6Al-4V) using PVD and CVD coated cutting tool. *Industrial Lubrication and Tribology*, 69(3), pp.363–370.
23. Yaqoob, S., Ghani, J.A., Jouini, N. and Juri, A.Z., 2024. Performance evaluation of PVD and CVD multilayer-coated tools in machining high-strength steel. *Coatings*, 14(7), pp.2079–6412.
24. Akgün, M., Özlü, B. and Kara, F., 2023. Effect of PVD-TiN and CVD-Al_2O_3 coatings on cutting force, surface roughness, cutting power, and temperature in hard turning of AISI H13 steel. *Journal of Materials Engineering and Performance*, 32(3), pp.1390–1401.
25. Gupta, M.K., Niesłony, P., Sarikaya, M., Korkmaz, M.E., Kuntoğlu, M. and Królczyk, G.M., 2023. Studies on geometrical features of tool wear and other important machining characteristics in sustainable turning of aluminium alloys. *International Journal of Precision Engineering and Manufacturing-Green Technology*, 10(4), pp.943–957.
26. Reddy Paresi, R., Arunachalam, N., Lou, Y. and Yoon, J.W., 2021. Machinability studies of AA2024-T351 alloy with uncoated carbide tool. *Materials Today: Proceedings*, 44(1), pp.1058–1064.
27. Haddag, B., Atlati, S., Nouari, M. and Moufki, A., 2016. Dry machining aeronautical aluminum alloy AA2024-T351: Analysis of cutting forces, chip segmentation and built-up edge formation. *Metals*, 6(9), p.197.
28. Singh, J., Gill, S.S., Dogra, M., Singh, R., Singh, M., Sharma, S., Singh, G., Li, C. and Rajkumar, S., 2022. State of the art review on the sustainable dry machining of advanced materials for multifaceted engineering applications: Progressive advancements and directions for future prospects. *Materials Research Express*, 9(6), p.064003.
29. Reis, D.D. and Abrao, A.M., 2005. The machining of aluminium alloy 6351. *Proceedings of the Institution of Mechanical Engineers, Part B: Journal of Engineering Manufacture*, 219(1), pp.27–33.
30. Singh, G., Bansal, A., Vasudev, H. and Mishra, V., 2024. Sliding wear study of the Inconel-625 clad deposits by microwave heating on SS–304. *Physica Scripta*, 99(6), p.065503.
31. Singh, J., Yadav, A., Gupta, S., Balu, R., Shukla, N., Singh, G. and Singh, S., 2024. Design and computational analysis of continuous wave fiber laser cutting of flat and conformal ITO PET sheets. *International Journal on Interactive Design and Manufacturing (IJIDeM)*, 1–9.
32. Singh, G., Singh, R. and Gul, J., 2023. Machinability behavior of human implant materials. *Journal of Electrochemical Science and Engineering*, 13(1), pp.99–114.
33. Sharanabasava, H., Prasad, C.D. and Ramesh, M.R., 2023. Microstructure, mechanical and wear properties of SiC and Mo reinforced NiCr microwave cladding. *Advances in Materials and Processing Technologies*, 10(4), pp.3620–3633. https://doi.org/10.1080/2374068X.2023.2257937.
34. Dutta, V., Thakur, L., Singh, B. and Vasudev, H., 2022. A study of erosion–corrosion behaviour of friction stir-processed chromium-reinforced NiAl bronze composite. *Materials*, 15(15), p.5401.

35. Poojari, M., Hanumanthappa, H., Prasad, C.D., Jathanna, H.M., Ksheerasagar, A.R., Shetty, P., Shanmugam, B.K. and Vasudev, H., 2024. Computational modelling for the manufacturing of solar-powered multifunctional agricultural robot. *International Journal on Interactive Design and Manufacturing (IJIDeM)*, 18(8), pp.5725–5736.
36. Vasudev, H., 2020. Wear characteristics of Ni-WC powder deposited by using a microwave route on mild steel: Microwave cladding of Ni-WC. *International Journal of Surface Engineering and Interdisciplinary Materials Science (IJSEIMS)*, 8(1), pp.44–54.
37. Prashar, G., Vasudev, H., Thakur, L. and Bansal, A., 2023. Performance of thermally sprayed hydroxyapatite coatings for biomedical implants: A comprehensive review. *Surface Review and Letters*, 30(1), p.2241001.
38. Mehta, A., Vasudev, H. and Jeyaprakash, N., 2024. Role of sustainable manufacturing approach: Microwave processing of materials. *International Journal on Interactive Design and Manufacturing (IJIDeM)*, 18(8), pp.5283–5299.
39. Singh, M., Vasudev, H. and Kumar, R., 2021. Corrosion and tribological behaviour of BN thin films deposited using magnetron sputtering. *International Journal of Surface Engineering and Interdisciplinary Materials Science (IJSEIMS)*, 9(2), pp.24–39.
40. Singh, M., Vasudev, H. and Singh, M., 2022. Surface protection of SS-316L with boron nitride based thin films using radio frequency magnetron sputtering technique. *Journal of Electrochemical Science and Engineering*, 12(5), pp.851–863.
41. Singh, G., Singh, H., Sharma, Y., Vasudev, H. and Prakash, C., 2024. Analysis and optimization of various process parameters and effect on the hardness of SS-304 stainless steel welded joints. *International Journal on Interactive Design and Manufacturing (IJIDeM)*, 18(5), pp.2889–2896.
42. Singh, G., Mehta, A. and Vasudev, H., 2024. Sustainability of additive manufacturing: A comprehensive review. *Progress in Additive Manufacturing*, 9, pp.2249–2272.
43. Prashar, G. and Vasudev, H., 2022. Surface topology analysis of plasma sprayed Inconel625-Al_2O_3 composite coating. *Materials Today: Proceedings*, 50, pp.607–611.
44. Yedida, V.S. and Vasudev, H., 2022. A review on the development of thermal barrier coatings by using thermal spray techniques. *Materials Today: Proceedings*, 50, pp.1458–1464.
45. Singh, J., Vasudev, H., Kumar, R. and Ubaidullah, M., 2024. Study on wear analysis of Ni-20Al_2O_3 HVOF micron layers using artificial neural network technique. *International Journal on Interactive Design and Manufacturing (IJIDeM)*, 18(8), pp.5925–5938.
46. Singh, J., Vasudev, H., Szala, M. and Gill, H.S., 2024. Neural computing for erosion assessment in Al-20TiO_2 HVOF thermal spray coating. *International Journal on Interactive Design and Manufacturing (IJIDeM)*, 18(4), pp.2321–2332.
47. Prakash, C., Gupta, L.R., Mehta, A., Vasudev, H., Tominov, R., Korman, E., Fedotov, A., Smirnov, V. and Kesari, K.K., 2023. Computing of neuromorphic materials: An emerging approach for bioengineering solutions. *Materials Advances*, 4(23), pp.5882–5919.
48. Prashar, G. and Vasudev, H., 2022. A review on the influence of process parameters and heat treatment on the corrosion performance of Ni-based thermal spray coatings. *Surface Review and Letters*, 29(1), p.2230001.
49. Prashar, G. and Vasudev, H., 2023. Understanding cold spray technology for hydroxyapatite deposition. *Journal of Electrochemical Science and Engineering*, 13(1), pp.41–62.
50. Prashar, G. and Vasudev, H., 2023. Understanding cold spray technology for hydroxyapatite deposition. *Journal of Electrochemical Science and Engineering*, 13(1), pp.41–62.
51. Parkash, J., Saggu, H.S. and Vasudev, H., 2022. A short review on the performance of high velocity oxy-fuel coatings in boiler steel applications. *Materials Today: Proceedings*, 50, pp.1442–1446.

52. Verma, P.K., Vasudev, H., Singh, G. and Singh, S., 2024. A review on the microstructural and biomedical properties of implants manufactured using additive manufacturing. *Journal of the Institution of Engineers (India): Series D*, pp.1–15.
53. Mehta, A., Singh, G. and Vasudev, H., 2024. Processing of shape memory alloys research, applications and opportunities: A review. *Physica Scripta*.
54. Singh, G., Bansal, A. and Vasudev, H., 2024. Corrosion and microstructural behaviour of Inconel 625 microwave clad deposited on mild steel. In *Thermal Claddings for Engineering Applications* (pp. 51–62). CRC Press.
55. Kapade, T.T., Shah, S.K., Shah, D.S., Kumar, N.S., Bhor, J.K., Panwar, V.S. and Singh, G., 2024. Thermal analysis of v-cut twisted tape fitted heat exchanger: A numerical approach. *International Journal on Interactive Design and Manufacturing (IJIDeM)*. https://link.springer.com/article/10.1007/s12008-024-01883-2
56. Bhowmik, A., Kumar, R., Beemkumar, N., Kumar, A.V., Singh, G., Kulshreshta, A. and Santhosh, A.J., 2024. Casting of particle reinforced metal matrix composite by liquid state fabrication method: A review. *Results in Engineering*, p.103152.
57. Kulkarni, S.D., Manjunatha, C.U., Manjunath, K.V., Prasad, C.D. and Vasudev, H., 2024. Design and optimization of polyvinyl-nitride rubber for tensile strength analysis. *International Journal on Interactive Design and Manufacturing (IJIDeM)*, 18(8), pp.5893–5908.
58. Kattimani, M.A., Venkatesh, P.R., Masum, H., Math, M.M., Bahadurdesai, V.N., Mustafkhadri, S., Prasad, C.D. and Vasudev, H., 2024. Design and numerical analysis of tensile deformation and fracture properties of induction hardened inconel 718 superalloy for gas turbine applications. *International Journal on Interactive Design and Manufacturing (IJIDeM)*, 18(8), pp.5939–5949.
59. Sunitha, K. and Vasudev, H., 2022. A short note on the various thermal spray coating processes and effect of post-treatment on Ni-based coatings. *Materials Today: Proceedings*, 50, pp.1452–1457.
60. Majji, B.G.R., Vasudev, H. and Bansal, A., 2022. A review on the oxidation and wear behavior of the thermally sprayed high-entropy alloys. *Materials Today: Proceedings*, 50, pp.1447–1451.
61. Rao, G.S., Mukkamala, U., Hanumanthappa, H., Prasad, C.D., Vasudev, H., Shanmugam, B. and KishoreKumar, K.C., 2024. Evaluating and optimizing surface roughness using genetic algorithm and artificial neural networks during turning of AISI 52100 steel. *International Journal on Interactive Design and Manufacturing (IJIDeM)*, 18(8), pp.6151–6160.
62. Singh, J., Gill, H.S. and Vasudev, H., 2023. Computational fluid dynamics analysis on role of particulate shape and size in erosion of pipe bends. *International Journal on Interactive Design and Manufacturing (IJIDeM)*, 17(5), pp.2631–2646.
63. Anjaneya, G., Sunil, S., Kakkeri, S., Math, M.M., Vaibhav, M.N., Solaimuthu, C., Durga Prasad, C. and Vasudev, H., 2023. Numerical simulation of microchannel heat exchanger using CFD. *International Journal on Interactive Design and Manufacturing (IJIDeM)*, pp.1–17.
64. Nagabhushana, P., Ramprasad, S., Prasad, C.D., Vasudev, H. and Prakash, C., 2024. Numerical investigation on heat transfer of a nano-fluid saturated vertical composite porous channel packed between two fluid layers. *International Journal on Interactive Design and Manufacturing (IJIDeM)*, 18(5), pp.2927–2944.
65. Singh, P., Vasudev, H. and Bansal, A., 2024. Effect of post-heat treatment on the microstructural, mechanical, and bioactivity behavior of the microwave-assisted alumina-reinforced hydroxyapatite cladding. *Proceedings of the Institution of Mechanical Engineers, Part E: Journal of Process Mechanical Engineering*, 238(3), pp.1060–1071.

66. Mahantesh, M.M., Rao, K.R., Chandra, A.P., Vijayakumar, M.N., Nandini, B., Prasad, C.D. and Vasudev, H., 2024. Design and modeling using finite element analysis for the sitting posture of computer users based on ergonomic perspective. *International Journal on Interactive Design and Manufacturing (IJIDeM)*, 18(8), pp.5875–5891.
67. Singh, J., Singh, R., Singh, S., Vasudev, H. and Kumar, S., 2023. Reducing scrap due to missed operations and machining defects in 90PS pistons. *International Journal on Interactive Design and Manufacturing (IJIDeM)*, 17(5), pp.2527–2539.
68. Arora, H., Kumar, V., Prakash, C., Pimenov, D., Singh, M., Vasudev, H. and Singh, V., 2021. Analysis of sensitization in austenitic stainless steel-welded joint. In *Advances in Metrology and Measurement of Engineering Surfaces: Select Proceedings of ICFMMP 2019* (pp. 13–23). Springer Singapore.

10 Hybrid Surface Processing Techniques to Enhance the Wear and Corrosion Properties of Inconel-800

Satish Kumar, Harvinder Singh, Gurbhej Singh, and Hitesh Vasudev

10.1 INTRODUCTION

The electric discharge machining (EDM) method is an unusual machining technique that is utilized globally in a variety of industrial sectors to produce a broad range of components on challenging-to-machine materials with greater surface finishes and regulated dimensions. Although the electric discharge machining method has a very long history dating back to 1770, it has seen amazing metamorphosis in the past 20 years in many different sectors, leading to a vast body of literature on the subject. Advanced technological industries have been requiring materials with high strength, high hardness, and temperature resistance. Old traditional machining techniques are being replaced by more widely used sophisticated non-conventional EDM for the machining of these kinds of materials. The researchers wanted to provide a critical explanation for the extensive study they conducted on the numerous machining aspects of the various grades of steels and alloys utilizing the EDM technique. Additionally, by using powder-suspended dielectric fluid in the EDM process, the researchers were able to enhance the machining characteristics and surface qualities of the workpiece. In addition to the EDM process, other thermo-electric methods have also been developed and are utilized in the manufacturing industries globally. Researchers have also examined the impact of cryogenic treatment on the machinability of various materials in an effort to increase machining efficiency. Moreover, nickel-based super alloys have a number of applications, and such hard material is not possible to machine with traditional EDM. Because of its enhanced mechanical and thermal qualities, nickel-based super alloy is employed in many common applications, including pump body components, jet engine casings, turbo chargers, and turbine blades for airplanes. These common applications necessitate precise product design and tight tolerances throughout production. Due to their common characteristics, these alloys cannot be

DOI: 10.1201/9781003546528-10

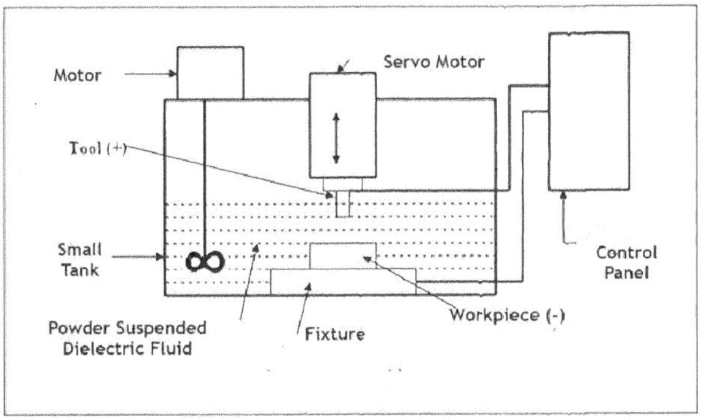

FIGURE 10.1 Schematic diagram of PMEDM setup.

machined using conventional production techniques [1]. The super alloy based on nickel has poor thermal characteristics and great toughness, making it challenging to process using conventional machining techniques. These constraints primarily lead to the employment of unconventional machining techniques, such as electric discharge machining, for processing Inconel [2]. EDM is a technical method that uses powder additives in the EDM oil to change the process variables [3, 4].

A schematic line diagram of a powder mixed electrical discharge machining (PMEDM) setup is shown in Figure 10.1 in which a separate tank has been designed with specific dimensions with a capacity of 10 liters EDM oil. To mix powder particles in the dielectric, a motorized stirrer was installed in the tank's left corner.

When the titanium alloy was machined using pure water and SiC powder in EDM oil, it was found that the use of both additives produced good conductivity and increased MRR and low electrode wear, respectively [5]. Many researchers carried out tests using three distinct titanium alloys, three types of electrodes, and a powder additive for the EDM fluid [6]. Studies using kerosene as the EDM oil revealed that the additions may lower the TWR and raise the MRR [7]. The study used two distinct dielectric fluids—kerosene and kerosene mixed with SiC powder—to compare the EDM technique. It was concluded that using kerosene with SiC powder, instead of plain EDM oil, increased both the material removal depth and the surface roughness (SR) [8]. When it comes to producing high-quality surfaces, the effectiveness of powder-mixed dielectric and traditional EDM were compared, and the results showed that using PMD-EMD encourages the decrease of surface roughness [9]. added powder to EDM oil and used the accretion procedure. The findings showed that powder suspended in the working fluid created two columns and a revolving disk electrode [10]. improved tungsten carbide alloy's surface properties by adding graphite powder to the EDM fluid. In comparison to other parameters, the findings showed that T_{on} is the primary factor [11]. suggested a technique that included adding tungsten carbide powder to EDM fluid. The results showed that the SR had significantly improved [12]. examined the workings of powder-mixed EDM. It was shown that adding abrasive grit to EDM oil increases the TWR and MRR while decreasing the oil's insulating power [13–16]. introduced a novel technique that uses μ-MoS_2 powder particles suspended in EDM oil. Findings showed that MoS_2 micro-powder was used as a dielectric. Caydas and Hascalik made an effort to simulate the output properties that resulted from machining Ti-6Al-4V. Throughout the experiment, five degrees of variation were made to three control factors: peak current (Ip), pulse off-time (T_{off}), and pulse on-time (T_{on}) [17–19]. Pardhan and Bhattacharyya documented the impact of polarity on Ti64 alloy micromachining [20]. Using grey relational analysis, Kao et al. improved the MRR, SR, and electrode wear ratio collectively in the EDM of Ti-6Al-4V [21]. Rahman et al. developed a mathematical model for MRR, TWR, and SR as well as determining the ideal input parameter values (Ip, T_{on}, and T_{off}) for Ti-6Al-4V machining [22, 23]. The impact of process variables Ip, T_{on}, and Toff on the machining of titanium alloy Ti-6Al-4V was examined by Kumar et al. [24]. When milling Ti-6Al-4V alloy, Priyadarshini and Pal investigated the impact of several machining parameters, including duty factor, peak current, gap voltage, and pulse duration [25]. The effect of cryogenic treatment on the most widely used HSS cutting tool material was investigated by Singh and Singh (2011). Their research revealed changes in the microstructure of the tool material that may have a major impact on tool life. In comparison to non-cryogenically treated tools, the test findings showed that cryogenically treated tools exhibit superior machining performance in terms of less tool wear, improved surface smoothness, and lower power consumption during machining operation [26]. According to Yildiz et al. (2011), the MRR of a beryllium copper alloy workpiece treated cold and cryogenically during the EDM process increased by 20%–30%. Workpieces exposed to low and cryogenic temperatures showed slight changes [27]. The findings showed that although MRR was comparable in the case of traditional EDM, EWR and SR in the UACEDM process were found to be much lower than in the conventional EDM method. Compared to the traditional EDM method, the UACEDM process produced better surface integrity [28]. When graphite powder with a concentration of 6 g/liter was utilized during the EDM process, Kumar et al. (2012) found that the DCT copper electrode greatly increased the SR of the workpiece (Inconel 718 super alloy) compared to the none-treated electrode [29]. Rotella et al. (2014) examined the outcomes of machining Ti-6Al-4 V alloy under various situations, including dry, low-quality lubrication, cryogenic cooling, and cutting speeds and feed rates utilizing coated tools. According to the experimental findings, cooling conditions of the cutting tools improve overall machining performance and have a significant impact on the surface integrity of the machined samples [30]. The EDM process is a novel machining process that is used internationally across various industrial fields to manufacture a wide variety of components on difficult-to-machine materials with improved surface finishes and controlled dimensions. Though the electric discharge machining process has an extremely long history since 1770, the last 20 years have witnessed tremendous transformation of the process in numerous industries, resulting in a sea of literature on the topic. High-strength, high-hardness, and temperature-resistant materials have been needed by advanced technology industries. Old conventional machining methods are being replaced by more commonly used advanced non-conventional EDM for machining these types of materials. The researchers aimed to present a critical justification for the extensive research they carried out on the many machining parameters of the different grades of steels and alloys using the EDM method. Moreover, with the EDM process employing powder-suspended dielectric fluid, the researchers could improve the workpiece's machining properties and surface finishes. Besides the EDM process, other thermo-electric techniques have also been established and are practiced in manufacturing sectors across the world. Researchers have also studied the effect of cryogenic treatment on the machinability of different materials with the aim of improving machining efficiency. In addition, nickel-based super alloys have a wide range of uses, and such a tough material is not possible to machine with conventional EDM. Due to its improved mechanical and thermal characteristics, nickel-based super alloy is used in numerous routine applications, which include pump body parts, jet engine casings, turbo chargers, and airplane turbine blades.

10.1.1 Preliminary Study to Assess the Process Parameters

In EDM, to identify the control factors and their impact on process performance, a cause and effect diagram has been developed, shown in Figure 10.2. The factors or process parameters that affect the machining characteristics can be classified in major categories as follows:

Electrical Parameters: Current, T_{on}, T_{off}, supply voltage, polarity, duty cycle, pulse wave form, and so on.

Non-Electrical Parameters: Flushing pressure, working time, lifetime, flushing method, jump time, electrode rotation, and so on.

Powder Additives and Dielectric Parameters: Properties of dielectric fluid, dielectric types, fluid viscosity, powder concentration, size and type of powder, additive types and powder conductivity, and so on.

Electrode Parameters: Type and size of electrodes, and so on.

A detailed discussion of each input parameter's impact on the EDM process may be found in Section 10.1. A fishbone cause-and-effect diagram is shown in Figure 10.2. Five process factors have been selected from the literature for further examination in relation

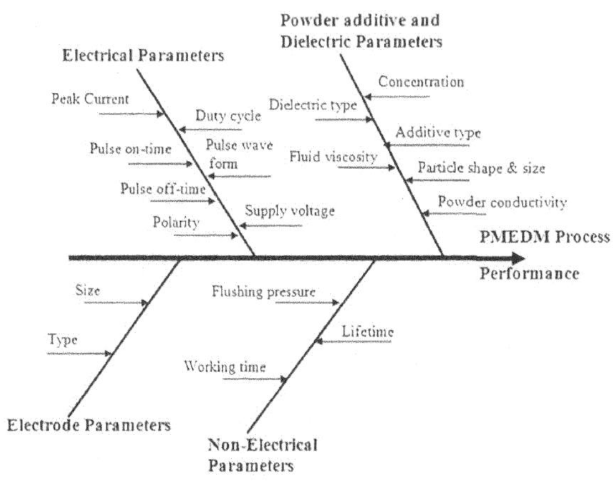

FIGURE 10.2 Fishbone cause-and-effect diagram.

to response parameters. The computerized control panel for this experiment allows for easy modification of the first three electrical input parameters. The selection of these parameters was based on the outcomes of a pilot study that varied one factor at a time.

10.1.2 Pilot Experimentation

For the initial experiments, Inconel-800, a super alloy based on nickel, was utilized as the work material, and a copper rod with a diameter of 12 mm was utilized as the tool electrode. During the experiment, positive polarity was employed. There is a lot of work available for different kinds of alloys, composites, die steels, and titanium alloys depending on these response factors in EDM. Using a "one factor at a time" approach, three factors—peak current, T_{on}, and T_{off}—were chosen for the first round of pilot tests in order to investigate their impacts on three surface roughness response characteristics. After analyzing the outcomes of pilot studies, the final values of these three electrical parameters and their ranges were determined. The following are the settings of the input process parameters selected for the pilot experiment based on the control panel and machine manual:

Current: 2, 4, 6, 8, 10, 12, 14, and 16 amperes
Pulse on-time: 30, 45, 60, 90, 120, 150, 200, and 300 μs
Pulse off-time: 15, 30, 45, 60, 90, 120, 150, and 200 μs

10.1.3 Effect on Surface Roughness

The effect of peak current, T_{on}, and T_{off} variations on SR in microns is depicted in Figures 10.3 to 10.6. The peak current first shifts from 2 to 16 A, while T_{on} = 120 μs and T_{off} = 45 μs remain constant. SR is seen to develop continuously as current and T_{on} rise in Figures 10.3 and 10.4.

The increased current caused more discharge energy to be accessible, which led to more material melting and the formation of large craters, which were the cause of the worse surface. The surface roughness also sharply rises at 14 and 16 A. The

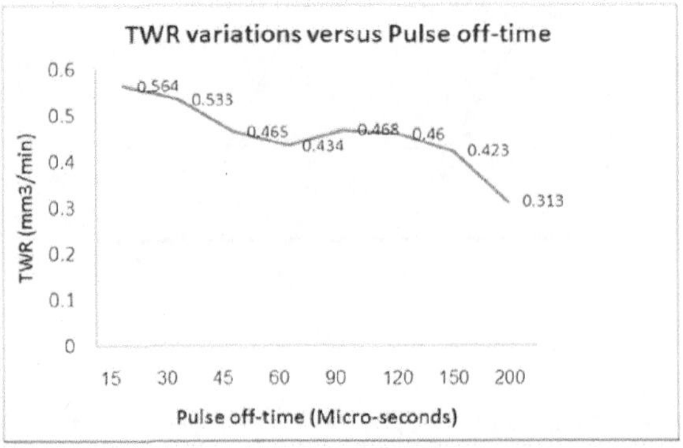

FIGURE 10.3 Impact of fluctuating pulse off-time on TWR at peak current of 8 amperes and pulse on-time of 90 μs.

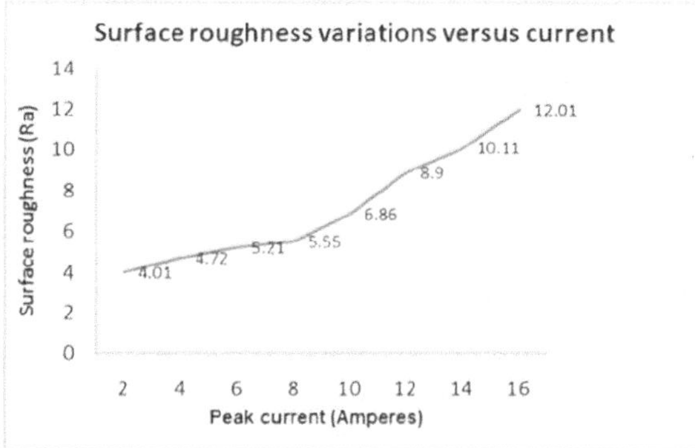

FIGURE 10.4 Effect of variation in peak current on SR at T_{on} = 120 μs and T_{off} = 60 μs.

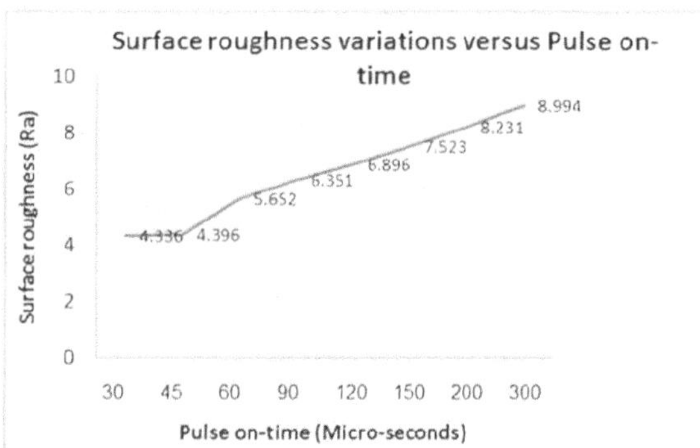

FIGURE 10.5 Effect of variation in pulse on-time on SR at pulse off-time = 45 μs and peak current = 12 amperes.

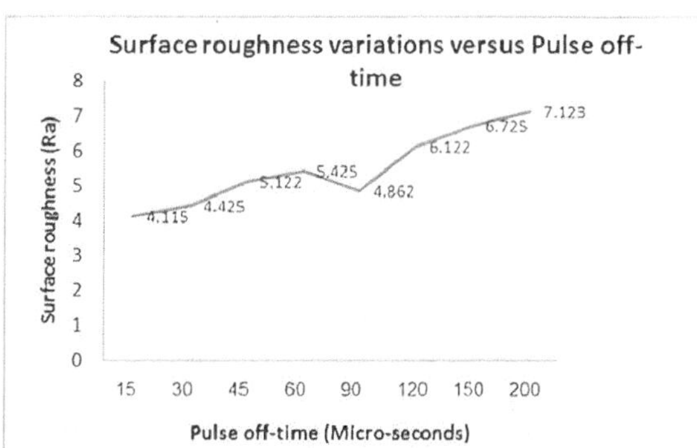

FIGURE 10.6 Effect of variation in pulse off-time on SR at pulse on-time = 90 μs and peak current = 8 amperes.

improved surface quality at low current and T_{on} is seen in Figures 10.4 and 10.5. Due to arcing on the machined surface, low T_{off} combined with high current and T_{on} had significant impacts on surface quality. The influence of $T_{\_off}$ change on surface roughness, with current and T_{on} held constant, is seen in Figure 10.6.

10.2 DESIGN OF EXPERIMENTS

10.2.1 Selection of Input Process Parameters

The quantities of the input parameters were established based on the experimental outcomes of the pilot experiment (Antony, 2003). Table 10.1 lists the process parameters and their levels for phase A and phase B experiments. The values of the input parameters for additional experiments are displayed in Table 10.2. It was demonstrated that at 6 g/l with powder concentration, SR improved. Consequently, for the main experiment, a powder concentration of 6 g/l is used. Pilot tests and published research are the sources from which the discharge current, T_{on}, and T_{off} ranges were chosen. These three primary EDM parameters, in addition to the tool material (copper) and powder particles (tungsten carbide), were chosen as input parameters for this investigation. Thus, a total of five input parameters were chosen, with three levels assigned to each parameter. Table 10.2 displays the input parameters' working range and level.

Inconel-800 was the material utilized in the trials. The dimensions of the rectangular samples were 150 × 15 × 6 mm. The chemical makeup of the tool electrode

TABLE 10.1
Input Parameters and Their Levels

Symbol	Factors	Level 1	Level 2	Level 3
A	Peak current	4	8	12
B	Pulse on-time	60	90	120
C	Pulse off-time	30	45	60

TABLE 10.2
Input Parameters and Their Levels

S. No	Symbols	Input Factors	Level			Units
			−1	0	1	
1	A	Current (I_p)	4	8	12	Amperes
2	B	Pulse on-time (T_{on})	60	90	120	μs
3	C	Pulse off-time (T_{off})	30	45	60	μs
4	D	Tool material	Copper Cu-Cr Graphite			_
5	E	Powder particles	WC Cobalt B4C			_

TABLE 10.3
Chemical Composition of Work Specimen and Tool Electrode Material

Work Material		Electrode Material	
Elements	**Inconel-800**	**Copper**	**Copper-Chromium**
Ni	Base material	0.0083	0.0104
Fe	39.5	0.109	0.0319
Cu	<0.75	99.7	98.4
Si	<1	<0.0050	<0.0050
Cr	20.5	0.0061	1.36
Al	–	<0.0020	<0.0020
S	<0.015	<0.0020	<0.0020
Bi	–	<0.0050	<0.0050
Sb	–	<0.0050	0.0072
Zn	–	0.0148	<0.0050
Pb	–	0.0206	0.0118
Sn	–	0.0356	<0.0050
Mn	–	0.005	0.006

TABLE 10.4
Mechanical Properties of Inconel–800

Density	Melting Point	Yield Strength	Tensile Strength	Elongation
7.94 g/cm^3	1385 °C	205 MPa	520 MPa	30%

and work specimen is shown in Table 10.3, whereas the mechanical properties of the Inconel-800 material are shown in Table 10.4. According to Furutani et al. [14], graphite's numerous industrial uses make it the material that is used most frequently. Thus, a tool electrode made of copper with a diameter of 12 mm and a length of 40 mm was chosen. Figure 10.7a depicts the copper tool electrodes, and Figure 10.7b shows the workpiece specimen following machining.

The EDM process requires careful consideration of a multitude of input process factors in order to identify the ideal process parameters. After taking into account the relevant research, it was determined that EDM performance is directly impacted by process variables such discharge current, T_{on}, T_{off}, electrode material, and powder concentration [13].

10.2.2 Experimental Setup

The powder mixing system and EDM machine configuration are depicted in Figures 10.8a and 10.8b. The work specimen was submerged in EDM oil and

FIGURE 10.7 Tool electrodes of (a) copper-chromium, (b) graphite, and (c) copper. (d) Inconel-800 work pieces after machining.

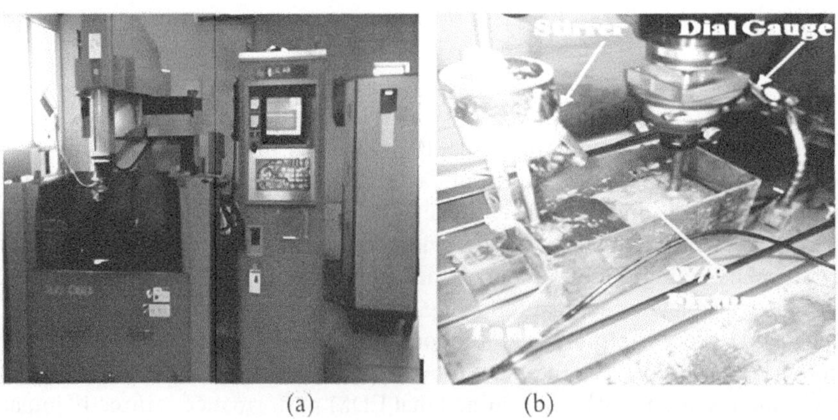

FIGURE 10.8 Experimental setup for (a) EDM machining and (b) powder mixing system.

securely fastened in the vice. Spark erosion 450 EDM oil was utilized as a dielectric fluid in this experimental investigation because of its high dielectric strength, high flash point, and low surface tension. Table 10.5 presents results from 46 trials. The RSM technique was followed in conducting the studies.

TABLE 10.5
Design of Experiments

Run	Current (A)	T_{on} (µs)	T_{off} (µs)	Tool Material	Powder
1	4	90	60	Cu-Cr	Co
2	8	90	45	Cu-Cr	Co
3	8	90	60	Gr	BC
4	8	90	45	Cu	WC
5	8	90	30	Cu-Cr	BC
6	8	120	30	Cu-Cr	Co
7	8	90	60	Cu-Cr	BC
8	8	90	60	Cu	Co
9	4	90	45	Cu	Co
10	8	90	45	Cu-Cr	Co
11	8	90	45	Cu-Cr	Co
12	8	90	45	Gr	WC
13	8	90	30	Gr	Co
14	12	90	30	Cu-Cr	Co
15	8	90	30	Cu	Co
16	12	90	45	Cu-Cr	BC
17	12	90	60	Cu-Cr	Co
18	8	120	45	Cu-Cr	WC
19	4	90	45	Gr	Co
20	12	90	45	Gr	Co
21	8	120	60	Cu-Cr	Co
22	4	90	45	Cu-Cr	BC
23	8	90	45	Gr	BC
24	8	120	45	Cu	Co
25	12	60	45	Cu-Cr	Co
26	8	90	45	Cu-Cr	Co
27	8	90	60	Cu-Cr	WC
28	4	90	30	Cu-Cr	Co
29	12	90	45	Cu	Co
30	4	120	45	Cu-Cr	Co
31	8	120	60	Cu-Cr	Co
32	12	90	45	Cu-Cr	WC
33	8	90	45	Cu	BC
34	8	90	45	Cu-Cr	Co
35	8	60	45	Cu-Cr	WC
36	12	120	45	Cu-Cr	Co
37	8	120	45	Gr	Co
38	8	120	45	Cu-Cr	BC
39	8	90	45	Cu-Cr	Co
40	8	90	30	Cu-Cr	WC
41	8	60	60	Cu-Cr	Co
42	4	60	45	Cu-Cr	Co
43	8	60	45	Cu-Cr	BC
44	8	60	45	Gr	Co
45	8	60	45	Cu	Co
46	4	90	45	Cu-Cr	WC

10.3 RESULTS AND DISCUSSION

10.3.1 Determination of Main Effects on Surface Roughness

The results of the trials carried out in accordance with the RSM approach are displayed in Table 10.6. Equations (1) and (2), respectively, illustrate the relationship between response in real form and machining parameters. Four factors, discharge current, $T_{\_off}$, tool material, powder particles, and two interactions (Ip × $T_{\_off}$, $T_{\_off}$ × powder particles) have significance for SR, according to analysis of variance (ANOVA), as indicated in Table 10.6 (prior to elimination. This model adequately fits the data, as indicated by the p-value of 0.2909, which suggests no significant lack of fit.

The adj R-squared of 0.96 and the pred R-squared of 0.93 are in appropriate contract. The signal-to-noise ratio is measured using adeq. precision. Backward elimination was used to exclude non-essential elements from the quadratic model in order to fit it for the SR. The SR reduced model is displayed in Table 10.7. The model is deemed verified after backward elimination, with R-squared and adjusted R-squared values of 97% and 96%, respectively. The primary effect plots based on Figure 10.9a show that the SR increases noticeably when the current is raised from 4 to 12 A. The difference in SR was from 4.23 to 12.57 μm. The largest effect on

TABLE 10.6
Surface Roughness of Samples after Experimentation

Run No.	**1**	**2**	**3**	**4**	**5**	**6**	**7**	**8**	**9**	**10**
R-1	3.50	9.92	6.45	8.89	9.16	9.16	5.90	6.88	3.01	8.72
R-2	4.59	9.92	6.48	7.89	11.21	9.45	6.11	6.78	3.91	8.72
R-3	3.51	7.17	5.08	7.29	9.36	9.67	5.11	6.14	4.78	8.94
Mean Value	3.87	9.00	6.00	8.02	9.91	9.43	5.71	6.60	3.90	8.79
Run no.	**11**	**12**	**13**	**14**	**15**	**16**	**17**	**18**	**19**	**20**
R-1	8.70	10.33	10.68	13.71	6.71	10.39	9.48	11.66	4.15	13.42
R-2	8.76	10.38	10.70	13.70	7.10	10.19	9.48	10.56	3.10	13.60
R-3	9.21	11.51	11.59	13.26	7.20	9.22	8.05	10.57	3.28	13.63
Mean Value	8.89	10.74	10.99	13.56	7.00	9.93	9.00	10.91	3.51	13.55
Run no.	**21**	**22**	**23**	**24**	**25**	**26**	**27**	**28**	**29**	**30**
R-1	9.16	3.54	7.89	9.07	12.06	10.32	8.71	3.25	10.52	4.33
R-2	9.10	3.70	7.88	8.42	13.50	10.32	8.84	3.07	9.70	4.56
R-3	8.15	3.71	7.89	8.63	13.66	9.34	8.87	2.68	10.80	4.57
Mean Value	8.80	3.65	7.89	8.71	13.07	9.99	8.81	3.00	10.34	4.49
Run no.	**31**	**32**	**33**	**34**	**35**	**36**	**37**	**38**	**39**	**40**
R-1	8.99	14.05	8.19	10.09	10.40	15.03	10.71	10.56	8.71	8.38
R-2	8.61	13.98	8.01	9.31	10.03	14.09	10.03	9.35	9.46	8.28
R-3	9.65	14.99	8.51	9.38	10.04	14.10	9.03	9.60	8.81	7.23
Mean Value	9.08	14.34	8.23	9.59	10.16	14.41	9.92	9.84	8.99	7.96
Run no.	**41**	**42**	**43**	**44**	**45**	**46**				
R-1	5.90	4.05	8.57	10.64	7.97	5.88				
R-2	6.01	3.80	7.49	9.60	7.09	5.01				
R-3	6.10	4.44	7.69	9.09	7.35	4.03				
Mean Value	6.00	4.09	7.92	9.78	7.47	4.97				

TABLE 10.7
Analysis of Variance for Surface Roughness (before Elimination)

Source	Sum of Squares	DF	Mean Square	F-Value	Prob > F	
Model	**370.04**	**20**	**18.5**	**65.45**	**<0.0001**	**Significant**
A	278.13	1	278.13	983.9	<0.0001	
B	3.59	1	3.59	12.72	0.0015	
C	16.28	1	16.28	57.59	<0.0001	
D	9.14	1	9.14	32.34	<0.0001	
E	10.31	1	10.31	36.47	<0.0001	
A2	5.75	1	5.75	20.36	0.0001	
B2	1.97	1	1.97	6.97	0.0141	
C2	9.4	1	9.4	33.28	<0.0001	
D2	2.78	1	2.78	9.84	0.0043	
E2	0.01	1	0.01	0.04	0.834	
AB	0.22	1	0.22	0.78	0.3851	
AC	7.35	1	7.35	26.01	<0.0001	
AD	3.24	1	3.24	11.48	0.0023	
AE	2.38	1	2.38	8.42	0.0076	
BC	2.78	1	2.78	9.85	0.0043	
BD	0.29	1	0.29	1.05	0.3143	
BE	0.33	1	0.33	1.19	0.2857	
CD	5.25	1	5.25	18.59	0.0002	
CE	6.36	1	6.36	22.5	<0.0001	
DE	2.34	1	2.34	8.3	0.008	
Residual	7.06	25	0.28			
Lack of Fit	5.9	19	0.31	1.6	0.2909	Not significant
Pure Error	1.16	6	0.1938806			
Cor Total	377.11	45				
R2	0.98		Pred. R^2	0.93		
Adj. R^2	0.96		Adeq. Precision	31.21		

TABLE 10.8
Analysis of Variance for Surface Roughness (after Elimination)

Source	Sum of Squares	DF	Mean Squares	F-Value	Prob > F	
Model	369.17	16	23.07	84.32	< 0.0001	Significant
A	278.13	1	278.13	1016.47	< 0.0001	
B	3.61	1	3.61	13.21	0.0011	
C	16.26	1	16.26	59.45	< 0.0001	
D	9.14	1	9.14	33.41	< 0.0001	
E	10.31	1	10.31	37.68	< 0.0001	
A2	6.06	1	6.06	22.16	< 0.0001	
B2	2.2	1	2.2	8.06	0.0082	

(Continued)

TABLE 10.8
(Continued)

Source	Sum of Squares	DF	Mean Squares	F-Value	Prob > F	
C2	9.91	1	9.91	36.22	< 0.0001	
D2	2.89	1	2.89	10.59	0.0029	
AC	7.35	1	7.35	26.87	< 0.0001	
AD	3.24	1	3.24	11.86	0.0018	
AE	2.38	1	2.38	8.7	0.0062	
BC	2.78	1	2.78	10.16	0.0034	
CD	5.25	1	5.25	19.2	0.0001	
CE	6.36	1	6.36	23.24	< 0.0001	
DE	2.34	1	2.34	8.57	0.0066	
Residual	7.93	29	0.27			
Lack of Fit	6.77	23	0.29	1.51	0.3166	Not significant
Pure Error	1.16	6	0.19			
Cor Total	377.11	45				
R^2	0.97		Pred. R^2	0.94		
Adj. R^2	0.96		Adeq. Precision	35.35		

SR, from 7.10 to 9.23 μm, was observed when T_{off} decreased from 60 to 30 μs, as Figure 10.3c illustrates. A similar pattern was noted by [22, 23]. Regarding pulse on-time, Figure 10.9b illustrates a small increase in surface roughness when pulse on-time was increased from 60 to 120 μs with 9.19 to 10.19 μm.

$$\begin{aligned} Ra = {} & +9.18+4.17\times A+0.50\times B-1.06\times C+0.76\times D-0.80\times A^2+0.49\times B^2-1.04\times C^2 \\ & -0.55\times D^2-1.36\times A\times E+0.90\times B\times C-1.15\times C\times D-1.26\times C\times E-0.77\times D\times E \end{aligned} \quad (1)$$

$$\begin{aligned} Ra = {} & -5.69+2.86\times current-0.17\times T_{on}+0.34\times T_{off}+2.39\times tool+4.52\times powder \\ & -0.50\times current^{2}+5.44E-4\times T_{on}{}^{2}-4.61E-3\times T_{off}{}^{2}-0.55\times tool^2- \\ & 0.022\times current\times T_{off}+0.22\times current\times tool-0.19\times current\times powder+1.99E- \\ & 3\times T_{on}\times T_{off}-0.77\times T_{off}\times tool \\ & -0.08\times T_{off}\times powder-0.76\times T_{off}\times tool-0.08\times T_{off}\times powder-0.76\times tool\times powder \end{aligned} \quad (2)$$

Figure 10.9d makes it evident that among the three tool electrode materials, copper has the most impact on SR, and Figure 10.9e shows that of the three powder particles, tungsten carbide has the greatest impact on SR. The surface roughness of cobalt powder particles was 9.98 μm, whereas that of boron carbide powder particles

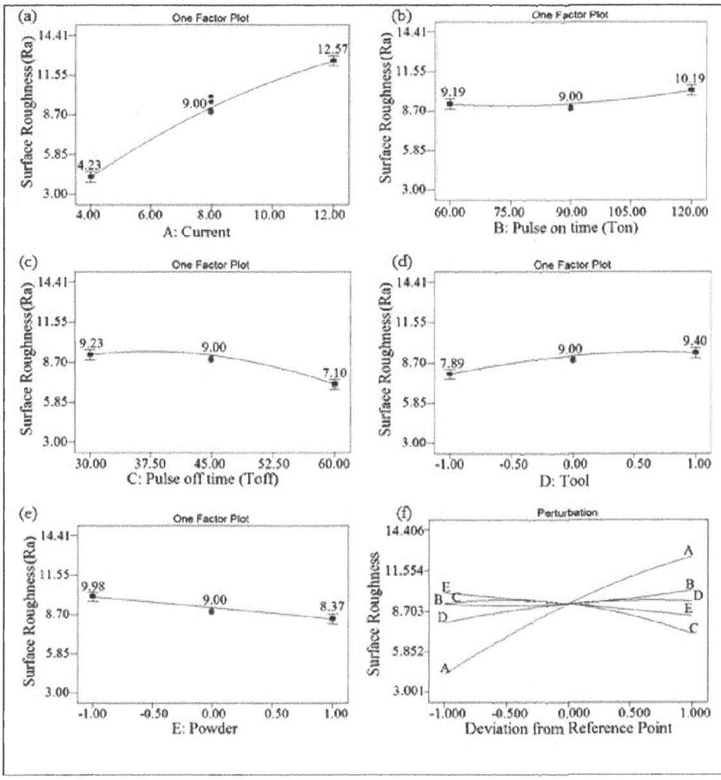

FIGURE 10.9 One-factor plots for surface roughness.

was 8.37 μm. More material is removed from the work specimen and the spark is sustained longer with increasing current. As a result, craters that emerge on the surface will be deeper, increasing SR. Apart from the standard residual plot, there are other plots depicting the residuals against the predictions. Figure 10.10a illustrates the normal distribution of the data. Additionally, Figure 10.10b shows that every experimental result is remarkably close to the predicted values.

10.3.2 Determination of Interaction Effects on Surface Roughness

Table 10.6 indicates that two interactions—Ip × $T_{\_off}$ and $T_{\_off}$ × powder particles—have occurred, as seen in Figures 10.4c and d. For both interactions, that is, Ip× $T_{\_off}$ and $T_{\_off}$ × powder particles, the Prob > F value is less than 0.0001. Figure 10.4c makes it evident that SR rose from 2.88 to 11.17 μm when current increased from 4 to 12 A, and $T_{\_off}$ varied from 30 to 60 μs. The surface roughness changed from 4.15 to 8.40 μm when the current was raised from 4 to 12 A with boron carbide powder additive and from 4.15 to 11.61 μm with tungsten carbide powder additive, as Figure 10.10d illustrates. Because the work specimen required more time to be machined, the SR grew as peak current climbed and pulse on-time continued to grow.

Figure 10.9e displays the interaction graph between the powder particles and the current. Figure 10.10 makes it evident that powder particles have a greater impact on SR at the same current value. The contour plot of current and powder, as seen in Figure 10.10f, exhibits a similar effect.

10.3.3 Microstructure Analysis

The surfaces of Inconel-800 undergo significant microstructural alterations during PMEDM procedures. Surface crack development and metallurgical alterations on the machined surface are indicative of the features of the surface. The primary variables that determine the surface quality of the machined specimen during EDM machining are current, tone, electrode polarity, electrode, workpiece material, and powder mixed dielectric medium [1, 24]. Three trials for the current setting—4, 8, and 12 A—as well as three distinct tool materials and three powder particles were thoroughly examined for the purpose of scrutiny.

Debris, craters, and pieces are depicted in Figure 10.10a. However, at greater magnification, deep fractures and subsurface cracks containing debris are visible on the surface, as seen in Figures 10.10d and g. The micrographs in Figure 10.10b, e, and h depict the machining conditions with graphite as the tool material; cobalt as a powder additive in the dielectric fluid; and 12 A current, 60 μs Ton, and 45 μs Toff. Figure 10.10b displays microcracks coupled with debris and a crater. As demonstrated in Figures 10.10e and h, surface fractures and deep cracks were identified by thorough evaluation of greatly enlarged pictures. The micrographs with the machining conditions at 4 A current, 90 μs Ton, 45 μs Toff, copper-chromium as the tool material, and tungsten carbide as a powder additive in the dielectric fluid are displayed in Figures 10.10c, f, and i. Figure 10.10c depicts microcracks, material pullout, and debris that were seen on the surface. However, Figures 10.10f and i reveal that deep fractures and a surface with increased magnification are visible. The surface may be created by an uneven fusing structure, microcracks, pock marks, craters, debris globules, subsurface cracks, and pieces, as seen by the micrographs shown in these pictures. When compared to Figures 10.10c, f, and i, it is seen from Figures 10.10a, b, d, e, g, and h that the PMEDM surface creates enormously uneven topography and defects, including debris, fragments, deep craters, pockmarks, surface cracks, sub-surface cracks, and micro-cracks.

The most important factor causing the surface texture to deteriorate is the peak current. This is because energy is transmitted to the machining region with an increase in current high discharge, causing melting and evaporation and the development of a tiny crater [15]. Moreover, the repeated electrical discharge vaporized the work specimen and left behind significant craters [24, 25]. The application presented in the current study can be utilized for various material and manufacturing techniques [31–37].

10.4 CONFIRMATION EXPERIMENT

For surface roughness, three confirmation experiments were carried out. Values of certain process variables for the confirmation test fall within the ranges from which the formula was generated [16].

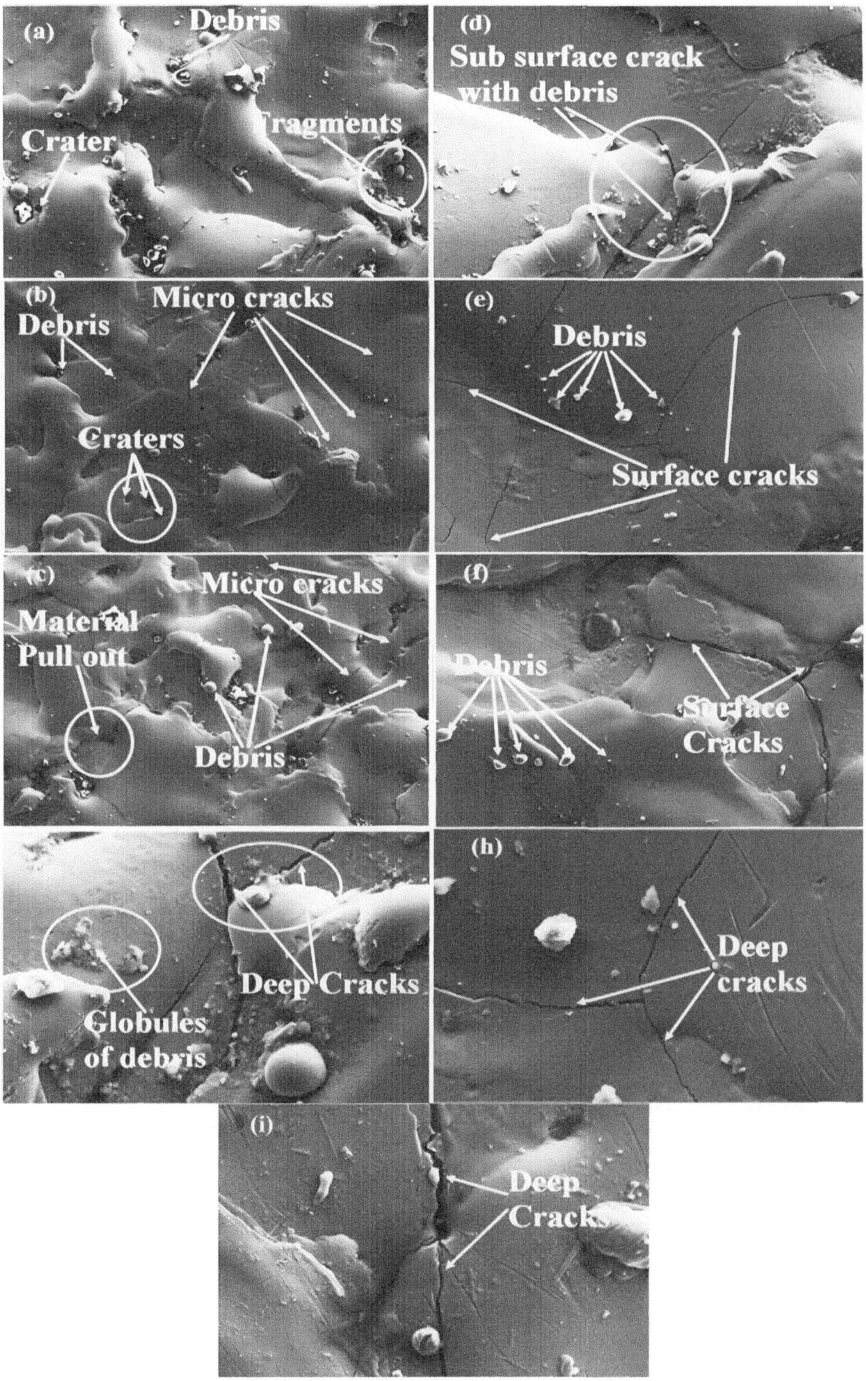

FIGURE 10.10 SEM micrographs of samples from run no. (a), (d) 3, (b), (e) 44, (c), (f) 46, (g)3,(h)44 and; (i)46.

Table 10.8 displays the results for the confirmation run and the anticipated design for SR. It was discovered that the estimated error was minimal. For SR, the range of error between experimental and predicted values was −8.55% to 5.1%, respectively. This clearly demonstrates how very reproducible the experimental results are.

TABLE 10.8
Conformation Test

	Machining Conditions					Surface Roughness (μm)		
Sample	Current (A)	T_{on}, B (μs)	T_{off}, C (μs)	Tool Material, D	Powder, E	Exp. Value	Pred. Value	Error (%)
1	4	60	45	Cu-Cr	Co	4.09	4.44	−8.55
2	8	90	45	Cu	BC	8.23	7.81	5.1
3	12	90	45	Cu-Cr	WC	14.34	14.1	1.67

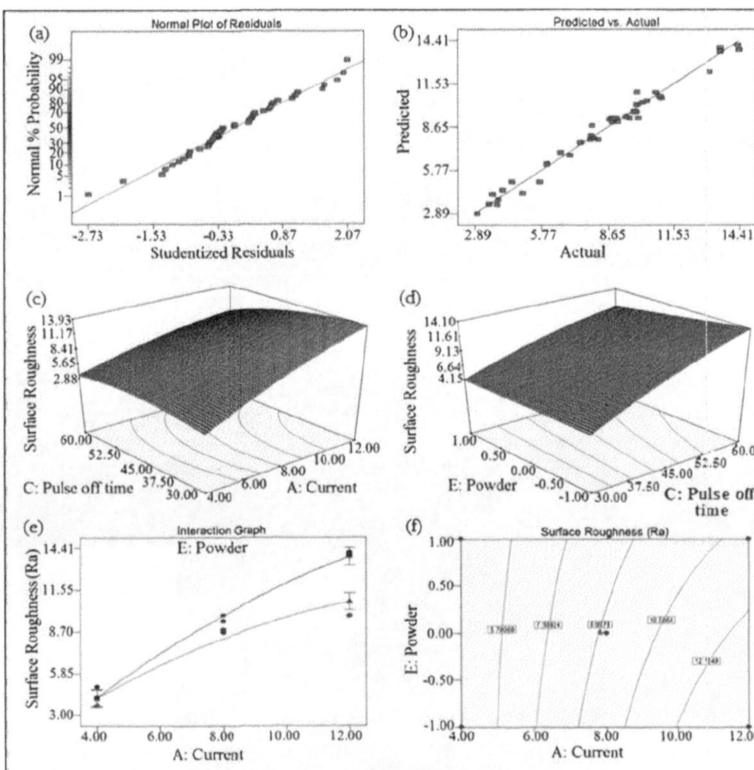

FIGURE 10.11 Response surface plots include the following: (a) plot of the normal probability residual; (b) plot of the predicted vs. actual; (c), (d), 3D surface plot for SR; (e) plot of the interaction for SR; and (f) plot of the contour for SR.

10.5 CONCLUSION

In order to link the machining process variables—current, T_{on}, $T_{off,}$ tool electrode, and powder particles—in the PMEDM process of Inconel-800, a quadratic model for SR was constructed. An RSM approach known as the Box-Behnken method was used for experimental work. The findings of the experimental investigation showed that the SR varied from 4.23 to 12.57 μm. The factors that had the most effect on the SR were the tool material, T_{off}, current, and powder particles. When the settings were adjusted to Ip = 4 amp, T_{on} = 60 μs, T_{off} = 60 μs, and tool material as a copper and boron-carbide powder particle, the lowest SR was achieved. With $T_{\_off}$, copper as a tool electrode, and boron carbide employed as a powder particle, surface roughness improves; however, surface roughness increases as current increases. T_{on} barely affects SR at all. For the produced surface roughness, the coefficients of regression R^2, adj. R^2, and projected R^2 are, respectively, 0.98, 0.96, and 0.93, and the latter two were acceptable. The results of the confirmation test indicated that there is a −8.55% to 5.1% discrepancy between the experimental and predicted values of SR. This attests to the results' validity for both experimental and forecasted values. SEM micrographs of machined surfaces revealed that peak current alters the surface structure of the materials, causing microcracks, craters, pockmarks, and debris to appear. It is possible to get improved surface properties with lower current and greater T_{off}.

REFERENCES

1. R. Rao and V. Yadava, "Multi objective optimization of Nd:YAG lasercutting of thin superalloy sheet using grey relational analysis with entropy measurement," Optic Laser Technology, 41:922–930, 2009.
2. S. Rajesha, A.K. Sharma and P. Kumar, "On electro discharge machining of Inconel718 with hollow tool," Journal of Materials Engineering and Performance, 21(6):882, 2012.
3. P. Pecas and E. Henriques, "Influence of silicon powder-mixed dielectric on conventional electrical discharge machining," International Journal of Machine Tools & Manufacture, 43:1465–1471, 2003.
4. P.P. Pande and R.K. Sambhe, "Artificial intelligence machining predictions by fuzzy GUI for Inconel 718 with PVD coated carbidecutting tool," International Journal of Mechanical and Production Engineering Research and Development (IJMPERD), 7(3):313–320, May–June 2017.
5. S. Singh and M.F. Yeh, "Optimization of abrasive powder mixed EDM of aluminum matrix composites with multiple responses using gray relational analysis," Journal of Materials Engineering and Performance, 21(4):481–491, 2012.
6. H.M. Chow, L.D. Yang, C.T. Lin and Y.F. Chen, "The use of SiC powder in water as dielectric for micro-slit EDM machining," Journal of Materials Processing Technology, 195:160–170, 2008.
7. S. Kumar, R. Singh, A. Batish and T.P. Singh, "Study the surface characteristics of cryogenically treated tool-electrodes in powder mixed electric discharge machining process," Materials Science Forum, 808:19–33, 2015.
8. Q.Y. Ming and L.U. He, "Powder-suspension dielectric fluid for EDM," Journal of Materials Processing Technology, 52:44–54, 1995.
9. H.M. Chow, B.H. Yan, F.U. Huang and J.C. Hung, "Study of added powder in kerosene for the micro-slit machining of titanium alloy using electro-discharge machining," Journal of Materials Processing Technology, 101:95–103, 2000.

10. P. Pecas and E. Henriques, "Electrical discharge machining using simple and powder-mixed dielectric: The effect of the electrode area in the surface roughness and topography," Journal of Materials Processing Technology, 200:250–258, 2008.
11. K. Furutani, A. Saneto, H. Takezawa, N. Mohri and H. Miyake, "Accretion of titanium carbide by electrical discharge machining with powder suspended in working fluid," Journal of the International Societies for Precision Engineering and Nanotechnology, 25:138–144, 2001.
12. E.A. Al-Ajaj, A. Abdul-Rassol Abdullah and A.A. Moosa, "Modeling and experimental studies of diffusion bonding of Inconel 600 to pyrolytic graphite," International Journal of Metallurgical & Materials Science and Engineering (IJMMSE), 3(1):9–30, January–March 2013.
13. J. Singh and R.K. Sharma, "Implementation of Taguchi method with hybrid decision making tools for prediction of surface characteristics for powder mixed EDM of WC," Engineering and Material Sciences, 8:455–458, 2016.
14. K. Furutani, A. Saneto, H. Takezawa, N. Mohri and H. Miyake, "Accretion of titanium carbide by electrical discharge machining with powder suspended in working fluid," Precision Engineering, 25(2):138–144, 2001.
15. H.K. Kansal, S. Singh and P. Kumar, "Technology and research developments in powder mixed electric discharge machining (PMEDM)," Journal of Materials Processing Technology, 184:32–41, 2007.
16. H. Singh, M. Kumar, R. Singh and S. Kumar, "Microstructure and cavitation erosion performance of cold-sprayed WC-12Co and WC-17Co coatings on hydraulic turbine steels," Journal of Materials Engineering and Performance, 2024. https://doi.org/10.1007/s11665-024-10097-2.
17. H. Singh, M. Kumar, R. Singh and S. Kumar, "Deposition behaviour of WC-17Co cold-sprayed composite coating on hydro turbine steels," Surface Review and Letters, 2024. https://doi.org/10.1142/S0218625X25500209.
18. H. Singh, M. Kumar and R. Singh, "Cavitation erosion behaviour of WC-Co coatings on turbine steel deposited by cold spray," Surface Review and Letters, 2024. https://doi.org/10.1142/S0218625X25500052.
19. H. Singh, M. Kumar and R. Singh, "Slurry erosion performance of cold-sprayed WC-12Co composite coatings," Surface Review and Letters, 2024. https://doi.org/10.1142/S0218625X2450050.
20. B.B. Pradhan and B. Bhattacharyya, "Improvement in micro hole machining accuracy by polarity changing technique for micro electrode discharge machining on Ti-6Al-4V," Proceedings of the Institution of Mechanical Engineers, Part: B, Journal of Engineering Manufacture, 222(2):163–173, 2008.
21. J.Y. Kao, C.C. Tsao, S.S. Wang and C.Y. Hsu, "Optimization of the EDM parameters on machining Ti–6Al–4V with multiple quality characteristics," International Journal of Advanced Manufacturing Technology, 47:395–402, 2010.
22. M.M. Rahman, M.A.R. Khan, K. Kadirgama, M.M. Noor and R.A. Bakar, "Modeling of material removal on machining of Ti-6Al-4V through EDM using copper tungsten electrode and positive polarity," International Journal of Mechanical and Materials Engineering, 1(3):135–140, 2010.
23. B. Jabbaripour, M.H. Sadeghi, S. Faridvand and M.R. Shabgard, "Investigating the effects of EDM parameters on surface integrity, MRR and TWR in machining of Ti–6Al–4V," Machining Science and Technology, 16:419–444, 2012.
24. M. Kumar, S. Datta and R. Kumar, "Electro-discharge machining performance of Ti–6Al–4V alloy: Studies onparametric effect and phenomenon of electrode wear," Arabian Journal for Science and Engineering, 44:1553–1568, 2019.
25. M. Priyadarshni and K. Pal, "A comparative study for machining of Ti–6Al–4V alloy for multi-criteria response," Journal of Advanced Manufacturing Systems, 17(4):515–531, 2018.

26. L.P. Singh and J. Singh, "Effects of cryogenic treatment on high speed steel tools," Journal of Engineering and Technology, 1(2):88–93, 2011.
27. G. Singh, H. Vasudev and H. Arora, "A short note on the processing of materials through microwave route," In *Advances in Materials Processing: Select Proceedings of ICFMMP 2019* (pp. 101–111). Springer Singapore, 2020.
28. V. Srivastava and P.M. Pandey, "Effect of process parameters on the performances of EDM process with ultrasonic assisted cryogenically cooled electrode," Journal of Manufacturing Processes, 14(3):393–402, 2012.
29. A. Kumar, S. Maheshwari and C. Sharma, "Effect of cryogenically treated electrode during aedm of nickel based super alloy 718 for surface quality improvement," Advanced Materials Research, 410:236–239, 2012.
30. G. Rotella, O. W. Dillon, D. Umbrello, L. Settineri and I. S. Jawahir, "The effects of cooling conditions on surface integrity in machining of Ti6Al4V alloy," International Journal of Advanced Manufacturing Technology, 71(1–4):47–55, 2014.
31. M. Singh, H. Vasudev and R. Kumar, "Corrosion and tribological behaviour of BN thin films deposited using magnetron sputtering," International Journal of Surface Engineering and Interdisciplinary Materials Science (IJSEIMS), 9(2):24–39, 2021.
32. M. Singh, H. Vasudev and M. Singh, "Surface protection of SS-316L with boron nitride based tin films using radio frequency magnetron sputtering technique," Journal of Electrochemical Science and Engineering, 12(5):851–863, 2022.
33. G. Singh, H. Singh, Y. Sharma, H. Vasudev and C. Prakash, "Analysis and optimization of various process parameters and effect on the hardness of SS-304 stainless steel welded joints," International Journal on Interactive Design and Manufacturing (IJIDeM), 18(5):2889–2896, 2024.
34. G. Singh, A. Mehta and H. Vasudev, "Sustainability of additive manufacturing: A comprehensive review," Progress in Additive Manufacturing:1–24, 2024.
35. G. Singh, A. Bansal, H. Vasudev and V. Mishra, "Sliding wear study of the Inconel-625 clad deposits by microwave heating on SS–304," Physica Scripta, 99(6):065503, 2024.
36. J. Singh, A. Yadav, S. Gupta, R. Balu, N. Shukla, G. Singh and S. Singh, "Design and computational analysis of continuous wave fiber laser cutting of flat and conformal ITO PET sheets," International Journal on Interactive Design and Manufacturing (IJIDeM):1–9, 2024.
37. G. Singh, R. Singh and J. Gul, "Machinability behavior of human implant materials," Journal of Electrochemical Science and Engineering, 13(1):99–114, 2023.
38. P.K. Verma, H. Vasudev, G. Singh and S. Singh, "A review on the microstructural and biomedical properties of implants manufactured using additive manufacturing," Journal of the Institution of Engineers (India): Series D:1–15, 2024.

11 Surface Engineering and Coatings to Enhance Wear and Corrosion Resistance Properties

Ranbir Singh Rooprai, Jagtar Singh, and Anuj Bansal

11.1 INTRODUCTION

Abrasive wear has posed significant challenges for tillage tools in agriculture and mining, especially as they engage with abrasive soil particles that degrade their surfaces over time [1, 2]. Chintha [3] describes abrasive wear occurring after tillage implements, such as diggers, rotary tillers, disc harrows, chisels, and ploughs, interacting with soil through mechanical action during cultivation. This degradation, caused by the continuous sliding and rolling of hard particles in the soil, diminishes tool performance and increases energy consumption and maintenance. Often, these tools are composed of high-carbon or low-alloy steels due to their cost effectiveness, durability, and machinability [4, 5]. However, frequent wear demands new solutions, as extending tool life and improving wear resistance are crucial to reducing operational costs [6]. Improving wear resistance in tillage tools includes surface treatments like thermal spray coatings, nitriding, cladding, and boronizing, each offering unique advantages [7, 8]. Among these, wire arc additive manufacturing (WAAM) is gaining recognition as an effective technique for its fast deposition rate, economic viability, and ability to produce intricate shapes [9]. WAAM is especially valuable for producing parts with minimal waste, making it an economical and sustainable option [10]. Unlike traditional methods, WAAM allows more efficient material usage and is well suited for use with various alloys, particularly for applications that require durability under abrasive conditions. Inconel 625, a nickel-based superalloy, is often used in cladding applications requiring durability owing to its structural integrity, surface and corrosion durability, and dimensional stability under stress [11]. Studies have shown that WAAM-fabricated parts with Inconel 625 offer better mechanical properties and wear resistance than conventionally produced parts, which is essential for extending the service life of agricultural tools [12, 13]. The unique composition of Inconel 625, with elements like columbium and moly in a nickel-chromium base, allows it to retain mechanical integrity at elevated temperatures, making it well suited for environments where high wear resistance is critical. In addition to agriculture, WAAM has gained

DOI: 10.1201/9781003546528-11

traction in aerospace, automotive, and construction, where parts with enhanced mechanical properties and high wear resistance are crucial [14]. Research highlights the importance of optimizing WAAM parameters such as current, voltage, wire feed speed, and scanning speed, as these aspects influence the microstructure and, consequently, the wear performance of the cladding [15]. By fine-tuning these parameters, WAAM can produce cladded parts with improved hardness and resilience.

In this study, Inconel 625 was cladded onto EN8 steel using WAAM with cold metal transfer (CMT) to boost the durability of agricultural tillage tools against wear. EN8 steel, a material widely used in tillage due to its balanced hardness, toughness, and cost effectiveness, is still susceptible to wear in abrasive conditions [16]. The aim was to develop a surface with increased durability capable of withstanding abrasive wear effects. An abrasive wear test was conducted to evaluate wear resistance, considering factors like applied load, particle size, and sand flow rate. Testing was carried out in longitudinal and transverse orientations to assess how directional loads affect performance [17]. To analyze the microstructure and surface characteristics, scanning electron microscopy (SEM), X-ray diffraction (XRD), and energy dispersive spectroscopy (EDS) were used. The tests revealed that Inconel 625 cladding on EN8 steel demonstrates a balance of ductile and brittle failure modes, highlighting the material's suitability for agricultural applications where abrasive wear resistance is crucial. This study advances WAAM as a solution for manufacturing wear-resistant agricultural tools, illustrating its potential for use in environments that demand high durability [18–20]. The EDM technique is an esoteric machining method that finds extensive application in a number of industry sectors throughout the world for making a large assortment of components on hard-to-cut materials with finer surface finish and controlled tolerances. Although the electric discharge machining method has a very long history dating back to 1770, it has seen amazing metamorphosis in the past 20 years in many different sectors, leading to a vast body of literature on the subject. Advanced technological industries have been requiring materials with high strength, high hardness, and temperature resistance. Aging conventional machining methods are giving place to more prevalent advanced non-conventional EDM for machining such materials. The researchers wished to present a critical justification for the widespread study they made on the many aspects of machining of the different grades of steels and alloys using the EDM method. Moreover, through the application of powder-suspended dielectric fluid during the EDM process, the researchers were able to increase the machining properties and surface qualities of the workpiece. Apart from the EDM process, other thermo-electric processes have been created and applied in the manufacturing sector across the world. Researchers have also investigated the effect of cryogenic treatment on the machinability of different materials in an attempt to enhance machining efficiency. In addition, nickel-based super alloys have a variety of uses, and such a hard material is not possible to machine with conventional EDM. Due to its superior mechanical and thermal properties, nickel-based super alloy is utilized in numerous universal applications, for example, pump body parts, jet engine casing, turbo chargers, and turbine blades for aircraft. Such universal applications required accurate product design and close tolerances during the production process. Because of the shared attributes, these alloys cannot be machined with the aid of standard production methods. Nickel-based super alloys possess weak thermal attributes and excellent toughness, which results in them being difficult to process utilizing standard machining methods. Such limitations mainly result in the usage of non-standard machining methods like electric discharge machining for the processing of Inconel. EDM is a technical process that employs powder additives in EDM oil to alter the process parameters.

11.1.1 Material Used

The base material selected for this experimental study was EN8 steel, sourced as rectangular bars from Punjab Steel, Ludhiana, Punjab. The specific elemental configuration of EN8 steel, detailed in Table 11.1, supports its suitability for cladding applications. These bars were machined to 300 mm in length, 50 mm in width, and 10 mm in thickness to prepare them for cladding using the wire arc additive manufacturing-cold metal transfer technique. In this process, Inconel 625 wire with a diameter of 1.2 mm served as the cladding material. The wire was obtained from the Caston Electrode Company, based in New Delhi, with its chemical composition outlined in Table 11.2. Natural sand was collected from numerous agricultural fields around the Sangrur district in Punjab, India, for the dry abrasive wear tests. This locally sourced sand provided a realistic simulation of abrasive conditions of farm applications, ensuring that the testing environment closely reflected practical wear scenarios for cladded EN8 steel surfaces. For the purpose of a complete analysis of wear resistance in agricultural uses, the experimental work adopted EN8 steel as the material base, with caution chosen due to its optimum mechanical properties and machinability. The material was obtained in the form of rectangular bars from Punjab Steel, Ludhiana, Punjab, whose elemental composition is specified in Table 11.1, confirming its fitness for cladding uses. These bars were accurately machined to dimensions of 300 mm in length, 50 mm in width, and 10 mm thickness to support the wire arc additive manufacturing-cold metal transfer process. For the cladding process, Inconel 625 wire—a high-performance nickel-based superalloy with excellent wear and corrosion resistance—was used as the material for deposition. The 1.2-mm-diameter wire was obtained from Caston Electrode Company, New Delhi, and its detailed chemical composition is presented in Table 11.2. Further,

TABLE 11.1
Elemental Composition of EN8 Steel

Base	Element	C	Si	P	Mn	S	Fe
EN8steel	Content	0.39	0.08	0.04	0.71	0.04	Rest

TABLE 11.2
Chemical Composition of Wire Inconel 625

Elements	Cr	Mo	Nb	C	Al	Fe	Ni
Content	15.5	4.5	4.7	18.3	0.29	0.6	Rest

to simulate actual agricultural wear conditions, natural sand was obtained from various agricultural fields of the Sangrur district of Punjab, India. This locally produced sand helped ensure that the dry abrasive wear tests closely represented the harsh working conditions of farm tools, closely simulating the abrasive interactions seen by cladded EN8 steel surfaces in real-world use.

11.2 EXPERIMENTAL DETAILS

11.2.1 WAMM-CMT Cladding

In this study, cladding was developed on EN8 steel using Inconel 625 wire through the wire arc additive manufacturing process. The WAAM system incorporated a six-axis Kawasaki robot (model 3BA006N-AC0, Japan), along with a cold metal transfer machine (Advanced 4000MVR, Fronius, Austria), a VR7000 wire feeder, and a system regulator. The experimental setup for WAAM-CMT employed in this research is illustrated in Figure 11.1. The Inconel 625 cladding process parameters, shown in Table 11.3, were selected based on preliminary trials and relevant studies [21, 22]. Following the cladding process, all samples, both cladded and uncladded,

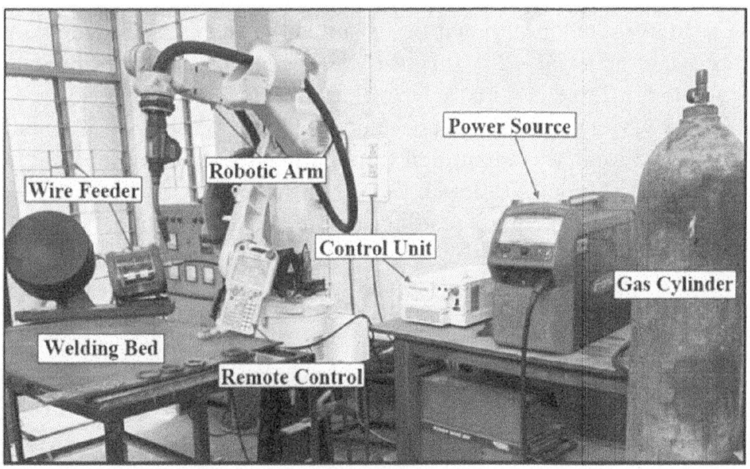

FIGURE 11.1 Experimental system of WAAM setup.

TABLE 11.3
Protocols for Cladding Operation

S. No.	Sample Name	Current (A)	Voltage (V)	Wire Speed (m/min)	Scanning Speed (cm/min)
1	Inc 1	146	16.0	7	10
2	Inc 2	156	16.5	8	15
3	Inc 3	166	17.0	10	20

were sectioned into dimensions of 50 × 25 mm. The upper surfaces were ground to ensure consistency in the dry sliding tests to create a uniform surface finish. Due to directional grain structures, certain microstructural variations were observed between the transverse and longitudinal orientations. Additionally, during wear testing, the interaction between the abrasive wheel and the number of cladding beads varied with orientation, potentially influencing the erosion patterns observed [23].

11.2.2 Microhardness Testing

A Mitutoyo tester (Model 500-196-20, Japan) was used for the microhardness test of all the samples under controlled conditions. Each measurement was conducted with a load of 300 g and a dwell time of 15 seconds. To ensure accuracy, the surface of each sample was tested, and an average microhardness value was calculated based on five individual measurements taken at separate points across each sample's surface.

11.2.3 Abrasive Wear Testing

To effectively assess the abrasion confrontation of cladded and uncladded samples, abrasive wear testing was conducted via a three-body rubber wheel abrasive testing apparatus (Model TR-50-M7, Ducom, India), as depicted in Figure 11.2. Adhering to the ASTM G65 standard, known for its suitability in examining medium to high abrasion-resistant materials [24], this test was performed under optimized parameters, as indicated in Table 11.4 and selected based on insights from prior research [21, 23]. The rubber wheel operated at a constant speed of 250 rpm for 15 minutes during the test. Abrasive wear was quantified through mass loss measurements, where an electronic balance (Model AB-250D, Denver Instrument, USA) with a precision of 0.0001 g provided accurate mass readings before and after testing.

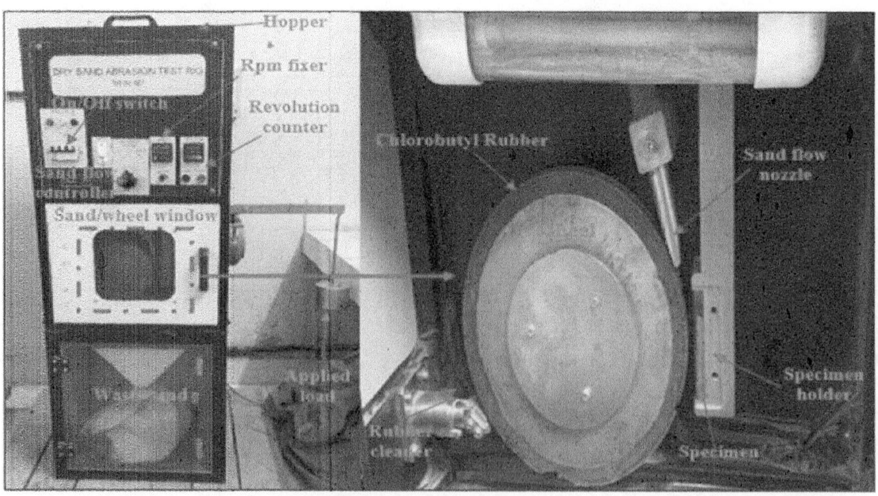

FIGURE 11.2 Abrasive wear testing machine.

TABLE 11.4
Chosen Parameters with Levels

S. No.	Parameters	Units	Levels		
			I	II	III
1	Applied load	N	60	80	100
2	Size of sand particle	µm	200	300	400
3	Flow rate of sand	gm/min	150	200	250

11.3 RESULTS

11.3.1 Microhardness Analysis

The sequence of assessing microhardness in different specimens, including both clad samples and the bare steel, unfolded as follows: EN8 < Inc3 < Inc2 < Incl. Their corresponding average microhardness values were measured at 205±5, 311±8, 338±10, and 385±12, as visually depicted in Figure 11.3. Remarkably, Incl exhibited the highest microhardness among all samples. This exceptional outcome can be attributed to the precise orchestration of optimal parameters encompassing current, voltage, wire feed speed, and scanning speed, set at 146 A, 16 V, 7 m/min, and 10 cm/min individually. The blend of these parameters played a vital role in granting the Inconel 625 material ample time to undergo solidification, setting it apart from alternative parameter combinations for the cladding process.

11.3.2 Abrasive Wear Testing Analysis

Using the specified process parameters, experimentation was conducted on both cladded and uncladded specimens (see Table 11.4). Figures 11.4a and b display bar graphs showing mass loss variations concerning sand particle size and sand flow rate. Among all claddings tested, the Incl sample demonstrated the lowest mass loss in both longitudinal and transverse orientations, which may be attributed to its higher hardness. Additionally, during wear testing, all developed clads exhibited improved wear resistance compared to EN8 steel. Longitudinal samples showed relatively low wear resistance compared to transverse samples, likely due to their more uniform yet weaker grain structure, which attacking particles can more easily disrupt, leading to cavity formation [21, 25]. Conversely, transverse samples contain diverse grain structures [25], making them more resistant to particle damage. The trend across tests indicates that abrasive parameters such as load, sand flow rate, and particle size significantly affect the mass loss of all samples in both cutting orientations. As depicted in Figures 11.4a and b, the highest mass loss for both sample types occurred at Run 8, characterized by the highest load (100 N), an intermediate sand particle size (300 µm), and the lowest sand flow rate (150 g/min). Under these conditions, the combination of high load and medium particle size, coupled with the reduced sand flow rate, likely allowed particles to adhere to the rubber wheel and strike the target

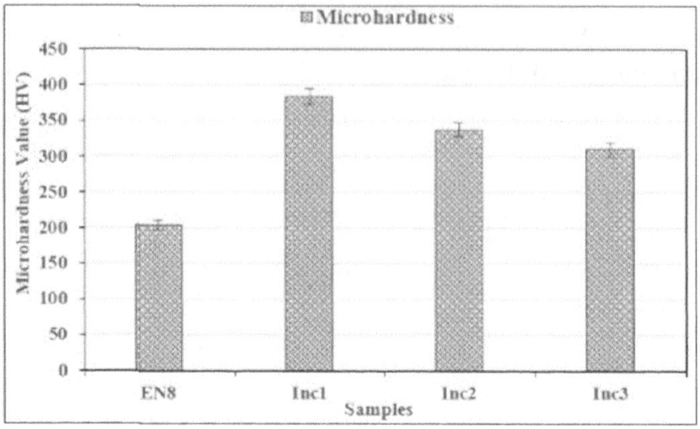

FIGURE 11.3 Microhardness values of uncladded and cladded specimens.

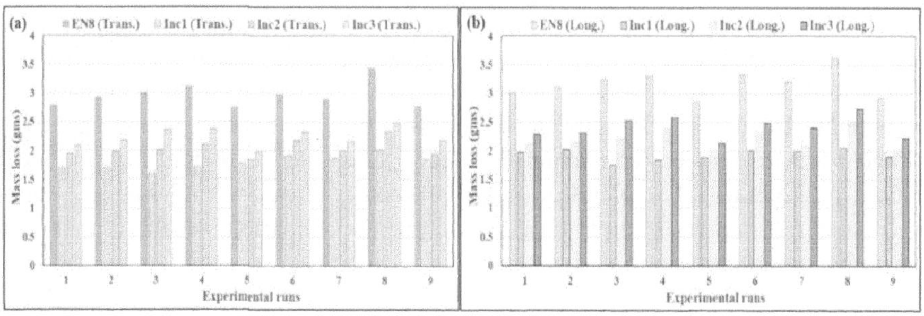

FIGURE 11.4 Mass loss of cladded and uncladded (a) transverse and (b) longitudinal cut samples.

area effectively, resulting in increased mass loss due to three-body abrasive wear. In contrast, the lowest mass loss for both longitudinal and transverse samples occurred at Run 5, with parameters including the highest sand flow rate (300 g/min), medium load (80 N), and medium sand particle size (300 μm). At this higher sand flow rate, mass loss is reduced, possibly because the sand particles are less effectively utilized in the abrasive process. A higher flow rate may lead to sand particles dispersing beyond the immediate abrasive zone or colliding with each other, which diminishes their impact on the specimen surface. These interactions likely result in less mass loss at elevated sand flow rates as some particles are deflected or wasted without significantly engaging in the abrasive wear process.

11.3.3 Analysis of Wear Mechanism

A field emission scanning electron microscope (FESEM, JSM-7610Fplus, JEOL, Japan) was utilized to examine the worn surfaces of both cladded and uncladded samples after dry abrasive wear tests. Based on SEM imagery, the samples were

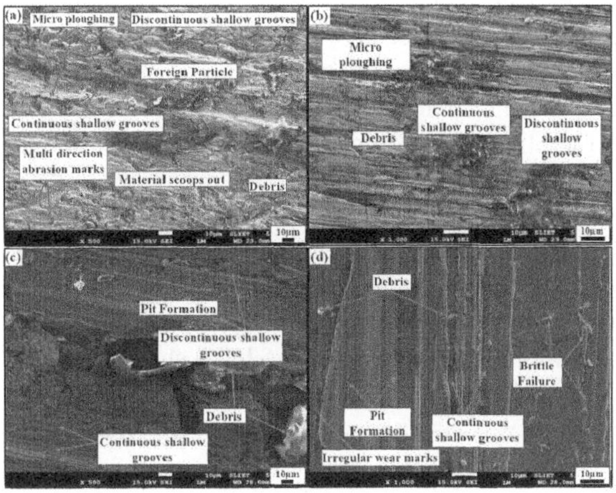

FIGURE 11.5 SEM micrographs of eroded surfaces of EN8 steel and Incl clad. (a) Long. (EN8), (b) Trans. (EN8) and (c) Long. (Incl), (d) Trans. (Incl).

categorized as either ductile or brittle. SEM micrographs for the eroded surfaces of EN8 steel in both longitudinal and transverse orientations are displayed in Figures 11.5a and b. The high pressure from the rubber wheel caused continuous and intermittent shallow grooves on both surfaces, with micro-ploughing observed particularly on the longitudinal sample (a). Due to the varied sizes of sand particles and the presence of grit within the soil, multidirectional abrasion marks developed, resulting in some material being displaced. Furthermore, debris from the eroded surfaces accumulated, showing characteristics of softening and suggesting a ductile failure mode for EN8 steel samples. Figures 11.5c and d present SEM images of the Incl clad samples in both longitudinal and transverse orientations. The longitudinal sample reveals continuous grooves and abraded material debris that may have resulted from the frequent ripping of the sample from the continuous attacking of sand particles at higher flow rates. This may be caused by the Incl clad's relatively increased hardness, mixed ductility, and brittleness characteristics when compared to EN8 steel. When moving from a soft to a hard surface area, abrasive particles change their direction, which are deflected and cause the formation of discontinuous grooves. The formation of pits and irregular wear marks are also visible on the critical examination of SEM images, which the large particle size and the increased effect of the applied load may have facilitated.

11.3.3.1 Integration of PMEDM for Wear-Resistant Surfaces

11.3.3.1.1 PMEDM as an Advanced Surface Modification Technique

Powder-mixed electrical discharge machining (PMEDM) has emerged as an effective technique for enhancing the wear resistance of engineering materials, including superalloys like Inconel 625. Unlike conventional EDM, which primarily focuses on material removal, PMEDM integrates fine powder particles such as boron carbide

(B_4C), silicon carbide (SiC), and aluminum oxide (Al_2O_3) into the dielectric fluid, significantly improving the surface characteristics of machined components. These powders contribute to the development of a modified recast layer with refined microstructures, reduced thermal defects, and enhanced hardness, making PMEDM an ideal candidate for applications requiring superior wear resistance.

11.3.3.1.2 Impact of PMEDM on Surface Hardening

PMEDM allows powder particles to be deposited on the machined surface, creating a hardened carbide and oxide compound-enriched layer. This improves the tribological characteristics of the workpiece by lowering friction and material degradation under abrasive wear conditions. Studies have shown that the addition of B_4C in PMEDM greatly reduces crater formation, enhances surface smoothness, and improves grain structures, thus enhancing the service life of wear-prone parts like agricultural tillage tools and turbine blades.

11.3.3.1.3 Process Optimization for Enhanced Performance

The success of PMEDM in creating wear-resistant surfaces is a function of some important process parameters, such as voltage, pulse duration, powder concentration, and dielectric medium choice. By optimizing these parameters:

- Higher powder concentrations produce more homogeneous recast layers with increased hardness but need careful control to avoid too much debris build-up.
- Tailored pulse durations guarantee effective energy transfer, minimizing micro-cracks and enhancing wear performance.
- Other dielectric media like oil-based fluid or deionized water may impact breakdown voltage as well as spark stability and thus overall machining efficiency.

11.3.3.1.4 Comparative Analysis with WAAM-CMT Cladding

While wire arc additive manufacturing with cold metal transfer enhances wear resistance well through metallurgical bonding, PMEDM presents a complementary methodology by improving surface properties at the microstructural level. The combination of both approaches could result in hybrid manufacturing techniques where WAAM-CMT is applied for bulk cladding and PMEDM is applied for post-processing to attain optimum surface integrity and hardness.

11.4 CONCLUSION AND FUTURE SCOPE

Some of the key points of this chapter are as follows:

- Results show that among all other clads, Incl shows better strength and ability to withstand harsh conditions due to its high hardness value (385 HV).
- Incl samples show better wear resistivity than other cladded and uncladded samples.

- Cladded samples show mixed ductile and brittle types of failure, whereas base steel shows a ductile mode of failure during SEM analysis.
- The optimum parameters (Incl) that are superlative in wear resistance are 146 (A) current, 16.0 (V) voltage, 7 (m/min) wire feed speed, and 10 (cm/min) scanning speed.

Abrasive wear has posed significant challenges for tillage tools in agriculture and mining.

11.4.1 Future Applications in Industrial Uses

The development of PMEDM and its synergy with additive manufacturing technology provide new possibilities for the development of high-performance wear-resistant materials. With further research in hybrid machining methods, aerospace, biomedical, and automotive industries can harness these technologies to improve the durability and efficiency of key components.

REFERENCES

1. Coronado, J. J., Caicedo, H. F., & Gómez, A. L. (2009). The effects of welding processes on abrasive wear resistance for hardfacing deposits. Tribology International, 42(5), 745–749.
2. Mulligan, C. P., Blanchet, T. A., & Gall, D. (2010). CrN–Ag nanocomposite coatings: High-temperature tribological response. Wear, 269(1–2), 125–131.
3. Singh, H., Singh, G., Singh, P., & Dixit, A. K. (2024, February). Effect of process parameters on the properties of stainless steel 304 and to optimize the parameters. In *AIP Conference Proceedings* (Vol. 2986, No. 1). AIP Publishing. https://pubs.aip.org/aip/acp/article-abstract/2986/1/030169/3266586/Effect-of-process-parameters-on-the-properties-of?redirectedFrom=fulltext#:~:text=MIG%20welding%20is%20an%20appropriate,parameters%20using%20the%20Taguchi%20method
4. Shmulevich, I. (2010). State of the art modeling of soil–tillage interaction using discrete element method. Soil and Tillage Research, 111(1), 41–53.
5. Yu, H. J., & Bhole, S. D. (1990). Development of a prototype abrasive wear tester for tillage tool materials. Tribology International, 23(5), 309–316.
6. Yu, H. J., & Bhole, S. D. (1990). Development of a prototype abrasive wear tester for tillage tool materials. Tribology International, 23(5), 309–316.
7. Natsis, A., Papadakis, G., & Pitsilis, J. (1999). The influence of soil type, soil water and share sharpness of a mouldboard plough on energy consumption, rate of work and tillage quality. Journal of Agricultural Engineering Research, 72(2), 171–176.
8. Kalácska, Á., De Baets, P., Fauconnier, D., Schramm, F., Frerichs, L., & Sukumaran, J. (2020). Abrasive wear behaviour of 27MnB5 steel used in agricultural tines. Wear, 442, 203107.
9. Kang, A. S., Cheema, G. S., & Singla, S. (2014). Wear behavior of hardfacings on rotary tiller blades. Procedia Engineering, 97, 1442–1451.
10. Khan, M. S., Ali, S., Westerbaan, D., Duley, W., Biro, E., & Zhou, Y. N. (2022). The effect of laser impingement angle on the optimization of melt pool geometry to improve process stability during high-speed laser welding of thin-gauge automotive steels. Journal of Manufacturing Processes, 78, 242–253.
11. Singh, G., Vasudev, H., Bansal, A., & Vardhan, S. (2021). Influence of heat treatment on the microstructure and corrosion properties of the Inconel-625 clad deposited by microwave heating. Surface Topography: Metrology and Properties, 9(2), 025019.

12. Raushan, H., Bansal, A., Singh, V., Singla, A. K., Singla, J., Omer, A., Singh, J., Goyal, D. K., Khanna, N., & Rooprai, R. S. (2023). Dry sliding and slurry abrasion behaviour of wire arc additive manufacturing–cold metal transfer (WAAM-CMT) cladded inconel 625 on EN8 steel. Tribology International, 179, 108176.
13. Xin, H., Yang, J., Zhang, W., Yang, J., Mao, J., Teng, C., Kong, X., Si, J., Xu, X., Wu, L., & Wu, X. (2022). Effect of Au ion irradiation on the surface morphology, microstructure and mechanical properties of AlNbTiZr medium-entropy alloy coatings with various Al content for ATF. Surface and Coatings Technology, 434, 128157.
14. Singh, V., Kunal, G., Rooprai, R. S., Bansal, A., Kaur, S., Vishnoi, M., & Singla, A. K. (2024). FEA-guided development of Inconel 625 clad on mild steel using WAAM and their cavitation erosion resistance. International Journal on Interactive Design and Manufacturing (IJIDeM), 1–15.
15. Li, W., Li, J., & Xu, Y. (2021). Optimization of corrosion wear resistance of the NiCrBSi laser-clad coatings fabricated on Ti6Al4V. Coatings, 11(8), 960.
16. Xing, Q., Feltrin, A. C., & Akhtar, F. (2022). High-temperature wear properties of CrFeHfMnTiTaV septenary complex concentrated alloy film produced by magnetron sputtering. Wear, 510, 204497.
17. Ennaceri, H., Erfurt, D., Wang, L., Köhler, T., Taleb, A., Khaldoun, A., El Kenz, A., Benyoussef, A., & Ennaoui, A. (2016). Deposition of multifunctional TiO_2 and ZnO top-protective coatings for CSP application. Surface and Coatings Technology, 298, 103–113.
18. Prashar, G., & Vasudev, H. (2021). A comprehensive review on sustainable cold spray additive manufacturing: State of the art, challenges and future challenges. Journal of Cleaner Production, 310, 127606.
19. Lazarevic, S., Lu, R. Y., Favede, C., Plint, G., Blau, P. J., & Qu, J. (2018). Investigating grid-to-rod fretting wear of nuclear fuel claddings using a unique autoclave fretting rig. Wear, 412, 30–37.
20. Thakur, L., Singh, J., & Vasudev, H. (Eds.). (2024). *Thermal Claddings for Engineering Applications*. CRC Press.
21. Rooprai, R. S., Bansal, A., & Singh, J. (2023). Influence of TiC powder content on wear behaviour of Inconel 625 clads developed by hybrid-mode wire arc additive manufacturing (WAAM) on EN-8 steel. Tribology International, 189, 108937.
22. Khanna, N., Patel, D., Raval, P., Airao, J., Badheka, V., & Rashid, R. A. R. (2024). Comparison of sustainable cooling/lubrication strategies for drilling of wire arc additively manufactured Inconel 625. Tribology International, 200, 110068.
23. Chigilipalli, B. K., & Veeramani, A. (2022). Investigations on dry sliding wear behavior of a wire arc additively manufactured nickel-based superalloy. Tribology Transactions, 65(5), 912–923.
24. Jayashree, P., Turani, S., &Straffelini, G. (2019). Effect of testing conditions on the dry sliding behavior of a Cu-Based refractory composite material. Tribology International, 140, 105850.
25. Rooprai, R. S., Singh, J., & Bansal, A. (2024). Slurry abrasion behaviour of ERNiCrMo-3 cladding on EN8 steel developed by WAAM. Materials Letters, 360, 136043.

12 Machine Learning for Predicting the Tribological Properties of Coatings

Jyotsna Barpute, Shubhangi Suryawanshi, Digvijay Bhosale, Amrut Bhosale, Umesh Ghorpade, Sunil B Shivarudraiah, and Hitesh Vasudev

12.1 INTRODUCTION

The significant influence of friction, lubrication, and wear on global energy consumption highlights the importance of tribology [1]. Understanding tribology principles can lead to new answers to future technical issues, supported by extensive experimental testing as well as advanced multiscale simulations [2]. Cutting-edge techniques in artificial intelligence (AI) and machine learning (ML) present excellent opportunities to explore the intricate workings of tribological systems, allowing effective, sometimes even instantaneous, behavior classification and quantification [2]. The advantages and possibilities of AI and ML strategies are most apparent in their capacity to deal with large-scale datasets along with issues and also in their ability to adjust to changing situations at a fair cost and effort [3].

These methods make it easier to find pertinent connections and causation, which improves what is already known using the data at hand. In the end, AI and ML can offer engineers, practitioners, and even intelligent and adaptive tribological systems clear and accurate action suggestions through analyses, predictions, and optimizations. Tribology's sophisticated experimental techniques set it apart from other fields. Experiments can be completed quickly and using a variety of materials, although they are very cost and resource intensive. Additionally, the difficulty of obtaining direct measurements of the contact interface makes it difficult to confirm experimental models. Although they can be utilized to improve some results, traditional methods such as completely randomized design, full factorial design, and response surface methodology do not accommodate the intricate and nonlinear nature of friction and wear. In dynamic situations, ML and AI prove to be highly effective in handling intricate problems and large datasets [4].

The area of machine learning is extremely varied and interdisciplinary, combining different data-driven designs with computational tools and learning algorithms.

DOI: 10.1201/9781003546528-12

These methods have been used for decades to train algorithms on empirical observations to examine big datasets. This makes it possible to create models that even in the face of ambiguity can predict future trends or other interesting outcomes [5, 6]. Tribology process optimization has garnered global interest due to ML techniques that can solve intricate, nonlinear issues. Sadegh et al. used artificial neural networks (ANNs) and genetic algorithms to observe and categorize the journal bearing's lubrication state [7]. Researchers analyzed surface wear with neural networks and pin-on-disc challenges. They applied ANNs to the wear particle morphology and were able to successfully control friction [8]. Using similar ML techniques, researchers identified in-flight particle characteristics during atmospheric plasma spray painting with a maximum correlation of 0.9996 [9].

Aleksendric et al. employed Bayesian regulation (BR) to optimize the friction properties of braking materials, and the BR neural model precisely predicted the optimal friction material's cold performance [10]. Li et al. used a hypersphere support vector machine to study mechanical seal conditions under different contact situations [11], and Markova et al. used the Mamdani algorithm and fuzzy logic to determine the status of the lubricant [12]. Other researchers investigated the effects of heat treatment on the tribological characteristics of AISI A8 tool steel by applying clustering techniques such as K-means and neural networks, which generated wear mechanism maps. The aim was to maximize the bearing's remaining functional life [13]. Wang et al. built a 3-D convolutional neural network (CNN) to determine a bearing's usable life [14].

12.2 MACHINE LEARNING'S FUNCTION IN TRIBOLOGY

Tribology benefits greatly from ML, which makes it easier to precisely model and simulate engineering processes. It is helpful for comprehending the interactions in tribological systems, which examine friction, lubrication, and wear, because it can manage complex and nonlinear correlations in data. ML can accurately identify relationships between inputs and outputs and does not impose restrictions on process parameters, in contrast to older approaches. In tribology, ML facilitates the decomposition of intricate design problems into more manageable components by highlighting significant connections among various aspects and procedures [15–17]. The following research methodology framework aids in understanding applications of ML in tribology: formulating research questions, locating and screening pertinent literature, examining the data to identify patterns and significant findings, and assessing the data to assess the usefulness of ML and recommend future lines of inquiry. This methodical approach offers a clear and thorough comprehension of ML applications in tribology (Figure 12.1).

12.2.1 Techniques for Machine Learning

Machine learning was introduced by a Cornell University research psychologist. Based on hypotheses regarding the functioning of the human nervous system, he created a computer system that was able to recognize alphabets [18]. Using both discrete and analog impulses, Rosenblatt's system—dubbed the perceptron—included

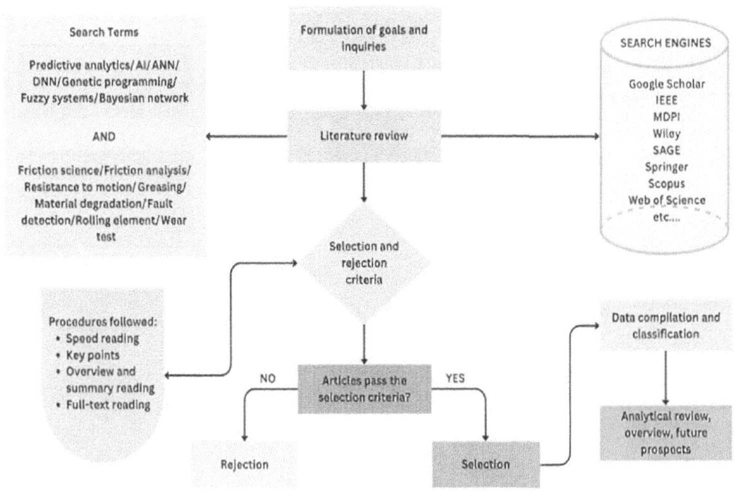

FIGURE 12.1 Tribology research framework flow chart.

a threshold element to transform analog signals into discrete ones. The first neural networks, which learned in the same way as people and animals, were prototyped by this technology [19]. Another key accomplishment was the Novikoff theorem, which defines how a perceptron learning algorithm will converge in a finite number of steps while using preexisting data.

Pattern recognition, data forecasting, and automation are among the functions carried out by ML algorithms that improve by "learning" from experience with particular tasks [20]. More accurate models are frequently produced by combining ML models with other methods. Generally, there are three types of ML algorithms: reinforcement, supervised, and unsupervised. Figure 12.2 shows categories of ML methods.

The goal of supervised learning is to map input variables (x) to desired output variables (y) in order to estimate an unknown function $y = (x)$ y = f(x). Regression analysis uses a database sample of function f to compute the approximation of a continuously unknown f. Classification is a kind of supervised learning in which the objective is to estimate a discretized function or identify numerical characters in a digital snapshot [21]. In contrast, unsupervised learning uses labeled input and output variables to identify patterns and clusters within a dataset. These systems are capable of clustering data without the need for human supervision [22]. One method that involves working with a reward function is reinforcement learning. Rather than being directed toward a predetermined set of activities, the learner must use trial and error to determine which actions result in the highest reward value [23]. Highly functional components must be designed with an understanding of the tribological behavior of materials. ML algorithms use tribology datasets to leverage examples or prior knowledge to improve task performance [24].

Tribological qualities depend on a number of elements, including the environment, the parameters of operation, material properties, and mechanical properties.

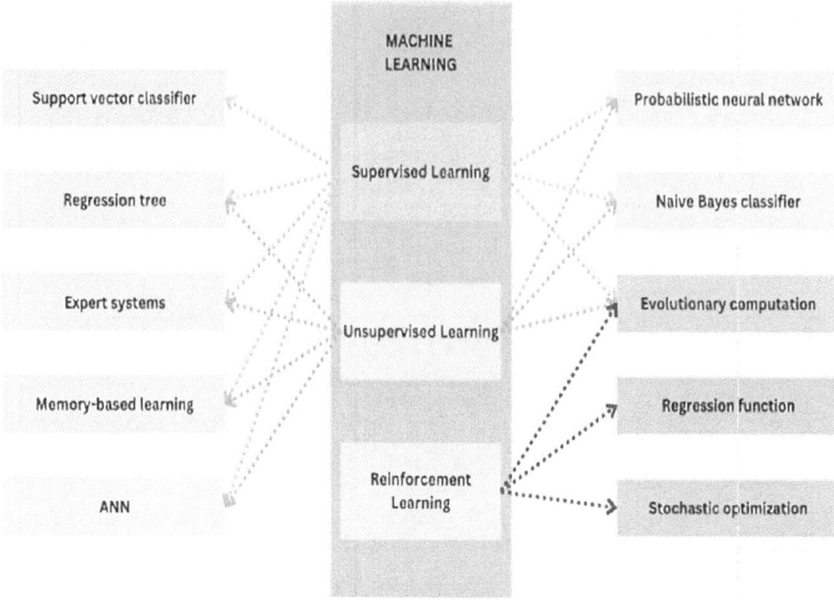

FIGURE 12.2 Common machine learning methods.

To properly characterize a product, one must comprehend the ways in which these components affect tribological performance. Tribological qualities can be adversely affected by sliding velocity, sliding speed, load, sliding distance, and temperature, among other factors. This complexity can be addressed in part by ML approaches. Researchers can better comprehend tribological phenomena and create useful links between process inputs and outputs by utilizing intelligent algorithms. Since the 1980s, tribology has used ML to improve friction, wear, and lubrication as well as condition monitoring and fault identification [25, 26].

12.2.1.1 Friction

Although friction reduces wear and energy consumption, it is necessary for daily living and energy saving. Friction has long been researched by scientists, and significant advances have been made. Adhesion and deformation caused by lubricants, coatings, and contaminants make up friction. Traditional approaches have difficulty reliably predicting friction coefficients because of the complexity and nonlinearity of the elements impacting friction; in contrast, ML algorithms are very good at predicting these nonlinear processes. Friction coefficients in a variety of materials have been accurately predicted by ML approaches [27].

12.2.1.2 Wear

Wear is caused by complex physical and chemical processes that occur over time. As the field of atomic force microscopy progressed in the 1980s, tribology began to focus on molecular mechanisms. This led to the identification of three wear

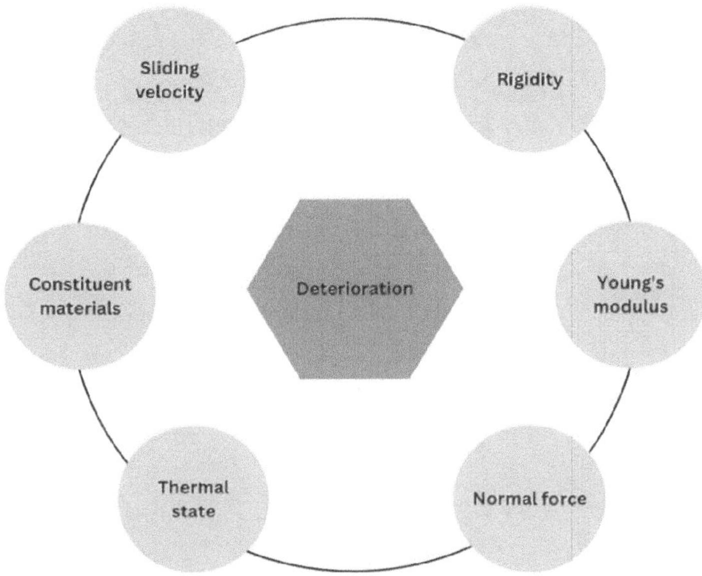

FIGURE 12.3 Factors that influence wear in tribological systems.

mechanisms at the asperity level: amorphization, smoothening by dislocation plasticity, and atomic-level attrition. Because of the difficulty in identifying wear cycles and mechanism heterogeneity, numerical modeling of wear processes has gained popularity. There is fresh promise for better wear models with ML [28]. Figure 12.3 shows factors affecting deterioration in tribology:

- Sliding velocity: the speed with which one surface glides over another
- Rigidity of the materials in contact
- Young's modulus, a measure of the material's elasticity
- Normal force: the perpendicular force exerted between surfaces
- Thermal state: the temperature conditions that influence the materials
- Constituent materials: the materials used including composition and qualities

12.2.1.3 Lubrication

In engineering and applied science, lubrication is essential because it directly affects component durability, energy efficiency, and financial gains. Lubrication, whether self-lubricating or external, is crucial for reducing wear and friction in tribological systems. In the nineteenth century, Reynolds designed empirical models to calculate lubricant flow dynamics. These days, computer simulations help to better comprehend complicated tribological interactions by examining variables such as surface geometry, sliding conditions, and the impact of the environment on wear and friction. In the process, researchers are increasingly turning to ML techniques because of the complexity and nonlinearity of lubrication models [29].

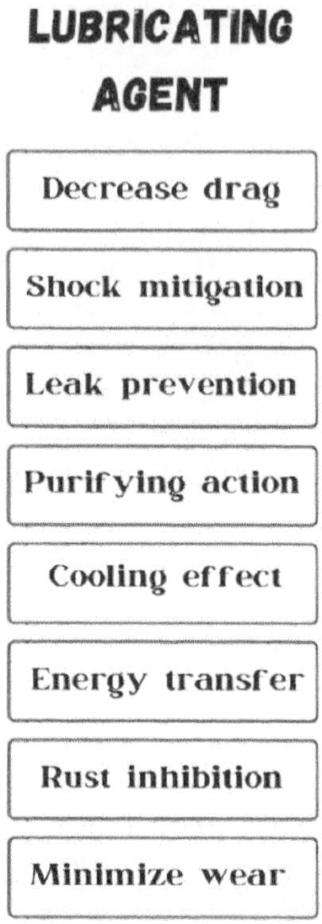

FIGURE 12.4 The roles of lubricating agents in tribological systems.

Figure 12.4 illustrates the many functions of lubricating agents, stressing their relevance in maintaining the efficiency and lifetime of tribological systems:

- **Decrease drag:** reduces the frictional resistance between moving surfaces
- **Shock mitigation:** absorbs and dampens shock loads to avoid harm
- **Leak prevention:** seals holes to prevent fluid leaking
- **Purifying action:** eliminates impurities and particulates from the system
- **Cooling effect:** removes heat produced by friction
- **Energy transfer:** enables the transfer of energy inside the system
- **Rust inhibition** protects metal surfaces against oxidation and corrosion
- **Minimize wear:** slows material loss and surface degradation over time

12.2.1.4 Surface Engineering

Surface engineering is the process of altering a component's surface properties to change how it behaves. For the purpose of improving mechanical, thermal, and tribological qualities, this involves changing the surface texture and using coatings. Researchers have developed coatings with characteristics like hardness, wear resistance, corrosion resistance, and self-lubrication, and precise process modeling and deposition parameter optimization are essential to achieving desired characteristics.

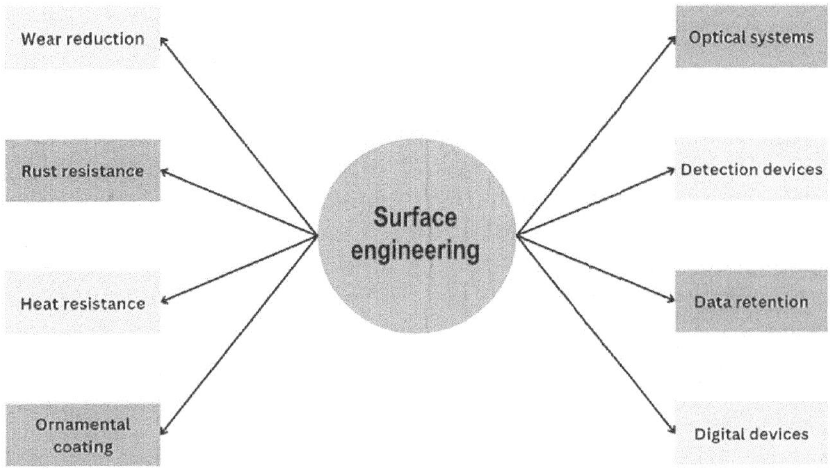

FIGURE 12.5 Roles of surface engineering in tribology.

Accurately predicting and characterizing coating qualities is difficult due to the complexity of coating technologies, which is a challenge for traditional approaches. To close this gap, ML is being increasingly applied to precisely simulate and optimize coating processes. Research indicates that ML strategies such as ANNs and CNNs can anticipate wear, friction, and coating characteristics more accurately, improving surface engineering skills [30].

Figure 12.5 shows the different applications and benefits of surface engineering:

- **Wear reduction:** minimizing friction and wear to extend the life of components
- **Rust resistance:** coatings and treatments that protect surfaces from corrosion
- **Heat resistance:** treatments that allow surfaces to tolerate extreme temperatures
- **Ornamental coatings:** decorative finishes applied to surfaces for aesthetic purposes
- **Optical systems:** used in the manufacture of lenses, mirrors, and other optical devices
- **Detection devices:** treatments to improve sensor and detector performance
- **Data retention**: treatments that improve the storage and preservation of data, such as in magnetic or optical storage media
- **Digital devices**: treatments that enhance the performance and durability of electronic components and devices

12.3 NEURAL NETWORK MODELS FOR WEAR PREDICTION

ANNs were used with characteristics such as input temperature, nozzle geometry, and particle size to forecast slurry erosion in thermally sprayed coatings. This method produced a more accurate erosion model that improved coating performance and design as well as helped anticipate and stop coating failure brought on by slurry erosion [31, 32].

12.3.1 ANNs

The computational models known as ANNs imitate the architecture and functions of genuine neural networks. They are widely used for tasks such as pattern recognition, regression analysis, and decision-making, and they are essential to machine learning [33]. The three primary layers of artificial neurons, or nodes, that make up ANNs are input, hidden, and output. Neurons use mathematical operations to process input signals and produce output signals that are transferred through connections represented by weights. Learning algorithms modify these weights during training to find patterns in the data [34]. Different ANNs include long short-term memory (LSTM), CNNs, recurrent neural networks (RNNs), and feed forward neural networks (FFNNs). FFNNs are best suited for simple input–output mappings, RNNs are best for sequential data, LSTM networks are best for long-term dependencies, and CNNs are best for image processing [35].

A recent ANN had error percentages of 0%–6%, making it more accurate than earlier iterations. The ANN gave fewer prediction errors than multiple linear, partial least squares, and principal component regression and a 125% higher R^2 than that from principal component regression [36]. The ANN also demonstrated superior fit to the observed data based on mean absolute error and root mean squared error as metrics and outperformed multiple linear regression with a 24-fold decrease in error. These results demonstrated how much better the ANN model is in predicting outcomes and capturing the relationships in the dataset [36]. Figure 12.6 shows the ANN's network configuration. The input layer (red circles) is where the first data input is received. The hidden layer (blue circles) applies activation functions and weights to the inputs received from the input layer via network nodes. The output layer (yellow circles) uses the information from the hidden layer that has been processed to create the final result.

12.3.2 Training, Testing, and Validating an ANN

It is standard procedure to create three sets of data to train ANNs. The validation set helps modify hyperparameters and minimize overfitting, the training set is used for

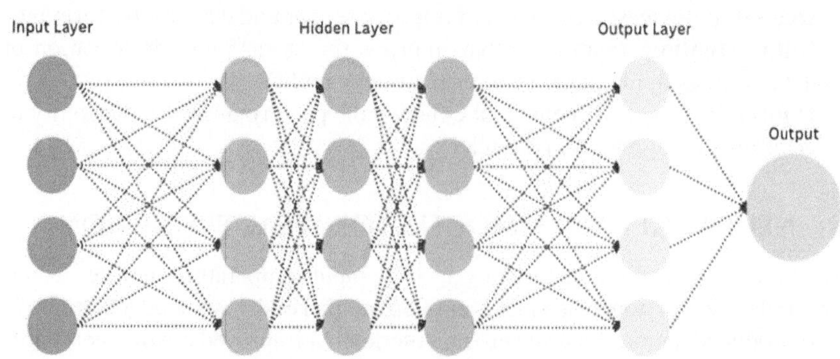

FIGURE 12.6 An ANN's network configuration.

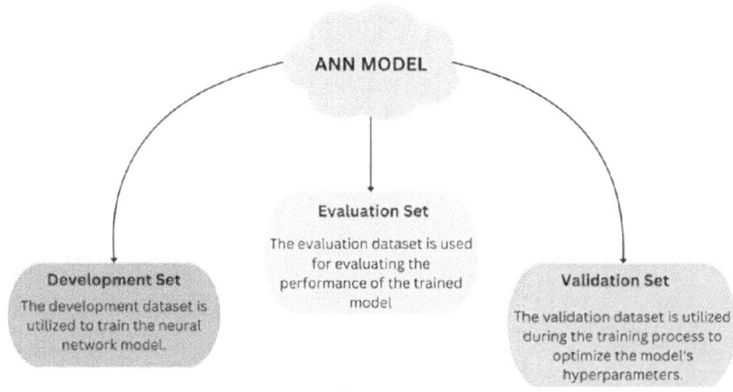

FIGURE 12.7 ANN datasets and their functions.

developing the model, and the testing set evaluates the model's performance on fresh information; see Figure 12.7. The figure includes most of the currently available data and is used to update the model's weights and biases throughout training [37].

Though these percentages can change, data is often divided into three categories: 70% for training, 15% for testing, and 15% for validation. The ANN model is thoroughly assessed and modified to improve its generalization to fresh data [38, 39]. Researchers used an ANN to design and optimize a thermal spray coating to improve qualities like wear resistance, corrosion resistance, or thermal insulation. The researchers employed ten neurons in the buried layer of the network and two 42 point datasets, one developed independently: 36 points were used for training and 6 for testing. The neural network was developed continuously for 52 iterations until the mean squared error decreased. The six testing data values were used to assess the network's performance, and the ANN demonstrated the ability to accurately predict outcomes from complicated datasets [40–42].

12.4 AI AND COMPUTATIONAL MODELS TO IDENTIFY WEAR BEHAVIOR

In mechanical engineering, ANNs are a common application of AI. Dr. Robert Hecht-Nielsen described an ANN as "a computing system composed of several simple, highly interconnected processing elements that process information through their dynamic state response to external inputs." ANNs are adaptable to a variety of tasks, including pattern recognition, prediction, classification, and optimization. They are very generalizable because of their flexibility, which enables them to model both linear and nonlinear connections. After being trained on specific datasets, ANNs are employed with a variety of techniques to anticipate output values [43, 44].

Figure 12.8 presents the wide variety of available AI models. Unsupervised learning entails identifying patterns in data without labeled outputs. Reinforcement learning teaches AI models by rewarding or penalizing them for their outcomes. In supervised learning, AI models learn from labeled training data. Deep learning is

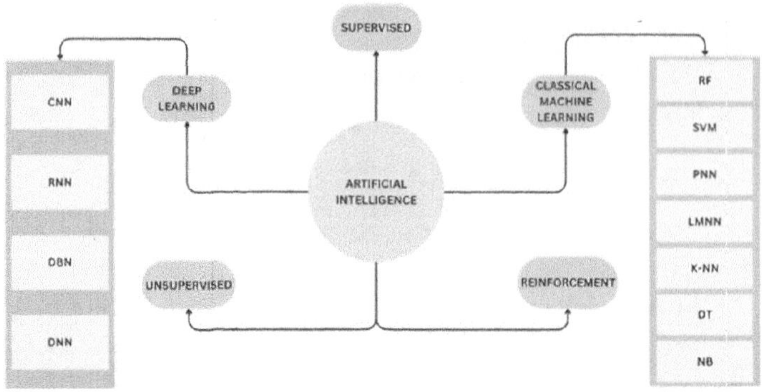

FIGURE 12.8 AI classification models.

a branch of neural network-based supervised learning that utilizes CNNs, RNNs, deep belief network, and deep neural networks. Classical machine learning is based on conventional supervised learning algorithms such as random forest, support vector machine, probabilistic neural network, large margin nearest neighbor, K-nearest neighbors, decision trees, and naive Bayes classifiers.

12.5 OUTSTANDING ISSUES

Researchers are increasingly using the Levenberg–Marquardt (LVM) algorithm with ANNs to create well-generalized, fault-tolerant models; however, the LVM is less effective for flat functions with few gradients because it usually just reaches a local optimum and can produce unwanted outcomes if the initial position is too far from the ideal. Takagi–Sugeno fuzzy inference was combined with ANN to create an adaptive neuro-fuzzy inference system that incorporated the benefits of both approaches; however, this model was not as effective in a large data environment where there were many inputs [45–47].

To anticipate wear and surface roughness, researchers employed RNNs, ANNs with memory; RNNs are better suited for dynamic situations like predicting tool wear. Investigators used random forests and multilayer perceptron (MLP) to predict surface wear. The latter required parameter adjustment, so random forests were preferred because the MLP result was nearly the same as that from random forest [48].

A number of factors influence a neural network's prediction accuracy with different coatings. Choosing the appropriate network model is essential because different networks respond differently to different data types; for instance, neural networks using backpropagation radial basis function can forecast and categorize electrochemical coating test results. CNNs and single-shot multibox detectors identify and classify faults in coated surface image data, and graph neural networks are applied to data that has multidimensional spatial features. Every day, new models are put out to advance prediction techniques and enhance outcomes. Accuracy can also be increased by combining neural networks with other prediction models. Additionally, as deep neural networks frequently need big datasets to function successfully on complicated tasks, transfer learning can help with the issue of insufficient training data.

To maximize prediction performance in neural network modeling, target variable selection is crucial. Stepwise regression, information gain computation, and statistical analysis are three efficient techniques. Stepwise regression adds variables one at a time and removes those that cease to be significant after multiple additions. Each input variable's importance to the result is determined by information gain, which gives higher weight to those variables. Statistical analysis can determine whether predictor factors have significant connections with the result by testing the links between them. By ensuring that neural networks only include significant predictors, these techniques improve the prediction accuracy of neural networks.

A research group used ANNs to predict coating life and degradation from electrochemical and salt-spray test data. The authors found greater prediction accuracy by combining data from modelling, simulation, image processing, and machine vision; specifically, they achieved 84.8% accuracy, indicating the model's generalizability and transferability. The researchers incorporated migrating learning networks that used pretrained models to extract common characteristics from test data sources to create universal life prediction models appropriate for different coatings [49].

Deep networks incorporate multiple input sources to diagnose faults, identify and classify targets, detect features, and assess performance. Researchers made extensive use of these networks to acquire precise and comprehensive coating data that allowed for simultaneously detecting the coating and the underlying substrate demonstrating the effectiveness of multisource deep networks for real-world coating defect detection in conjunction with non-destructive testing and other approaches [50, 51].

12.6 FUTURE STUDIES

AI-based wear analysis is a novel field of study, so far garnering much less research attention that materials like aluminum composites. Wear-resistant materials are being developed with AI, but more research is needed to fully realize its potential. Future researchers could concentrate on analyzing the wear behavior of different tool metals to potentially improve wear and performance. With less effort and more accuracy than conventional techniques, AI can assist in determining optimal coating materials for better wear resistance. The primary advantage of AI is its capacity to evaluate massive datasets and generate accurate predictions, despite its current limitations. To fully utilize this technology and enable industries to anticipate material wear ahead of time and run uninterrupted, more testing is required [52, 53].

The task of choosing and refining the best algorithms for particular situations is at the heart of ML in tribology. Although ANNs have been the main tool employed, there is value in investigating alternative ML techniques that might work better for specific problems. It is imperative that results be reproducible and comparable, but this is currently hampered by imprecise databases and uneven implementation details. Future researchers should publish underlying information and models and develop inverse models and ML techniques influenced by physics. The secret to maximizing the value of current information in the field will be to automate and optimize data processing and collection [54].

Tribology prioritizes the use of AI in place of conventional statistical techniques for variable optimization and modeling. Although ANNs are an excellent choice for intricate, nonlinear tribology issues, additional ML algorithms including ensemble learning, fuzzy logic, genetic algorithm, RNN, decision trees, random forest, and SVM should also be considered. Additionally, the use of nanoparticles in the investigation of friction, wear, and lubrication has broadened the discipline and given rise to novel phenomena; AI algorithms can predict how different combinations of base oils and nanoparticles will perform under varying test conditions. However, the lack of adequate and comparable datasets and inadequate uncertainty management are major obstacles; researchers should include appendices or data repositories with the underlying datasets and models. To optimize both ML utilization and tribology skills across industries, automation and optimization of data collection and processing are crucial. Future developments in physics, chemistry, materials science, mechanical engineering, and computer science will require extensive multidisciplinary collaborations. AI and ML algorithms offer insights into the important field of tribology and can point scholars toward a new, more energy-efficient, environmentally sustainable society. Machine learning has emerged as a powerful tool for predicting tribological properties of coatings, offering significant advantages in terms of accuracy and efficiency [55–62].

Recent researchers have applied ANNs and SVMs to model and predict coating wear and corrosion [63–72]. For example, researchers used ML models to extensively analyze and predict the tribological behavior and real-world performance of boron-nitride thin films deposited using magnetron sputtering. These models assisted in understanding the influence of process parameters on wear resistance and provided insights into the optimization of coatings for specific industrial applications, such as corrosion protection of stainless steel [73–81].

ML is also being employed to analyze the sliding wear properties of Inconel-625 clad deposits on stainless steel, specifically to predict tribological performance under different operating conditions. By integrating ML techniques with experimental data, researchers can identify key parameters affecting the wear resistance of coatings and optimize coating designs and processes, leading to more durable and efficient materials for industries like aerospace, biomedical, and energy [82–93].

12.7 CONCLUSION

The purpose of this chapter is to steer future directions and convince scientists that computational intelligence is a better option for modeling and variable optimization than traditional statistical methods. The most suitable course for tribology research in the future was determined by analyzing two decades of data. Although ANNs are attractive for tribology because they work particularly well for highly complicated, non-linear fundamental and applied problems, many other ML algorithms. To fully leverage existing ML and tribology expertise across industries, substantial data must be automated and optimized, much of which is still handled manually. Because tribology is an interdisciplinary field, it requires collaboration between researchers in physics, chemistry, materials science, mechanical engineering, and computer science, among other fields. However, ML does not guarantee accurate outcomes for every problem, so choosing and refining an appropriate algorithm is crucial. Indeed, many different strategies have already been effectively used to address tribological research issues.

REFERENCES

1. Holmberg K, Erdemir A. Influence of tribology on global energy consumption, costs and emissions. Friction 2017, 5, 263–284.
2. Vakis A, Yastrebov V, Scheibert J, Nicola L, Dini D, Minfray C, Almqvist A, Paggi M, Lee S, Limbert G, Molinari JF. Modeling and simulation in tribology across scales: An overview. Tribol. Int. 2018, 125, 169–199.
3. Rosenkranz A, Marian M, Profito FJ, Aragon N, Shah R. The use of artificial intelligence in tribology: A perspective. Lubricants 2021, 9, 2.
4. Wuest T, Weimer D, Irgens C, Thoben K-D. Machine learning in manufacturing: Advantages, challenges, and applications. Prod. Manuf. Res. 2016, 4, 23–45.
5. Wuest T, Weimer D, Irgens C, Thoben KD. Machine learning in manufacturing: Advantages, challenges, and applications. Prod. Manuf. Res. 2016, 4, 23–45. https://doi.org/10.1080/21693277.2016.1192517.
6. Barber D. Latent linear models. Bayesian Reason Mach. Learn. 2012. https://doi.org/10.1017/cbo9780511804779.026.
7. Jackson AH. Machine learning: A probabilistic perspective. 1988. https://doi.org/10.1111/j.1468–0394.1988.tb00341.x.
8. Sadegh H, Mehdi AN, Mehdi A. Classification of acoustic emission signals generated from journal bearing at different lubrication conditions based on wavelet analysis in combination with artificial neural network and genetic algorithm. Tribol. Int. 2016, 95, 426–434. https://doi.org/10.1016/j.triboint.2015.11.045.
9. Jaza R, Mollon G, Descartes S, Paquet A, Berthier Y. Lessons learned using machine learning to link third body particles morphology to interface rheology. Tribol. Int. 2021, 153, 106630. https://doi.org/10.1016/j.triboint.2020.106630.
10. Choudhury TA, Hosseinzadeh N, Berndt CC. Improving the generalization ability of an artificial neural network in predicting in-flight particle characteristics of an atmospheric plasma spray process. J. Therm. Spray Technol. 2012, 21, 935–949. https://doi.org/10.1007/s11666-012-9775-9.
11. Aleksendrić D, Duboka Č, Mariotti GV. Neural modelling of friction material cold performance. Proc. Inst. Mech. Eng. Part D. 2008, 222, 1201–1209. https://doi.org/10.1243/09544070JAUTO583.
12. Li X, Fu P, Chen K, Lin Z, Zhang E. The contact state monitoring for seal end faces based on acoustic emission detection. Shock Vib. 2016. https://doi.org/10.1155/2016/8726781.
13. Markova LV. Intelligent method for monitoring the state of lubricating oil. J. Friction Wear. 2016, 37, 308–314. https://doi.org/10.3103/S1068366616040115.
14. Pillai N, Karthikeyan R, Davim JP. Heat treatment effects on tribological characteristics for AISI A8 tool steel and development of wear mechanism maps using K means clustering and neural networks. Tribology. 2018, 12, 44–56. https://doi.org/10.1080/17515831.2018.1437335.
15. Wang X, Wang T, Ming A, Han Q, Chu F, Zhang W, Li A. Deep spatiotemporal convolutional-neural-network-based remaining useful life estimation of bearings. Chin. J. Mech. Eng. (English Edition). 2021, 34, 1–15. https://doi.org/10.1186/s10033-021-00576-1.
16. Sangwa NR, Sangwan KS. Leanness assessment of organizational performance: A systematic literature review. J. Manuf. Technol. Manag. 2018, 29, 768–788. https://doi.org/10.1108/JMTM-09-2017-0196.
17. Snyder H. Literature review as a research methodology: An overview and guidelines. J. Bus. Res. 2019, 104, 333–339. https://doi.org/10.1016/j.jbusr es.2019.07.039.
18. Sihag N, Sangwan KS. A systematic literature review on machine tool energy consumption. J. Clean. Prod. 2020, 275, 123125. https://doi.org/10.1016/j.jclepro.2020.123125.
19. Rosenblatt F. The perceptron: A perceiving and recognizing automaton. Report 85, Cornell Aeronautical Laboratory, 1957, 460–461.

20. Bush RR, Mosteller F. A mathematical model for simple learning. Psychol. Rev. 1951, 58, 313–323. https://doi.org/10.1037/h0054388.
21. Albert BJ, Novikoff. On convergence proofs for perceptrons. Station AHAU Arlington Hau Station Unclassified. Stanford Research Institute SRI Project No. 3605, 1963.
22. Alpaydin E. Neural networks and deep learning. Mach. Learn. 2021. https://doi.org/10.7551/mitpress/13811.003.0007.
23. Alpaydin E. Neural networks and deep learning. Mach. Learn. 2021. https://doi.org/10.7551/mitpress/13811.003.0007.
24. Prezelj J, Murovec J, Huemer-Kals S, Häsler K, Fischer P. Identification of different manifestations of nonlinear stick–slip phenomena during creep groan braking noise by using the unsupervised learning algorithms k-means and self-organizing map. Mech. Syst. Signal Process. 2022. https://doi.org/10.1016/j.ymssp.2021.108349.
25. Schmidt J, Marques MRG, Botti S, Marques MAL. Recent advances and applications of machine learning in solid-statematerials science. NPJ Comput. Mater. 2019. https://doi.org/10.1038/s41524-019-0221-0.
26. Tallian TE. Tribological design decisions using computerized databases. 1986, 109, 381–386. https://doi.org/10.1115/1.3261450.
27. Tallian TE. A computerized expert system for tribological failure diagnosis. 1988, 111. https://doi.org/10.1115/1.3261896.
28. Bowden FP, Tabor D, Palmer F. The friction and lubrication of solids. Am. J. Phys. 1951, 19, 428–429. https://doi.org/10.1119/1.1933017.
29. Binnig G, Quate CF, Gerber C. Atomic force microscope. Phys. Rev. Lett. 1986, 56, 930–933. https://doi.org/10.1103/PhysRevLett.56.930.
30. Reynolds O. On the theory of lubrication and its application to Mr. Beauchamp tower's mental determination of the viscosity of olive oil. Philos. Trans. R Exp. Soc. Lond. 1983, 177, 135–217. https://doi.org/10.1098/rstl.1886.0005.
31. Zhang G, Guessasma S, Liao H, Coddet C, Bordes JM. Investigation of friction and wear behaviour of SiC-filled PEEK coating using artificial neural network. Surf. Coat. Technol. 2006, 200, 2610–2617. https://doi.org/10.1016/j.surfcoat.2004.12.026.
32. Jagota V, Sharma RK. Wear volume prediction of AISI H13 die steel using response surface methodology and artificial neural network. J. Mech. Eng. Sci. 2020, 14, 6789–6800.
33. Szala M, Łatka L, Awtoniuk M, Winnicki M, Michalak MJP. Neural modelling of APS thermal spray process parameters for optimizing the hardness, porosity and cavitation erosion resistance of Al_2O_3–13 wt% TiO2 coatings. Processes. 2020, 8, 1544.
34. Ye D, Xu Z, Pan J, Yin C, Hu D, Wu Y, Li R, Li Z. Prediction and analysis of the grit blasting process on the corrosion resistance of thermal spray coatings using a hybrid artificial neural network. Coatings. 2021, 11, 1274.
35. Paturi UMR, Palakurthy ST, Reddy NS. The role of machine learning in tribology: A systematic review. Arch. Comput. Methods Eng. 2023, 30, 1345–1397.
36. Sahu SP, Satapathy A, Mishra D, Patnaik A, Sreekumar KJTT. Tribo-performance analysis of fly ash–aluminum coatings using experimental design and ANN. Tribol. Trans. 2010, 53, 533–542.
37. Pati PR, Satapathy AJTT. Triboperformance analysis of coatings of LD slag premixed with TiO_2 using experimental design and ANN. Tribol. Trans. 2015, 58, 349–356.
38. Razavipour M, Legoux J-G, Poirier D, Guerreiro B, Giallonardo JD, Jodoin B. Artificial neural networks approach for hardness prediction of copper cold spray laser heat treated coatings. J. Therm. Spray Technol. 2022, 31, 525–544.
39. Kumar S, Chandra SK, Dixit S, Kumar K, Kumar S, Murali G, Vatin NI, Sabri Sabri MM. Neural network prediction of slurry erosion wear of Ni-WC coated stainless steel. Metals. 2022, 420(12), 706.
40. Meerasri J, Sothornvit RJ. Artificial neural networks (ANNs) and multiple linear regression (MLR) for prediction of moisture content for coated pineapple cubes. Case Stud. Therm. Eng. 2022, 33, 101942.

41. Singh JJIL. Tribology: Analysis on suitability of HVOF sprayed Ni-20Al, Ni-20Cr and Al-20Ti coatings in coal-ash slurry conditions using artificial neural network model. Ind. Lubr. Tribol. 2019, 71, 972–982.
42. Mahendru P, Tembely M, Dolatabadi AJJTST. Artificial intelligence models for analyzing thermally. Spray. Funct. Coat. 2023, 32, 388–400.
43. Paturi UMR, Vanga DG, Cheruku S, Palakurthy ST, Jha NKJMTP. Estimation of abrasive wear of nanostructured WC-10Co-4Cr TIG weld cladding using neural network and fuzzy logic approach. Mater. Today Proc. 2023, 78, 449–457.
44. Karthik RM, Malghan RL, Kara F, Shettigar A, Rao SS, Herbert MA. Influence of support vector regression (SVR) on cryogenic face milling. Adv. Mater. Sci. Eng. 2021, Article ID 9984369, 18.
45. Eser A, Akar Ayyıldız E, Ayyıldız M, Kara F. Artificial intelligence-based surface roughness estimation modelling for milling of AA6061 alloy. Adv. Mater. Sci. Eng. 2021, Article ID 5576600, 10.
46. Leyland A, Matthews A. On the significance of the H/E ratio in wear control: A nanocomposite coating approach to optimised tribological behavior. Wear. 2000, 246(1–2), 1–11.
47. Pattnaik SK, Nayak A, Parida A, Kumar SS. +e study of surface roughness and tool wear analysis in turning of aluminum using different advanced cutting tools (January 16, 2021). In Proceedings of the International Artificial Intelligence in Manufacturing & Renewable Energy Conference on (ICAIMRE), Bhubaneswar, India, October 2019.
48. Sivaprasad DK, Babu SK, Natarajan S, Narayanasamy R, Kumar BA, Dinesh G. Study on abrasive and erosive wear behavior of Al 6063/TiB2 in situ composites. Mater. Sci. Eng. A. 2008, 498(1–2), 495–500.
49. Hermann G. Artificial intelligence in monitoring and the mechanics of machining. Comput. Ind. 1990, 14(1–3), 131–135.
50. Tian W, Meng F, Liu L, Li Y, Wang F. Lifetime prediction for organic coating under alternating hydrostatic pressure by artificial neural network. Sci. Rep. 2017, 7, 12. https://doi.org/10.1038/srep40827.
51. Wang S, Li W. GeoAI in terrain analysis: Enabling multi-source deep learning and data fusion for natural feature detection. Comput. Environ. Urban Syst. 2021, 90, 1715.
52. Bu K, Liu Y, Wang F. Operating performance assessment based on multi-source heterogeneous information with deep learning for smelting process of electro-fused magnesium furnace. ISA Trans. 2022, 128, 357–371. https://doi.org/10.1016/j.isatra.2021.10.024.
53. Selvaraj SK, Srinivasan K, Chadha U, Mishra R, Arpit K, Apurb K, Hu YC. Contemporary progresses in ultrasonic welding of aluminium metal matrix composites. Front. Mater. 2021, 8, Article ID 647112.
54. Sharma A, Chouhan A, Pavithran L, Chadha U, Selvaraj SK. Implementation of LSS framework in auto motive component manufacturing: A review, current scenario and future directions. Mater. Today Proc. 2021, 46.
55. Almqvist A. Fundamentals of physics-informed neural networks applied to solve the Reynolds boundary value problem. Lubricants. 2021, 9, 82.
56. Singh M, Vasudev H, Kumar R. Corrosion and tribological behaviour of BN thin films deposited using magnetron sputtering. Int. J. Surf. Eng. Interdiscipl. Mater. Sci. IJSEIMS. 2021 Jul 1, 9(2), 24–39.
57. Singh M, Vasudev H, Singh M. Surface protection of SS-316L with boron nitride based thin films using radio frequency magnetron sputtering technique. J. Electrochem. Sci. Eng. 2022, 12(5), 851–863.
58. Singh G, Singh H, Sharma Y, Vasudev H, Prakash C. Analysis and optimization of various process parameters and effect on the hardness of SS-304 stainless steel welded joints. Int. J. Interact. Des. Manuf. IJIDeM 2024, 18(5), 2889–2896.
59. Singh G, Mehta A, Vasudev H. Sustainability of additive manufacturing: A comprehensive review. Progr. Addit. Manuf. 2024, 1–24.

60. Singh G, Bansal A, Vasudev H, Mishra V. Sliding wear study of the Inconel-625 clad deposits by microwave heating on SS–304. Phys. Script. 2024, 99(6), 065503.
61. Singh J, Yadav A, Gupta S, Balu R, Shukla N, Singh G, Singh S. Design and computational analysis of continuous wave fiber laser cutting of flat and conformal ITO PET sheets. Int. J. Interact. Des. Manuf. IJIDeM. 2024, 1–9.
62. Singh G, Singh R, Gul J. Machinability behavior of human implant materials. J. Electrochem. Sci. Eng. 2023, 13(1), 99–114.
63. Verma PK, Vasudev H, Singh G, Singh S. A review on the microstructural and biomedical properties of implants manufactured using additive manufacturing. J. Inst. Eng. India Ser. D. 2024, 1–15.
64. Mehta A, Singh G, Vasudev H. Processing of shape memory alloys research, applications and opportunities: A review. Phys. Scr. 2024. https://iopscience.iop.org/article/10.1088/1402-4896/ad48cb
65. Singh G, Bansal A, Vasudev H. Corrosion and microstructural behaviour of Inconel 625 microwave clad deposited on mild steel. In Thermal Claddings for Engineering Applications (pp. 51–62). CRC Press, 2024.
66. Kapade TT, Shah SK, Shah DS, Kumar NS, Bhor JK, Panwar VS, Singh G. Thermal analysis of v-cut twisted tape fitted heat exchanger: A numerical approach. Int. J. Interact. Des. Manuf. IJIDeM. 2024. https://link.springer.com/article/10.1007/s12008-024-01883-2
67. Bhowmik A, Kumar R, Beemkumar N, Kumar AV, Singh G, Kulshreshta A, Mann VS, Santhosh AJ. Casting of particle reinforced metal matrix composite by liquid state fabrication method: A review. Res. Eng. 2024, 103152.
68. Kulkarni SD, Manjunatha, Chandrasekhar U, Manjunath KV, Prasad CD, Vasudev H. Design and optimization of polyvinyl-nitride rubber for tensile strength analysis. Int. J. Interact. Des. Manuf. IJIDeM. 2024, 18(8), 5893–5908.
69. Prashar G, Vasudev H. Surface topology analysis of plasma sprayed Inconel625-Al_2O_3 composite coating. Mater. Today Proc. 2022, 50, 607–611.
70. Yedida VS, Vasudev H. A review on the development of thermal barrier coatings by using thermal spray techniques. Mater. Today Proc. 2022, 50, 1458–1464.
71. Sharanabasava H, Raviprakash M, Prasad CD, Ramesh MR, Phanibhushana MV, Vasudev H, Kumar S. Microstructure, mechanical and wear properties of SiC and Mo reinforced NiCr microwave cladding. Adv. Mater. Process. Technol. 2023, 10(4), 3620–3633. https://doi.org/10.1080/2374068X.2023.2257937.
72. Dutta V, Thakur L, Singh B, Vasudev H. A study of erosion–corrosion behaviour of friction stir-processed chromium-reinforced NiAl bronze composite. Materials. 2022, 15(15), 5401.
73. Poojari M, Hanumanthappa H, Prasad CD, Jathanna HM, Ksheerasagar AR, Shetty P, Shanmugam BK, Vasudev H. Computational modelling for the manufacturing of solar-powered multifunctional agricultural robot. Int. J. Interact. Des. Manuf. IJIDeM. 2024, 18(8), 5725–5736.
74. Prashar G, Vasudev H. A review on the influence of process parameters and heat treatment on the corrosion performance of Ni-based thermal spray coatings. Surf. Rev. Lett. 2022, 29(1), 2230001.
75. Kattimani MA, Venkatesh PR, Masum H, Math MM, Bahadurdesai VN, Mustafkhadri S, Prasad CD, Vasudev H. Design and numerical analysis of tensile deformation and fracture properties of induction hardened inconel 718 superalloy for gas turbine applications. Int. J. Interact. Des. Manuf. IJIDeM. 2024, 18(8), 5939–5949.
76. Sunitha K, Vasudev H. A short note on the various thermal spray coating processes and effect of post-treatment on Ni-based coatings. Mater. Today Proc. 2022 Jan 1, 50, 1452–1457.
77. Majji BGR, Vasudev H, Bansal A. A review on the oxidation and wear behavior of the thermally sprayed high-entropy alloys. Mater. Today Proc. 2022, 50, 1447–1451.

78. Rao GS, Mukkamala U, Hanumanthappa H, Prasad CD, Vasudev H, Shanmugam B, KishoreKumar KC. Evaluating and optimizing surface roughness using genetic algorithm and artificial neural networks during turning of AISI 52100 steel. Int. J. Interact. Des. Manuf. IJIDeM. 2024, 18(8), 6151–6160.
79. Singh J, Gill HS, Vasudev H. Computational fluid dynamics analysis on role of particulate shape and size in erosion of pipe bends. Int. J. Interact. Des. Manuf. IJIDeM. 2023, 17(5), 2631–2646.
80. Anjaneya G, Sunil S, Kakkeri S, Math MM, Vaibhav MN, Solaimuthu C, Durga Prasad C, Vasudev H. Numerical simulation of microchannel heat exchanger using CFD. Int. J. Interact. Des. Manuf. IJIDeM. 2023, 1–17.
81. Nagabhushana P, Ramprasad S, Prasad CD, Vasudev H, Prakash C. Numerical investigation on heat transfer of a nano-fluid saturated vertical composite porous channel packed between two fluid layers. Int. J. Interact. Des. Manuf. IJIDeM. 2024, 18(5), 2927–2944.
82. Singh P, Vasudev H, Bansal A. Effect of post-heat treatment on the microstructural, mechanical, and bioactivity behavior of the microwave-assisted alumina-reinforced hydroxyapatite cladding. Proc. Inst. Mech. Eng. Part E J. Process Mech. Eng. 2024, 238(3), 1060–1071.
83. Mahantesh MM, Rao KR, Chandra AP, Vijayakumar MN, Nandini B, Prasad CD, Vasudev H. Design and modeling using finite element analysis for the sitting posture of computer users based on ergonomic perspective. Int. J. Interact. Des. Manuf. IJIDeM. 2024, 18(8), 5875–5891.
84. Singh J, Singh R, Singh S, Vasudev H, Kumar S. Reducing scrap due to missed operations and machining defects in 90PS pistons. Int. J. Interact. Des. Manuf. IJIDeM. 2023, 17(5), 2527–2539.
85. Arora H, Kumar V, Prakash C, Pimenov D, Singh M, Vasudev H, Singh V. Analysis of sensitization in austenitic stainless steel-welded joint. In Advances in Metrology and Measurement of Engineering Surfaces: Select Proceedings of ICFMMP 2019 (pp. 13–23). Springer Singapore, 2021.
86. Vasudev H. Wear characteristics of Ni-WC powder deposited by using a microwave route on mild steel: Microwave cladding of Ni-WC. Int. J. Surf. Eng. Interdiscip. Mater. Sci. IJSEIMS. 2020, 8(1), 44–54.
87. Prashar G, Vasudev H, Thakur L, Bansal A. Performance of thermally sprayed hydroxyapatite coatings for biomedical implants: A comprehensive review. Surf. Rev. Lett. 2023, 30(1), 2241001.
88. Mehta A, Vasudev H, Jeyaprakash N. Role of sustainable manufacturing approach: Microwave processing of materials. Int. J. Interact. Des. Manuf. IJIDeM. 2024, 18(8), 5283–5299.
89. Prashar G, Vasudev H. Understanding cold spray technology for hydroxyapatite deposition. J. Electrochem. Sci. Eng. 2023, 13(1), 41–62.
90. Parkash J, Saggu HS, Vasudev H. A short review on the performance of high velocity oxy-fuel coatings in boiler steel applications. Mater. Today Proc. 2022, 50, 1442–1446.
91. Singh J, Vasudev H, Kumar R, Ubaidullah M. Study on wear analysis of Ni-20Al_2O_3 HVOF micron layers using artificial neural network technique. Int. J. Interact. Des. Manuf. IJIDeM. 2024, 18(8), 5925–5938.
92. Singh J, Vasudev H, Szala M, Gill HS. Neural computing for erosion assessment in Al-20TiO_2 HVOF thermal spray coating. Int. J. Interact. Des. Manuf. IJIDeM. 2024, 18(4), 2321–2332.
93. Prakash C, Gupta LR, Mehta A, Vasudev H, Tominov R, Korman E, Fedotov A, Smirnov V, Kesari KK. Computing of neuromorphic materials: An emerging approach for bioengineering solutions. Mater. Adv. 2023, 4(23), 5882–5919.

13 Effect of Different Hard-Facing Welds on Erosion Wear Behavior of AISI 310SS

Ravi Kumar, Karthikeyan A, Dhirendra Nath, Abhinav Kumar, Vladimir Ivanovich Velkin, and Rajeev Rathi

13.1 INTRODUCTION

Solid particle erosion is a type of wear that occurs from repeated impressions of solid particles on a metal surface due to the mechanical interaction of the metal surface and solid particles.[1,2] Boiler nozzle erosion due to impingement of pulverized coal particles is one of the significant problems that is generally faced in coal-fired boilers.[3] Stainless steels (SS) are commonly used in boiler components working under a combined erosive and corrosive environment.[3–5] Normally, austenitic stainless steels are preferred over martensitic stainless steels due to their high corrosion and wear-resistant properties.[6] In the current investigation, erosion studies were carried out on coated and uncoated American Iron and Steel Institute 310 Stainless Steel (AISI 310SS) at room temperature. Hard facing was applied using three different hard-faced welding electrodes of different chemical compositions using shield metal arc welding on the surface of base metal (AISI 310SS). Welding parameters were kept constant during deposition. Erosion tests were performed according to American Society for Testing and Materials (ASTM) G76–95 standards. Erosion experiments were carried out on an erosion testing rig at different impact angles. This work focused on how to repair the already eroded surfaces of components that could be reused by depositing hard-facing weld metal (WM), which allows for repair instead of replacement. In this experimental work, the authors analyzed behavior and performance of the erosion of uncoated and coated hard-faced materials is analyzed. [7,8] Laguna-Camacho et al.[9] conducted a study of solid particle erosion of AISI 304, 310, and 420 stainless steels at 300, 450, 600, and 900 using silicon carbide (Si-C) abrasive particles with a particle velocity of 24 m/s. It was observed that AISI 420 stainless steel was the material with higher erosion resistance at all impact angles compared to the materials the current study, whereas Patel et al.[10] carried out investigation of the solid particle erosion behavior of AISI 304SS at the surrounding temperature.

DOI: 10.1201/9781003546528-13

It was observed that erosion takes place via plastic deformation of material. It was also observed that the erosion rate of AISI 304SS at an impingement angle of 300 was little higher than at an impingement angle of 900. Morrison et al.[11] conducted an erosion test on 304SS. In this investigation, SEM findings shows similar morphologies for low impact angles as well as high impact angles, which can be observed in ductile materials. Indentation, ploughing, and cutting actions were observed when high-velocity particles impacted a material surface.[12,13] In the present study, the performance and behavior of base metal against erosive wear are analyzed. It is characterized by brittle fracture of large fragments; a few fragments were embedded on the surface of the specimen, and pits were formed.[14–16] The surface of the specimen after erosion testing shows different features like wear debris, ploughing action, and irregular indentations.[17–20] Erosive wear is a significant concern in industries where there is high-speed particle impingement, including boiler equipment, heat treatment units, and high-temperature furnaces. Among other materials, AISI 310 stainless steel (SS) is used widely in applications where high-temperature strength, oxidation resistance, and mechanical stability are needed. Nevertheless, even with such benefits, exposure to erosive conditions over extended periods causes material deterioration, enormously cutting component performance and life.

To reverse such wear-induced problems, hard-facing methods have evolved as a surface alteration method to offer improved erosion resistance. Hard-facing welds, built through shielded metal arc welding processes, provide a surface protection layer above the base metal, enhancing hardness, toughness, and solid particle erosion resistance. The performance of such coatings is based on parameters like weld chemistry, microstructure, and processing, which affect their capacity to endure impact by abrasive particles at varying angles and speeds. In this work, AISI 310SS samples were exposed to hard-facing by various welding electrodes of differing chemical composition. The erosion response of coated and uncoated samples was evaluated by employing solid particle erosion tests under laboratory-controlled conditions. The effects of significant parameters, such as impact angle, particle velocity, and microhardness, were also studied to identify the optimum hard-facing material for severe wear applications. The research also discusses failure modes like pitting, ploughing, and wear debris formation, with the results showing the erosion resistance of various weld metals.

The outcomes of this research help to create high-performance surface coatings for industrial applications demanding outstanding wear resistance. By tuning the hard-facing weld composition and deposition conditions, the objective of this work is to maximize the longevity of AISI 310SS components, decrease maintenance expenditure, and optimize operation efficiency under erosive conditions.

13.1.1 Testing Information

- **Specimen Details:** AISI 310 stainless steel was used in the test as a base material whose chemical composition is given in Table 13.1. The microhardness value of the base material was 214 VHN when measured using Vickers hardness tester. Seven different points were measured over the surface to calculate microhardness.

TABLE 13.1
Chemical Composition of Base Metal (%)

Material	Form	C	Cr	Ni	Si	Mn	P	S	Mo
AISI 310SS	Base Metal	0.067	21.5	13	0.5	1.3	0.043	0.003	0.02

TABLE 13.2
Chemical Composition of Hard-Facing Electrodes (%)

Hard-Facing Electrode	Form	C	Cr	Ni	Mn	Si	Mo	Fe
Electrode 1	WM 1	5	33	_	_	_	_	Rest
Electrode 2	WM 2	0.06	26.5	9.5	1.0	1.10	0.20	Rest
Electrode 3	WM 3	0.10	29	9.5	0.60	1.0	0.5	Rest

TABLE 13.3
Welding Parameters

Form	Current; I (Amp)	Voltage; V (volts)	Welding Speed; S (mm/min)	Heat Input; H H=I*V/S (kj)
WM 1	135	22	12.5	14.27
WM 2	133	21	13	12.88
WM 3	136	22	12.5	14.34

Alumina powder was used to boost the erosion wear process and to investigate the performance of these materials under different erosive conditions. The size of the alumina powder particle was 100 microns. The specimen was of square shape with dimensions of 15 × 15 × 5 mm.

- **Specimen Preparation:** Three hard-facing electrodes of distinct composition, as shown in Table 13.2, were used to deposit weld metal on the surface of base metal using shield metal arc welding. While depositing electrode material on the base metal, the welding parameters remained constant, as shown in Table 13.3, where welding speed was manual.
- Three layers of weld metal were deposited for every specimen. After depositing the first layer or weld bead on the surface of the base metal, subsequent beads were deposited by overlapping the previous ones by approximately 25%. Welding was done horizontally in all cases.
- **Test Procedure:** Solid particle erosion tests were performed on the erosion test rig, which is represented in the ASTM G76–95 test standards. A schematic diagram of the erosion testing equipment is shown in Figure 13.1.

In this machine, small abrasive particles were fed through the nozzle using compressed air, which caused the impingement of particles on the

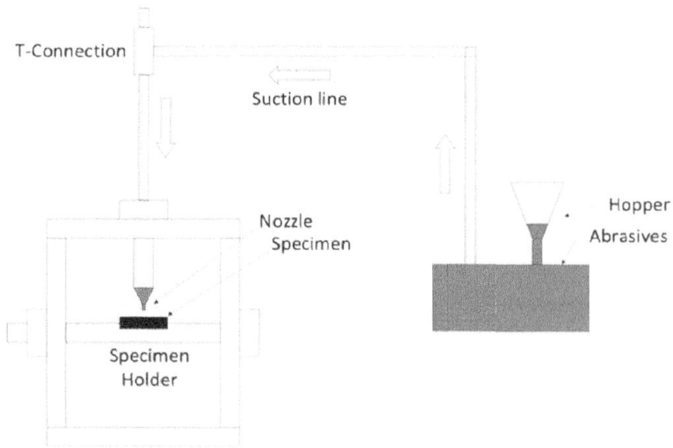

FIGURE 13.1 Schematic of erosion testing rig.

surface of the specimens. The weld metal and base metal were eroded in a period of 10 min, whereas each specimen was removed in a cycle of 2 min to measure the weight loss. Specimens were weighed with an accuracy of 0.0001 gm before the start of each test and removed after every 2 min to measure the weight.

- Tests were carried out at a temperature of approximately 350 °C. Tests were conducted using different impingement angles of 450 and 600 at a flow rate of 50 gm/min. The stand-off distance between the nozzle tip and the surface of the specimen to be tested was 10 mm. Microhardness tests were performed using a Vickers hardness tester at seven different points over the surface and normal to the surface from top to bottom.

13.2 RESULTS AND DISCUSSION

Erosion Test: The erosion test gives the results by measuring the weight loss after the surface is eroded by the impingement of alumina particles. The results were obtained in terms of weight loss of samples at a particular angle upon which the erodent strikes. The initial weights of the specimen and their process parameter/variables are shown in Table 13.4 and Table 13.5.

The degree of damage in erosion is usually a function of the mass of particle impingement acting on a surface, their velocity, the angle of impingement, and the particle sharpness. This damage leads to weight loss of the specimen. While weighing the specimens after every 2 min, the specimen was precleaned to remove wear debris from the surface. Progressive values of weight loss are given in Table 13.6. This depicts the weight loss of the eroded specimen after every 2 min at 60°. The base metal shows higher weight loss, whereas weld metal 1 presents minimum weight loss, as mentioned in the table. Metals at 60° showed debris of large eroded fragments detached by the action of alumina particles. Surface damage by irregular scratches

TABLE 13.4
Process Variables Used for Testing and Initial Weight of Samples at 60°

Material	Initial Weight (gm)	Angle of Impingement	Particle Velocity (m/s)
Base Metal	15.2442	60°	100
WM 1	14.5685	60°	100
WM 2	15.9318	60°	100
WM 3	14.1866	60°	100

TABLE 13.5
Process Variables used for Testing and Initial Weight of Samples at 45°

Material	Initial Weight (gm)	Angle of Impingement	Particle Velocity (m/s)
Base Metal	16.1119	45°	100
WM 1	15.1625	45°	100
WM 2	15.9133	45°	100
WM 3	15.9288	45°	100

TABLE 13.6
Weight Loss of Samples after Every 2 Min at 60° (gm)

Material	Initial Weight	Weight Loss	Weight Loss	Weight Loss	Weight Loss	Weight Loss	Total Weight Loss
Base Metal	15.2442	0.0021	0.0029	0.0024	0.0020	0.0023	0.0117
Weld Metal 1	14.5685	0.0008	0.0006	0.0003	0.0002	0.0002	0.0021
Weld Metal 2	15.9318	0.0013	0.0010	0.0012	0.0009	0.0006	0.0050
Weld Metal 3	14.1866	0.0011	0.0008	0.0009	0.0006	0.0004	0.0038

in random places is observed. At 60°, some craters are found on the tested surface along with wear debris of abrasives in the case of all specimens. Figure 13.2 shows a graph of the weight loss versus the time for all test specimens. It shows trends in weight loss at 60° with respect to time. It is possible to conclude that weld metal 1 is best in terms of erosion resistance. It can be clearly observed from Figure 13.2 that weld metal has better erosion resistance in severe erosive conditions. Table 13.7 presents the progressive weights of different test specimens under severe erosive conditions at an impact angle of 45°. Wear mechanisms like pitting action, ploughing, and wear debris are clearly seen on the damaged surface. The base metal undergoes plastic deformation followed by micro-cutting due to hard alumina abrasive particles. The continued impact of alumina particles on the specimen resulted in the formation of craters on the periphery.

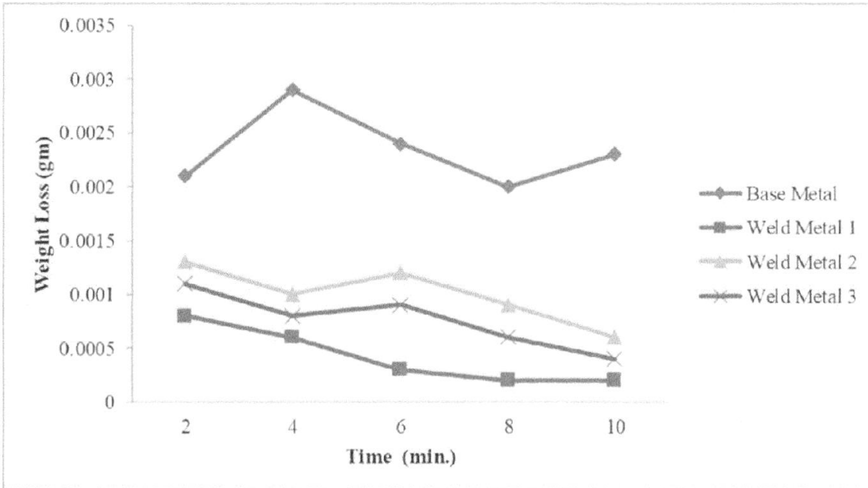

FIGURE 13.2 Weight loss with respect to time at 60°.

TABLE 13.7
Weight Loss of Samples after Every 2 Min at 45° (gm)

Material	Initial Weight			Weight Loss			Total Weight Loss
Base Metal	16.1119	0.0023	0.0025	0.0023	0.0021	0.0017	0.0109
WM 1	15.1625	0.0008	0.0003	0.0003	0.0002	0.0001	0.0017
WM 2	15.9133	0.0011	0.0008	0.0009	0.0006	0.0003	0.0037
WM 3	15.9288	0.0009	0.0008	0.0008	0.0006	0.0003	0.0034

It can be observed from the plot shown in Figure 13.3 that the Weld metal 1 has minimum weight loss at 45°. Here it is possible to see that weld metal (WM) 1 performed better in severe erosive conditions.

It can be clearly observed from Figure 13.4 that the base metal performs the worst against both impact angles, whereas weld metal 1 shows the least weight loss against both impact angles. It is possible to conclude that weld metal 1 at a 45° impact angle performs the best in severe erosive conditions.

- **Microhardness Tests:** Microhardness tests were conducted on an erosion test specimen. A Vickers hardness tester was used to evaluate microhardness results. Further, these results were correlated with erosion tests. Weld metal 1 shows the highest microhardness values, followed by weld metal 3 and weld metal 2 measured on the surface of specimen. The base metal shows the minimum microhardness value measured on the surface of the specimen. Average microhardness values are shown in Figure 13.5.

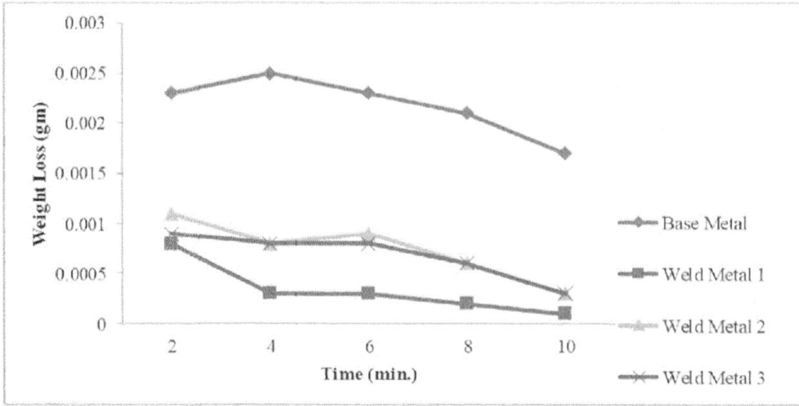

FIGURE 13.3 Weight loss with respect to time at 45°.

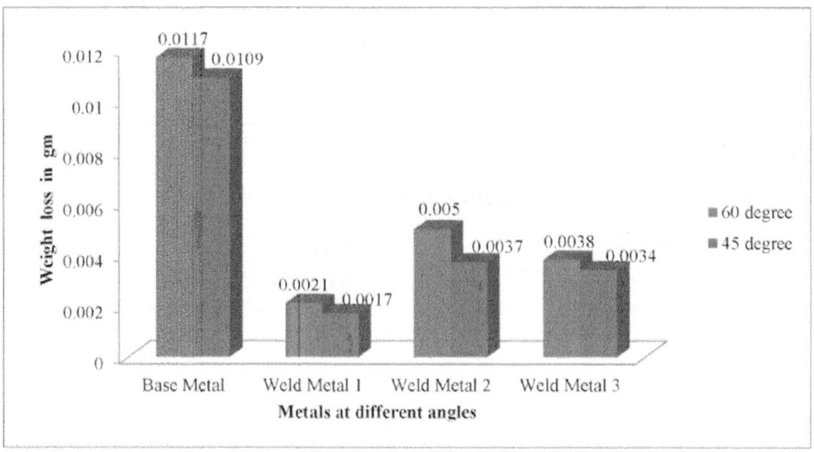

FIGURE 13.4 Interaction of weight losses at 60° and 45°.

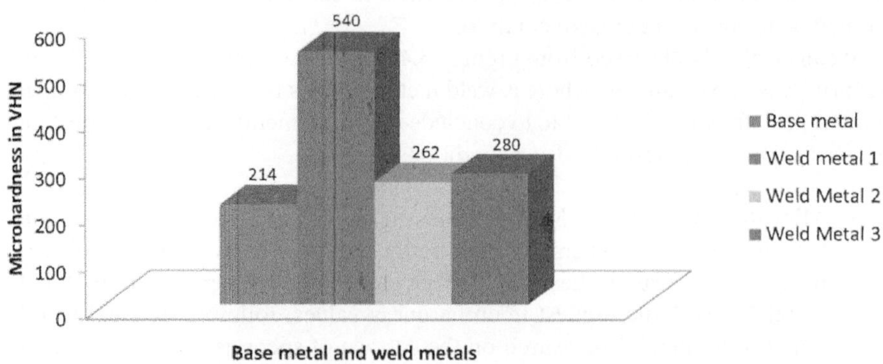

FIGURE 13.5 Average microhardness results.

TABLE 13.8
Microhardness Values (VHN) Normal to the Specimen from Weld Top to Bottom

S. No.	Distance (mm)	Microhardness		
		WM 1	WM 2	WM 3
1.	1	598.6	241.3	258.2
2.	2	428.2	236.7	237.5
3.	3	373.0	216.1	218.5
4.	4	338.0	201.5	206.3
5.	5	190.5	188.3	189.7
6.	6	178.5	180.0	179.0
7.	7	170.1	172.3	173.3

In this present study, there is a specific interest in the weld and heat-affected zone (HAZ) in order to observe the microhardness values of hard-faced specimens using a diamond indenter. The microhardness of the region is determined in the longitudinal direction of the weld. It has been discovered that the microhardness is lowest at the bottom of the sample, highest at the top of the weld, and moderate in the middle of the weld. This is consistent with the idea that the grain size alters in the weld's depth direction relative to the surface. Here, the results show that the weld has the highest microhardness, followed by the HAZ and base metal normal to the surface. Table 13.8 depicts microhardness at seven different points on the specimen normal to the surface.

13.2.1 Discussion and Future Scope

The current research has given a complete assessment of the erosion wear behavior of AISI 310SS with various hard-facing welds. The experimental outcomes clearly show that weld metal 1 has the maximum erosion resistance among all test materials and shows the least weight loss and maximum microhardness. The greater erosion resistance of weld metal 1 is due to its high hardness, which minimizes material removal due to particle impact. In comparison, the base metal exhibited serious erosion, demonstrating the need for surface alterations in enhancing wear resistance under high-erosion conditions.

The results of this investigation concur with the findings of past studies on erosion behavior of stainless steels in solid particle impact conditions. Nonetheless, this research goes a step further in the understanding by examining the role of various hard-facing materials and welding conditions in erosion resistance. The findings verify that higher microhardness materials in general have greater erosion wear resistance, especially under low impact angles. This is due to harder surfaces deforming less upon abrasive contact, hence decreasing the degree of material removal.

The mechanisms of erosion studied here are largely pitting action, ploughing, and crater formation resulting from repetitive particle contact. At an impingement angle of 60°, greater debris and deeper indentations were observed, while at an angle of 45°, material loss was more progressive, with a higher tendency for surface ploughing. These results indicate that the adjustment of welding conditions and electrode chemistry will substantially increase component life under erosive conditions. Although the experiment has effectively defined the microhardness vs. erosion resistance relation, there are some drawbacks in the experimental results that have to be taken care of in subsequent work. Initially, tests were performed under room temperature, whereas practical cases frequently encounter increased temperatures, possibly changing the behavior of wear. Future research needs to examine erosion resistance at elevated temperatures of various hard-facing welds to more closely represent field conditions. The research also utilized alumina particles as the erodent material, but particle type, shape, and velocity variations can have a strong influence on the rate of erosion. Exploring the effect of various abrasive materials would offer better insight into the wear behavior of AISI 310SS under varied operation conditions.

The microstructural evolution of hard-faced materials over long periods of usage is another important area for future work. Long-term erosion exposure has the potential to induce microstructure alterations like grain refinement, phase transformation, or work hardening, which may affect the material's wear resistance. Advanced characterization tools like scanning electron microscopy (SEM) and energy-dispersive spectroscopy (EDS) need to be used to determine microstructural transformations prior to and subsequent to erosion testing. Application-wise, this research can help design more aggressive and wear-proof coatings for industrially used parts that are subjected to erosive conditions like boiler nozzles, heat exchangers, and turbine blades. Additional refinement of welding methods and compositions of alloys may result in cost-effective means for prolonging the service life of key engineering components.

13.2.1.1 Conclusion

The main aim of this investigation was to assess the mechanical properties and erosion wear resistance of hard-faced material. This study aimed for surface modification of the AISI 310SS. The following results are obtained:

- Weld metal 1 was found to be best erosion-resistant material at all impact angle as compared to other tested materials. Weld metal 1 shows minimum weight loss and maximum average microhardness when measured over the surface, followed by weld metal 3 and weld metal 2. Microhardness value is also maximum when investigated normal to the surface of specimen. Results shows that hardness affects the erosion resistance of the material.
- Weld metal 1 presents the highest erosion resistance at an impact angle of 45°, followed by an impact angle of 60°, where results also show minimum weight loss in severe erosion conditions. Erosion wear resistance was considered as a function of weight loss.
- It is possible to conclude that the erosion wear mechanisms found in this work were pitting action, irregular scratches, craters, ploughing action, and wear debris. Weight loss was characterized by the detachment of large eroded fragments from the surface, along with the formation of pits caused by a combination of ploughing, irregular scratches, and wear debris.

13.2.1.1.1 Acknowledgments

The research funding from the Ministry of Science and Higher Education of the Russian Federation (Ural Federal University Program of Development within the Priority-2030 Program) is gratefully acknowledged. The Youth Laboratory was established with grant number FEUZ-2022–0031.

REFERENCES

1. E. Chantziara, K. Lentzaris, A.G. Lekatou, and A.E. Karantzalis, "Sliding wear and solid particle erosion response of aluminium reinforced with tungsten carbide nanoparticles and aluminide particles," Fatigue Fract Eng Mater Struct 42(7), 1548–1562 (2019).
2. M.S. Parancheerivilakkathil, S. Parapurath, S. Ainane, Y.F. Yap, and P. Rostron, "Flow velocity and sand loading effect on erosion–corrosion during liquid-solid impingement on mild steel," Appl Sci 12(5), 2530 (2022).
3. P. Singh, P. Kumar, and R.L. Virdi, "Investigation of solid particle erosion behaviour of SS-304 under different conditions," Mater Today Proc 48, 1147–1152 (2022).
4. G. Singh, S. Singh, and J.S. Grewal, "Erosion wear characterisation of DLC and AlCrN-based coated AISI-304/316 steels," Surf Eng 35(4), 304–316 (2019).
5. S. Tiwari, R. K. Yadav, N. Kumar, and G. Singh, "Investigation of stress and wear analysis for aluminum-based metal matrix composite reinforced silicon carbide using ANSYS software package," Recent Pat Mech Eng 18(3), 248–258 (2025).
6. G. Singh, A. Bansal, and H. Vasudev. "Corrosion and microstructural behaviour of Inconel 625 microwave clad deposited on mild steel," In *Thermal Claddings for Engineering Applications* (pp. 51–62). CRC Press (2024).
7. G. Singh, H. Vasudev, and Arora, H. "A short note on the processing of materials through microwave route," In *Advances in Materials Processing: Select Proceedings of ICFMMP 2019* (pp. 101–111). Springer Singapore (2020).
8. J.R. Laguna-Camacho, S.M. Sánchez-Yáñez, C. Calderón-Ramón, J.E. Escalante-Martínez, G. Juárez-Morales, M.I. Cruz-Orduña, L.M. Ramos-González, and A. Galicia-Badillo, "Erosion of metals using angular silicon carbide particles," Tribol Mater Surf Interfaces 13(4), 215–222 (2019).
9. J.R. Laguna-Camacho, A. Marquina-Chávez, J.V. Mendez-Mendez, M. Vite-Torres, and E.A. Gallardo-Hernandez, "Solid particle erosion of AISI 304, 316 and 420 stainless steels," Wear 301(1–2), 398–405 (2013).
10. M. Patel, D. Patel, S. Sekar, P.B. Tailor, and P. V Ramana, "Study of solid particle erosion behaviour of SS 304 at room temperature," Proc Technol 23, 288–295 (2016).
11. C.T. Morrison, R.O. Scattergood, and J.L. Routbort, "Erosion of 304 stainless steel," Wear 111(1), 1–13 (1986).
12. T.A. Nagaraj, S.P.K. Babu, M. Rajkumar, and R. Mahendran, "High temperature erosion behavior of cast nickel base 73Ni–13Cr–4Sn–4Bi–Mo alloy," Mater Res Express 6(8), 0865f6 (2019).
13. S.A. Alidokht, J. Lengaigne, J.E. Klemberg-Sapieha, S. Yue, and R.R. Chromik, "Effect of microstructure and properties of Ni-WC composite coatings on their solid particle erosion behavior," J Mater Eng Perform 28, 1532–1543 (2019).
14. A. Evstifeev, N. Kazarinov, Y. Petrov, L. Witek, and A. Bednarz, "Experimental and theoretical analysis of solid particle erosion of a steel compressor blade based on incubation time concept," Eng Fail Anal 87, 15–21 (2018).
15. H.S. Nithin, V. Desai, and M.R. Ramesh, "Elevated temperature solid particle erosion behaviour of carbide reinforced CoCrAlY composite coatings," Mater Res Express 5(6), 066529 (2018).

16. S.G. Sapate and A.V. RamaRao, "Erosive wear behaviour of weld hardfacing high chromium cast irons: Effect of erodent particles," Tribol Int 39(3), 206–212 (2006).
17. A. Patnaik, A. Satapathy, N. Chand, N.M. Barkoula, and S. Biswas, "Solid particle erosion wear characteristics of fiber and particulate filled polymer composites: A review," Wear 268(1–2), 249–263 (2010).
18. R.B. Nair, K. Selvam, H.S. Arora, S. Mukherjee, H. Singh, and H.S. Grewal, "Slurry erosion behavior of high entropy alloys," Wear 386, 230–238 (2017).
19. P. Mukhopadhyay, M. Srinivas, and M. Roy, "Microstructural developments during erosion of tribological steels," Mater Charact 113, 43–51 (2016).
20. V. Sharma, M. Kaur, and S. Bhandari, "Slurry erosion performance study of high velocity flame sprayed Ni–Al_2O_3 coating under hydro accelerated conditions," Mater Res Express 6(7), 076436 (2019).

14 Optimization of Plasma Spray Coatings Using the Taguchi Method and Machine Learning

S. Maharajan, F. Michael Thomas Rex, and S. Valaiganesh

14.1 INTRODUCTION

Sliding wear is caused by the slow degradation of one or both relatively moving surfaces. Wear is the forced removal of material caused by mechanical activity. When wear cannot be eliminated, it can be diminished. It is typically crucial to develop ceramic coatings with low wear and friction characteristics compared to those of bare materials used in various applications. Austenitic stainless steel is widely used in environments where sliding wear occurs, such as in aircraft missile bearings, engine missile casings, and missile structures. The coated material enhances the wear resistance behavior of the components. Hence, it is imperative to investigate the sliding wear characteristics of coated materials in this critical environment [1]. Ceramic materials based on carbides show exceptional resistance to deterioration due to their extreme hardness and chemical stability under these conditions. Coating with these ceramic particles may be accomplished most effectively by the use of the thermal spray technique. The process parameters must be optimized for each coating composition for high-density coatings. The amount of current, the distance between the stand-offs, and the powder feed rate are the key factors that impact the coating quality [2]. Several investigations have demonstrated that tungsten carbide (WC) and chromium carbide (Cr_3C_2) are ceramic powders that provide significant wear and corrosion resistance. These powders also enhance the lifespan of components when used as coatings. Furthermore, these studies have examined the sliding wear behavior and mechanism of materials coated with ceramics. The inclusion of a hard tungsten (W), carbide (C), and cobalt (Co) matrix in WC coating significantly improves its wear resistance. The soft Co matrix forms strong bonds with the hard phases, resulting in a considerable increase in the microhardness of ceramic coatings. The superior wear resistance of these coated materials could be attributed to their low porosity, dense coatings, and high microhardness [3, 4]. The abrasive wear characteristics of the thermally sprayed WC- and Cr_3C_2-based coatings were assessed by Murthy and Venkataraman [5], who determined that the Co

DOI: 10.1201/9781003546528-14

and NiCr matrices provide strength to the WC hard ceramic phases and Cr_3C_2 hard phases, respectively. As a result, ceramic coatings composed of WC and Cr_3C_2 are suitable alternatives to hard chromium coatings on furnace steels due to the presence of finely dispersed carbides that possess greater hardness [6–9]. The composite coating of WC+Cr_3C_2 outperformed WC and Cr_3C_2 coatings in terms of microhardness, with an increased mean microhardness of 1350.2 HV compared to 966.96 and 869.82 HV, respectively [10].

The literature discusses the sliding wear mechanism of Cr_3C_2 coatings. The primary causes of increased material removal during sliding include carbide pull outs, grooves, and loose particles [11–14]. Koutsomichalis et al. [15] studied the sliding wear mechanism of WC ceramic coatings on aluminum substrates and revealed that the primary wear failure mechanisms include surface microcracking, flattening coating layer fragmentation, and progressive carbide particle pullout. Tabrizi et al. [16] discovered that the wear characteristics of coatings are influenced by surface roughness. The presence of textures and steeper peaks in sliding wear increases the vulnerability of surfaces to fracture through abrasion.

The necessary spray parameters for process optimization have been determined to ensure coating quality. The coating performance depends on the current, powder feed rate, and standoff distance. Effective melting without the presence of partially melted particles with higher density can only be achieved by ensuring sufficient particle flight duration, adequate current, and an appropriate stand-off distance [17–21]. Li et al. [22] and Thirumalaikumarasamy et al. [23] performed thermal spray optimization work on the process variables to increase the efficiency of the deposition of coated powder and substantially improve microhardness while minimizing the porosity of 8YSZ coatings. The deposition and temperature of the molten particles increase, resulting in improving the degree to which the particle clings to the substrate. This leads to enhanced splat formation and improved surface characteristics. The coating porosity was greatly influenced by the power input. Powder feeding rate and stand-off distance considerably influenced coating surface qualities. When powder particle supply is low, splats flatten and disperse poorly.

In recent years, researchers have focused more on machine learning algorithms for the prediction of results and have overcome the shortcomings of conventional optimization techniques such as the Taguchi method and Response Surface Methodology (RSM). Random Forest (RF) is a prevalent supervised learning method for machine learning. Machine learning can be applied to both regression and classification problems. Additionally, novel materials with desirable qualities, such as strength, resilience, and various surface properties, can be predicted and developed using machine learning [24, 25]. Sharma et al. [26] compared the effectiveness of machine learning algorithms with surface roughness methods for tensile strength prediction. Machine learning algorithms provide reliable predictions and outperform optimization techniques. Most notably, materials scientists employ linear regression, random forest regression, and gradient boosting regression in their research. The machine learning algorithm is a cutting-edge technology that can precisely illustrate the relationship between different process factors and the tribological characteristics of materials [27–29]. Although the process parameters have a significant influence on the longevity of a coating, few studies have reported on optimizing these parameters for coated austenitic stainless steel with the objective of improving the sliding wear resistance. Therefore, the current study aimed to determine the most effective process parameters, including the current, powder feed rate, and standoff distance, to enhance the performance of the coating. Using the Taguchi design, the coating performance is ensured by determining the sliding wear loss by varying the process parameters. This work also focuses on the quantitative comparison of computational modeling methods versus traditional Taguchi optimization for enhancing process efficiency.

Surface Modification Strategies for Wear and Corrosion Resistance: The need for higher-performance materials with better corrosion and wear resistance has prompted a thorough investigation of surface processing methods. Of these, thermal spray coatings, laser cladding, and plasma nitriding have become leading candidates for use in increasing the longevity of engineering parts subjected to harsh conditions. Thermal spray coatings like high-velocity oxy-fuel (HVOF) and plasma spray are commonly employed for depositing ceramic and metal-ceramic composite films on metallic substrates with outstanding wear resistance under high-temperature and abrasion conditions. Laser cladding, however, allows for localized deposition of wear-resistant alloys through the application of high-power lasers, which results in good metallurgical bonding and low dilution with the substrate material. Equally, plasma nitriding is a potent diffusion-based process that improves corrosion resistance and surface hardness through nitrogen introduction into the metal lattice in applications such as stainless steels and titanium alloys. These methods not only prolong the lifespan of vital components in aerospace, automotive, and energy industries but also provide cost-effective options to standard material replacement techniques. The fusion of machine learning and optimization methods, including the Taguchi approach, also maximizes the processes' efficiency through the forecasting of optimal parameters, minimization of experimental trials, and enhanced overall coating performance. With industries still looking for improved solutions to wear and corrosion issues, hybrid surface modification processes integrating thermal spraying, laser treatment, and electrochemical methods will shape the future of materials engineering.

14.2 MATERIALS AND METHODS

14.2.1 Substrates and Coatings

Austenitic stainless steel (SS316) is used as a substrate material in this study. The morphology of the individual particle configurations is very clearly visible in the composite mixed powders and EDAX analysis, as illustrated in Figure 14.1. The principal components of the composite powder for $WC+Cr_3C_2$ were W, Cr, C, Ni, and a minimal amount of Cobalt (Co). Ni and Co are strong binder phase particles. The W and Cr grains are strongly connected with binder elements such as Ni and Co in the composite powder [6, 10].

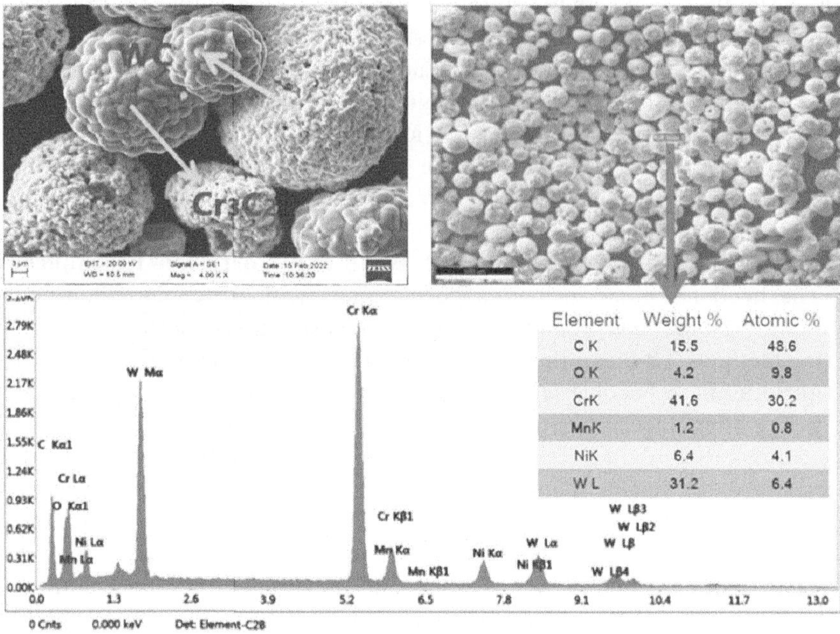

FIGURE 14.1 Surface morphology and EDAX analysis of the WC+Cr_3C_2 composite powder.

14.2.2 Coatings

The ceramic powders underwent mechanical ball milling in a powder mixer for 30 minutes to ensure thorough homogeneity of the composite powder with a distinctive size and shape without any alteration. The nickel–chromium bond coating has a thickness of 50 µm, and the primary coating deposition has a thickness of 200 µm. Before coating, surface irregularities were generated with shot peening to improve adherence. The 3MB is a multipurpose sprayer that creates plasma-sprayed coatings of the highest quality, greatest consistency, and greatest repeatability by adjusting the process parameters.

14.2.3 Sliding Wear Test

A pin-on disc (Ducom, model No. ED-201, Bangalore) was used to perform the sliding wear test on the WC+ Cr_3C_2 composite coating samples according to ASTM G99–05 [10]. The length and diameter of the specimen were 30 mm and 10 mm, respectively. The coating was at the top of the pin. Acetone was applied to both the disc and specimen to ensure cleanliness. The coated region was placed against the spinning disc plate made up of EN31 hardened steel of the testing machine. The wear lane radius was 50 mm. The disc wheel was spun at a rotating velocity of 500 rpm with a sliding rate of 2.61 m/s. The mass of the pin was measured prior to and following the experiment using a digital balance, and the loss of material was calculated. Furthermore, the wear mechanism was studied by scanning electron microscopy.

14.2.4 Taguchi Analysis

Taguchi's orthogonal array L9 (3^4) estimates which factors influence the performance criteria. The objectives of this study were accomplished through the use of the signal-to-noise ratio (S/N) and analysis of variance (ANOVA). ANOVA was performed to calculate the percentage contribution of each independent variable. Currently, the standoff distance and powder feed rate are process variables used in this study and are optimized by measuring dry sliding wear.

14.3 MACHINE LEARNING MODELING APPROACH

Linear benchmark (linear regression), nonlinear tree-based (decision tree, random forest), and kernelized (support vector regression) models provide a diverse set of flexible regression approaches for predictive modeling that can capture linear, nonlinear, and complex relationships in the data. The comparative analysis will reveal the suitable level of model complexity needed for optimizing the process variables considering the amount of available experimental data. RF builds an ensemble of decision trees, usually improving the predictive performance over that of a single tree. Averaging multiple trees reduces overfitting and improves generalizability.

14.4 RESULTS AND DISCUSSION

14.4.1 Taguchi Experimental Design

The Taguchi method is employed to determine the optimum conditions for enhancing performance, particularly sliding wear and corrosive wear loss behavior. The coating density is calculated using the mass difference before and after coating and the volume occupied by the coating. The specific sliding wear rate (Ks) and sliding distance are determined using Equations (1) and (2), respectively.

$$KS = \frac{\Delta m}{L\rho F} \tag{1}$$

$$\mathrm{L} = \mathrm{V} \times \mathrm{t} \tag{2}$$

Ks = specific sliding wear rate (mm^3/N-km), Δm = weight loss (gm), = coating density (gm/mm^3), L = pin and disc contact distance (m), V = velocity (m/s), t = duration of the test (sec), F = applied normal force (N).

Three input factors of plasma spray coating are correlated with the specific wear rate of the coated materials through experiments. The signal-to-noise ratio is determined by evaluating the quality attributes, where a lower value indicates greater quality. The specific wear rate and S/N ratio are presented for each experimental trial in Table 14.1. Using the Minitab 17 application, the mean plot for the calculated signal-to-noise ratios of a specific wear rate were constructed. All factors exhibited great statistical significance, since their individual P values were below 0.05, as seen in Table 14.2. Notably, the factor current makes the greatest contribution. The model built for a given sliding wear rate has R^2 of 99.67%, as seen in Table 14.3. The generated model best matches the experimental data, given its high R^2.

TABLE 14.1
S/N Ratio with Input and Output Response of the Plasma-Sprayed $WC+Cr_3C_2$ Coating for a Specific Wear Rate

Exp. Trials	Current (amp)	Powder Feed Rate (gm/min)	Standoff Distance (inches)	Specific Sliding Wear Rate (Ks), (mm^3/N. km)	Taguchi S/N Ratio
1	450	40	2	0.29800	10.5157
2	450	45	3	0.07820	22.1359
3	450	50	4	0.07956	21.9861
4	475	40	3	0.22320	13.0261
5	475	45	4	0.03980	28.0023
6	475	50	2	0.19850	14.0448
7	500	40	4	0.03600	28.8739
8	500	45	2	0.03680	28.6830
9	500	50	3	0.03913	28.1498

TABLE 14.2
ANOVA for S/N Ratios for Specific Wear Rates

Source	DF	Seq SS	Adj SS	Adj MS	F	P	Contribution Ratio
Current (amp)	2	211.544	211.544	105.772	143.30	0.007	47.99%
Powder Feed Rate (gm/min)	2	116.668	116.668	58.334	79.03	0.012	26.48%
Stand-Off Distance (inches)	2	111.058	111.058	55.529	75.23	0.013	25.20%
Residual Error	2	1.476	1.476	0.738	–	–	0.33%
Total	8	440.746	–	–	–	–	100%

TABLE 14.3
Model Summary for the Specific Wear Rate

S	R-Sq	R-Sq (adj)
0.8591	99.67%	98.66%

14.4.2 Process Parameter Optimization

The primary effect plots of the specific sliding wear rates are used to determine the optimal process parameters, as shown in Figure 14.2. The optimal value of the process parameter is denoted by a greater S/N ratio, as illustrated in Figure 14.2a. The optimal values for the key factors were identified as follows: current at 500 amps,

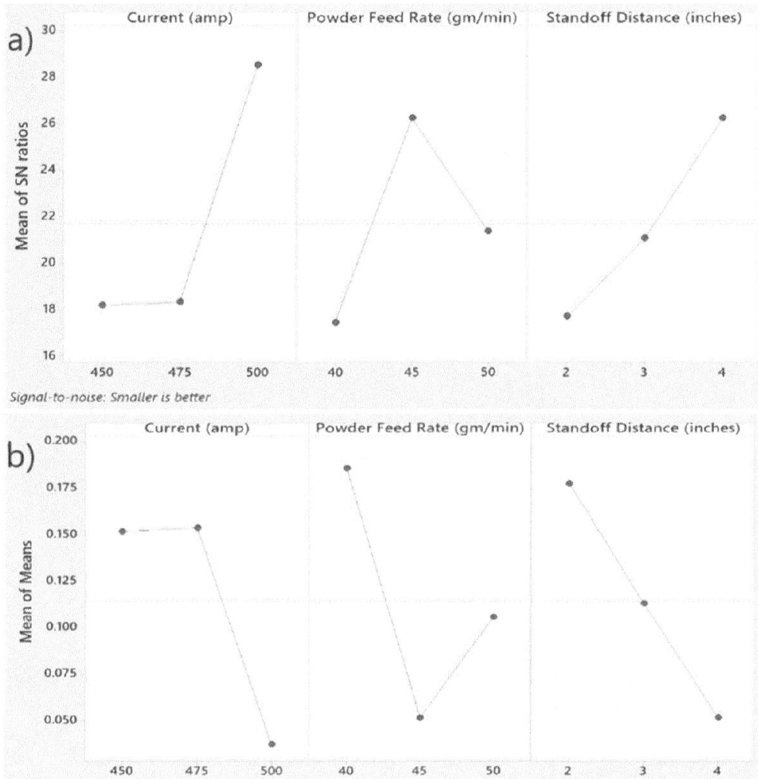

FIGURE 14.2 Main effect plot for (a) S/N ratios on specific wear rate, (b) means on specific wear rate.

powder feed rate at 45 gm/min, and standoff distance at 4 inches. A confirmation assessment investigation was conducted to validate the effectiveness of the approach with the optimum process parameter, which yielded a lower specific wear rate of 0.02800 $mm^3/N \cdot km$.

14.4.3 Implications of Process Variables on Specific Sliding Wear Rate

The most significant impact plot for the mean specific sliding wear rate, as shown in Figure 14.2b, demonstrates the impact of each parameter on the wear rate. It is inferred from the graph that the response is almost constant as the current increases from 450 to 475 amp. Despite this, an increase in current from 475 to 500 amp results in a substantial decrease in the rate of wear.

Additionally, it is predicted that a lower wear rate will be obtained as the powder feed rate approaches the middle value of 45 gm/min. This phenomena can be explained by the combination and fusion of particles at high temperatures, which in turn increases the effectiveness of deposition. A linear correlation is evident between the standoff distance and wear rate, as depicted in the graph. The wear rate decreases

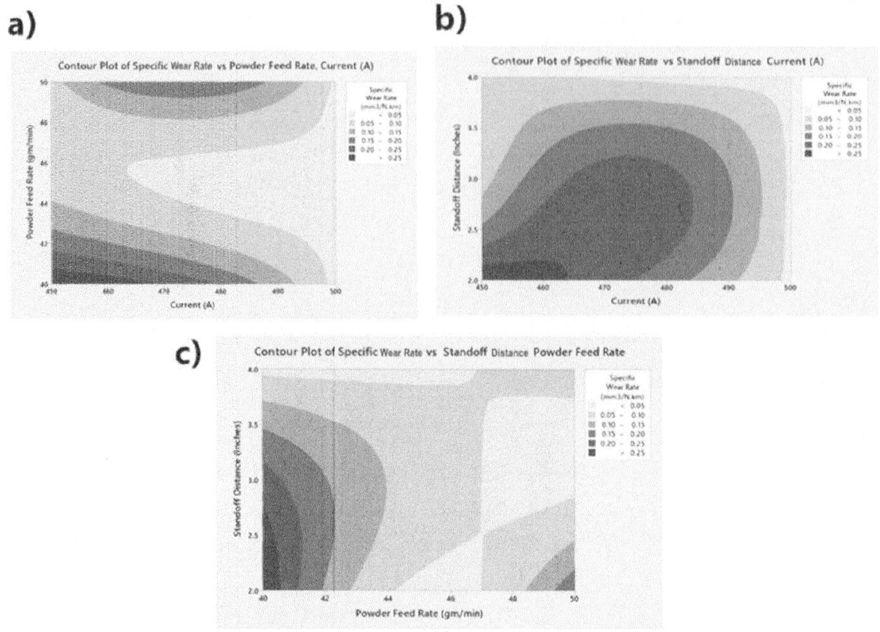

FIGURE 14.3 Contour plots of (a) current vs powder feed rate, (b) current vs stand-off distance, and (c) powder feed rate vs stand-off distance.

with increasing standoff distance. This phenomenon is ascribed to the fact that as the standoff distance increases, the powder particles have ample time to travel with a hot plasma jet, consequently augmenting the coating deposition. A drop in the specific wear rate results from increasing the current from 450 to 500 amp with a mid-feed rate ranging between 44 and 47 gm/min. An increase in current at the mid-powder feed rate provides a lower spectral wear rate, as shown in Figure 14.3a. The specific wear rate is very low at a high current of 500 amp and a powder feed rate of 45 gm/min. A lower current and standoff distance are associated with a greater spectral wear rate, as shown in Figure 14.3b. An increase in the mid-powder feed rate with increasing stand-off distance results in a reduction in the specific wear rate, as shown in Figure 14.3c.

14.4.4 Performance Analysis of the Machine Learning Algorithms

Table 14.4 provides the details of the prediction accuracy of various models. Table 14.5 illustrates the performance of the model for predicting the wear rate. The machine learning models demonstrate varying degrees of prediction accuracy for the wear rate, as evidenced by the mean absolute error (MAE) and R^2 values. The linear regression model has a MAE of 0.1284, indicating that the average absolute error in the predictions is 0.1284. However, the poor negative R^2 score suggests that this baseline model fails to account for the relationship between the process

TABLE 14.4
Prediction Accuracy of the Models

Model	Mean Absolute Error	Root Mean Square Error	R-squared (R^2)
Linear Regression	0.1284	0.1284	−37.67
Decision Tree Regressor	0.2031	0.2112	−103.15
Random Forest Regressor	0.1104	0.1306	−29.12
SVR (Support Vector Regression)	0.085	0.0856	−16.11

TABLE 14.5
Model for Predicting the Wear Rate

Model	Current (amp)	Powder Feed Rate	Standoff Distance (inches)	Predicted Specific Sliding Wear Rate
Linear Regression	500	40	4	0.0428
Decision Tree Regressor	500	40	4	0.036
Random Forest Regressor	488.9	45.6	3.6	0.041
SVR (Support Vector Regression)	500	40	4	0.1359

parameters and the wear rate. Similarly, the decision tree regressor with a MAE of 0.2031 and a highly negative R^2 is unable to adequately fit the input data to the output variable. This could be due to the inability to model complex nonlinear relationships. Hyperparameter tuning may improve performance. In comparison, the random forest model with its ensemble of decision trees provides a better MAE of 0.1104. The bagging and averaging of multiple trees help reduce overfitting evident in a single decision tree outcome. However, R^2 is still negative, implying suboptimal modeling of the wear rate data. Finally, the support vector regression model demonstrated a superior MAE of 0.085 and the least negative R^2 among the tested models. This indicates SVR's effectiveness in handling small complex datasets and identifying patterns for accurate predictions through its kernel functions. In conclusion, SVR achieves the best overall performance, followed by RF. The nonlinear modeling capability provided by both of these techniques makes them suitable for developing accurate predictive models even when working with small experimental datasets, as in the case of specialized plasma spray coating processes. However, the RF regressor model provides a value that is comparable to that of the Taguchi optimal process parameter and close to the predicted sliding wear rate. Linear regression and single decision trees cannot be generalized well in this scenario.

14.4.5 Random Forest Regression Model

The random forest prediction model for the predicted sliding wear rate vs a) current, b) powder feed rate, and c) stand-off distance is shown in Figure 14.4. The predicted sliding wear rate in the range between 0.075 and 0.09 mm^3/N.km for two trial runs at the current 500 amp. The lowest predicted sliding wear rate is recorded below 0.075 mm^3/N.km at a powder feed rate of 45 gm/min. Another two trials at a 45 gm/min powder feed rate resulted in a predicted sliding wear rate ranging from 0.150 to 0.175 mm^3/N.km. For all three experimental trials, the prediction was less than 0.09 mm^3/N.

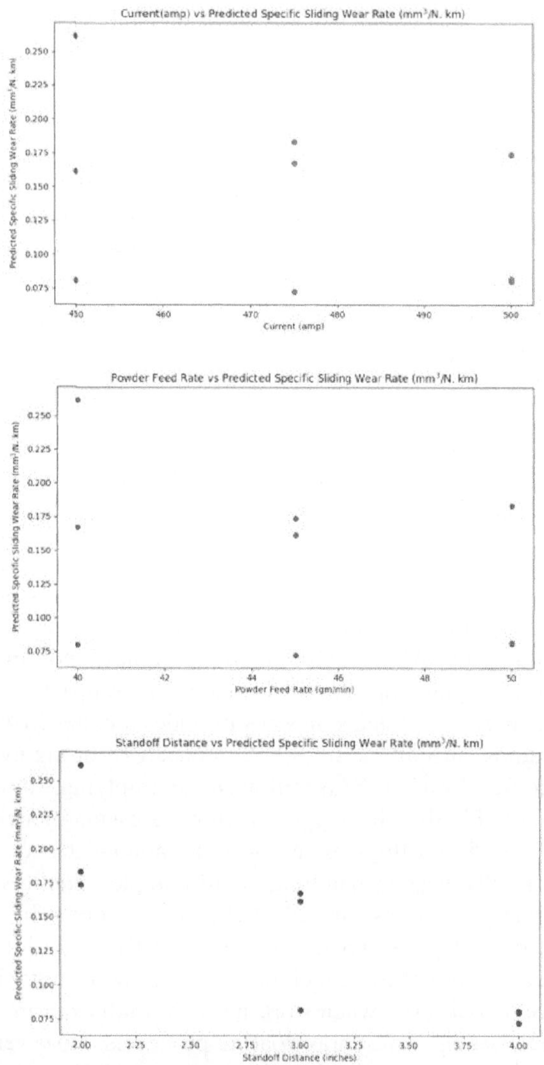

FIGURE 14.4 Random forest prediction model for the predicted sliding wear rate vs (a) current, (b) powder feed rate, and (c) stand-off distance.

km at a stand-off distance of 4 inches. The optimized process parameters included a current of 488.9 amp, a powder feed rate of 45.6 gm/min, and a stand-off distance of 3.6 inches, as shown in Table 14.6. The random forest model provides the closest prediction to the actual confirmation test output of the Taguchi technique. While decision trees have slightly lower errors, random forests enhance predictions by averaging multiple trees. This demonstrates that random forest regression, with its ensemble modeling capability, can more accurately capture the complex relationships between process parameters and the wear rate. The overall lower errors and proximity of predictions to experimental confirmation tests make it the best performing machine learning approach among the compared models. The inclusion of more training data can further improve the results, as expected. However, even with the current small dataset, random forest exhibits superior predictive performance over the other regression techniques.

The nine actual trial experiments were given and trained. A comparative study on the actual experimental value with the random forest model–predicted value of the sliding wear rate after the trained model with the nine trial experiments was performed.

Figure 14.5 shows a line plot comparing the actual experimentally measured specific sliding wear rate values (mm^3/N.km) to the predicted values from the RF regression model over the nine trial experiments. The actual wear rate values vary significantly over the experiments, ranging from approximately 0.03 to 0.3 mm^3/N.km. Experiment 1 has the highest wear rate of 0.298 mm^3/N.km. Experiments 5 and 7 have the lowest values of approximately 0.036–0.04 mm^3/N.km. The RF-predicted values follow the general trend of the actual values quite well. The difference between the actual and predicted values is lowest for trials 5, 7, 8, and 9 in the 0.036–0.045

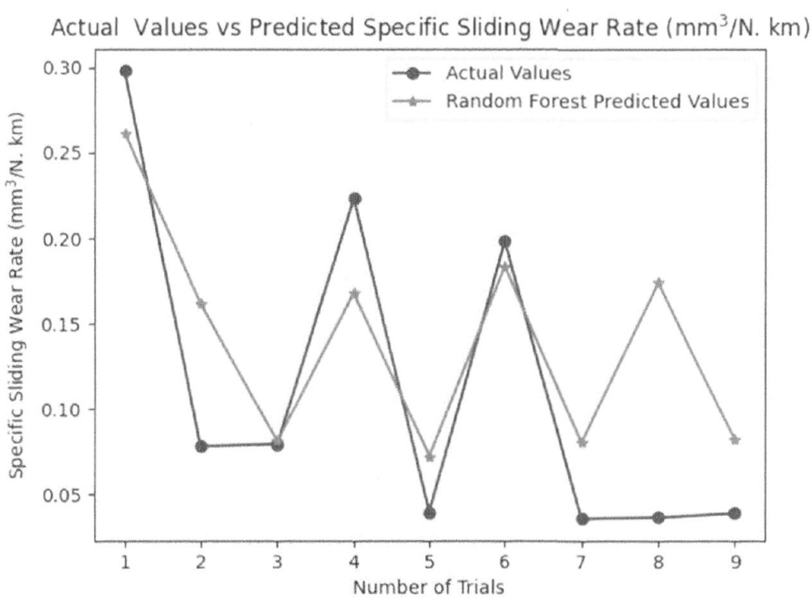

FIGURE 14.5 Comparison of the actual values and random order–predicted specific wear rates in the experimental trial.

mm^3/N.km range. The greatest difference is for trial 1, where RF underpredicted the very high experimental wear rate. The prediction is also somewhat inaccurate for the mid-range experimental value of trial 6.

The RF regression model provides good predictive capability in mapping the plasma spray process parameters to the wear rate, despite limited data. It tracks the experimental values quite well, especially at lower wear rates. Some deviation occurs at very high or medium wear rates. Overall, RF shows promise for predictive optimization of the coating process, as shown in Figure 14.5.

14.4.6 Wear Surface Topography on the Confirmation Test Sample

Figure 14.6a shows the SEM image of the worn-out coated sample of trial 1. The wear fragments, which are produced as solid particles, function as abrasive materials at the boundary between the coated sample and the rotating disc. As a result, the bonding strength of the coating steadily weakens over time, ultimately resulting in failure by splat delamination. This leads to significant fracture and the extraction of particles from the worn debris. The specimen loses stability when subjected to high loads and contact speeds over long distances. An SEM image of the worn sample from the optimized confirmation test is shown in Figure 14.6b. The wear surface picture shows significant resistance to sliding wear from the pin, as the coated surface particles have fewer fractures and pullouts than other wear samples.

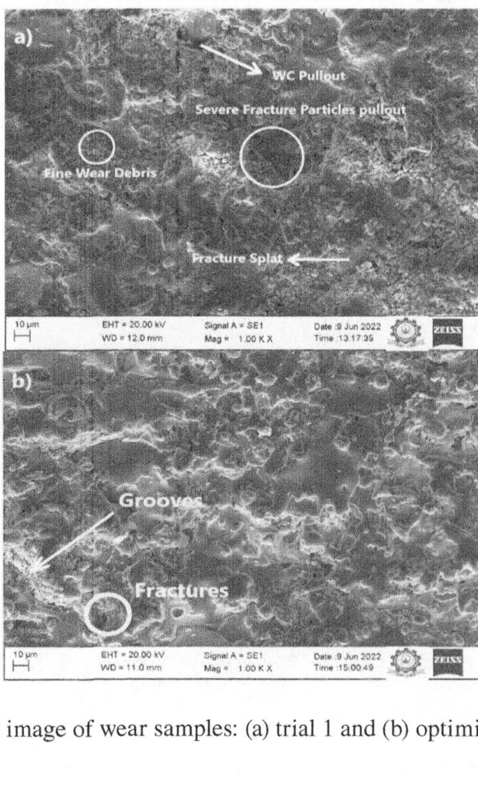

FIGURE 14.6 SEM image of wear samples: (a) trial 1 and (b) optimized coating confirmation test.

14.5 CONCLUSION

An austenitic stainless steel substrate was coated with a ceramic composite coating comprising 50 wt.% WC and remaining Cr_3C_2 by plasma spraying. The parameters of the plasma spraying process were systematically varied using the Taguchi technique and compared with those of machine learning algorithms for predicting the sliding wear rate. Compared with other coatings, the coating developed with optimized process parameters exhibited superior sliding wear resistance. This superiority was evident in its lower wear rate, highlighting the effectiveness of the proposed approach with a minimum number of experiments. The profound effect of the process parameters on the sliding wear resistance was analyzed, and the following conclusions were drawn.

1. The current in the plasma spray process is the most influential factor affecting the sliding wear rate. The particles are melted to a solid state and strongly adhere to the substrate when the current increases to a higher temperature. As a result, the wear resistance is greatly enhanced.
2. Nonlinear behavior is observed for the powder feed rate. Achieving complete melting is difficult if the powder feed rate is high. At the same time, a lower feed rate does not result in sufficient coating density on the substrate. Therefore, the parameter between the lower and upper limits leads to the best coating density without partial melting.
3. Most fractures occurred at lower current, feed rate, and standoff distance settings. Conversely, optimal combinations of factors and levels led to improved fusion of powder particles as a result of adequate duration of particle flight and increased temperature of the plasma jet. As a result, the applied coating solidified, creating a durable connection on the surface of the substrate.
4. The random forest model provided the most accurate predictions (0.041 $mm^3/N.km$ wear rate) and used Taguchi's optimal parameters, which were closest to those of the confirmation run. The optimal value produced by the RF algorithm is nearly equivalent to the optimal setting determined by the Taguchi approach.

14.6 FUTURE SCOPE

Plasma spray coating optimization in wear and corrosion resistance advance hand in hand with breakthroughs in material science, computational simulation, and machine intelligence. Research for the future can advance along the following directions for greater coating efficiency and process effectiveness.

1. Advanced Coating Materials: Novel composite coatings with nanostructured materials, functionally graded materials (FGMs), and high-entropy alloys (HEAs) may further enhance wear resistance and mechanical strength. Alternative binder compositions can also improve adhesion and durability of coatings.

2. Hybrid Coating Methods: The integration of plasma spray with other surface engineering processes like laser cladding, electrochemical deposition, or chemical vapor deposition (CVD) can result in better coating characteristics. Hybrid methods can offer higher hardness, lower porosity, and better bonding strength.
3. High-Temperature and Corrosive Conditions: Examining the behavior of plasma-sprayed coatings in harsh operating conditions, such as high-temperature oxidation, marine conditions, and exposure to aggressive chemicals, will enable the development of coatings for specific industrial uses.
4. Process Parameter Optimization Using AI: Combining artificial intelligence and machine learning for real-time process optimization can greatly minimize trial-and-error experimentation. Predictive models based on AI, in conjunction with sensor-based monitoring, can provide consistent coating quality with minimal material wastage and production costs.
5. Tribological and Mechanical Characterization: Extensive investigations employing in-situ monitoring methods, including scanning electron microscopy (SEM), X-ray diffraction (XRD), and Raman spectroscopy, can yield more detailed information on wear mechanisms, phase transformations, and microstructural evolution of coatings.
6. Sustainability and Eco-Friendly Coatings: Future studies must investigate eco-friendly coatings that minimize the use of toxic materials such as chromium and cobalt. The creation of sustainable spray methods with reduced energy usage and minimal environmental footprint is an important area of research.

14.6.1 Acknowledgments

The authors wish to acknowledge Spraymet Technologies, Bangalore, for their provision of the coating facility, as well as Ramco Institute of Technology, Rajapalayam, Kalasalingam Academy of Research and Education, and Krishnan Koil for their assistance in conducting the experimental work and microstructural characterization studies.

REFERENCES

1. Arul Raj K, Arunkumar MP, Kanikapula PKC, et al. Tribological and vibrational characteristics of AISI 316 L tested at elevated temperature and 600 Torr vacuum. *Defence Technol.* 2019; 15(1): 58–64.
2. Rubino F, Merino D, Silvestri AT, et al. Mechanical properties optimization of Cr_3C_2-NiCr coatings produced by compact plasma spray process. *Surf. Coat. Technol.* 2023; 465: 129570. https://doi.org/10.1016/j.surfcoat.2023.129570.
3. Wood RJK, Herd S, Thakare MR. A critical review of the tribocorrosion of cemented and thermal sprayed tungsten carbide. *Tribol. Int.* 2018; 119: 491–509.
4. Guo H, Li B, Lu C, et al. Effect of WC-Co content on the microstructure and properties of NiCrBSi composite coatings fabricated by supersonic plasma spraying. *J. Alloys Compd.* 2019; 789: 966–975.
5. Murthy JKN, Venkataraman B. Abrasive wear behaviour of WC-CoCr and Cr_3C_2-20(NiCr) deposited by HVOF and detonation spray processes. *Surf. Coat. Technol.* 2006; 200(8): 2642–2652.

6. Sidhu HS, Sidhu BS, Prakash S. Mechanical and microstructural properties of HVOF sprayed WC-Co and Cr_3C_2–NiCr coatings on the boiler tube steels using LPG as the fuel gas. *J. Mater. Process. Technol.* 2006; 171(1): 77–82.
7. Bhosale DG, Prabhu TR, Rathod WS. Sliding and erosion wear behaviour of thermal sprayed WC-Cr_3C_2-Ni coatings. *Surf. Coat. Technol.* 2020; 400: 126192.
8. Ahmed R, Ali O, Berndt CC, et al. Sliding wear of conventional and suspension sprayed nanocomposite WC-Co coatings: An invited review. *J. Therm. Spray Technol.* 2021; 30: 800–861.
9. Wang Q, Luo S, Wang S. 2019 Wear, erosion and corrosion resistance of HVOF-sprayed WC and Cr_3C_2 based coatings for electrolytic hard chrome replacement. *Int. J. Refract. Met. Hard Mater.* 2019; 81: 242–252.
10. Maharajan S, Michael Thomas Rex F, Ravindran D, et al. Erosive and corrosive wear performance and characterization studies of plasma-sprayed WC/Cr_3C_2 coating on SS316'. *Int. J. Appl. Ceram.* 2022; 19(5): 2845–2861.
11. Bolelli G, Berger LM, Borner T, et al. Sliding and abrasive wear behaviour of HVOF- and HVAF-sprayed Cr_3C_2-NiCr hardmetal coatings. *Wear.* 2016; 358–359: 32–50.
12. Torkashvand K, Selpol VK, Gupta M, et al. Influence of test conditions on sliding wear performance of high velocity air fuel-sprayed WC–CoCr coatings. *Materials.* 2021; 14(11): 3074.
13. Rajinikanth V, Venkateswarlu K. An investigation of sliding wear behaviour of WC-Co coating. *Tribol. Int.* 2011; 44(12): 1711–1719.
14. Ishikawa Y, Kuroda S, Kawakita J, et al. Sliding wear properties of HVOF sprayed WC-20%Cr_3C_2–7%Ni cermet coatings. *Surf. Coat. Technol.* 2007; 201(8): 4718–4727.
15. Koutsomichalis A, Vardavoulias M, Vaxevanidis N. HVOF sprayed WC-CoCr coatings on aluminum: Tensile and tribological properties. In *13th International Conference on Tribology, ROTRIB'16, IOB Conference Series, Materials Science and Engineering.* 2017; 174: Article No. 012062.
16. Tabrizi AT, Aghajani H, Laleh FF. Tribological study of thin-electroplated chromium: Evaluation of wear rate as a function of surface roughness. *Exp. Tech.* 2023; 47: 369–379.
17. Patel GCM, Pradeep NB, Girisha L, et al. Experimental analysis and optimization of plasma spray parameters on microhardness and wear loss of Mo-Ni-Cr coated super duplex stainless steel. *Aust. J. Mech. Eng.* 2022; 20(5): 1426–1438.
18. Pakseresht AH, Javadi AH, Nejati M, et al. Statistical analysis and multiobjective optimization of process parameters in plasma spraying of partially stabilized zirconia. *Int. J. Adv. Manuf. Technol.* 2014; 75: 739–753.
19. Thao DX, Got HV, Cuong PD. Optimization of plasma spraying parameters with respect to shear adhesion strength of Cr_3C_2-NiCr coating on 16Mn steel. *Tribol. Ind.* 2022; 44(2): 221–229.
20. Ramaiah K, Eswaregowda NB, Tambrallimath V, et al. Chapter 8: Optimization of deposition parameters in plasma spray coatings. In *Modeling and Optimization in Manufacturing: Toward Greener Production by Integrating Computer Simulation*, eds. CI Pruncu, J Jiang. Wiley-VCH GmbH, 2021: 217–235.
21. Azarmi F, Coyle T, Mostaghimi J. Optimization of atmospheric plasma spray process parameters using a design of experiment for alloy 625 coatings. *J. Therm. Spray Technol.* 2008; 17: 144–155.
22. Li JF, Liao HL, Ding CX, et al. Optimizing the plasma spray process parameters of yttria stabilized zirconia coatings using a uniform design of experiments. *J. Mater. Process. Technol.* 2005; 160(1): 34–42.
23. Thirumalaikumarasamy D, Shanmugam K, Balasubramanian V. Influences of atmospheric plasma spraying parameters on the porosity level of alumina coating on AZ31B magnesium alloy using response surface methodology. *Prog. Nat. Sci. Mater. Int.* 2012; 22(5): 468–479.

24. Xu X, Wang X, Wu S, et al. Design of super hard high-entropy ceramics coatings via machine learning. *Ceram. Int.* 2022; 48(21): 32064–32072. https://doi.org/10.1016/j.ceramint.2022.07.145.
25. Wu S, Xu X, Yang S, et al. Data-driven optimization of hardness and toughness of high-entropy nitride coatings. *Ceram. Int.* 2023; 49(13): 21561–21569. https://doi.org/10.1016/j.ceramint.2023.03.292.
26. Sharma S, Gupta V, Mudgal D. Response surface methodology and machine learning based tensile strength prediction in ultrasonic assisted coating of poly lactic acid bone plates manufactured using fused deposition modelling. *Ultrasonics.* 2024; 137: 107204. https://doi.org/10.1016/j.ultras.2023.107204.
27. Wang G, Song J, Zhao G, et al. Tribological performance prediction of WS2 coating under different conditions by machine learning. *Wear.* 2023; 532–533: 205092. https://doi.org/10.1016/j.wear.2023.205092.
28. Jia B, Wan Q, Yan L, et al. Tribological properties and machine learning prediction of FeCoCrNiAlN high entropy coatings. *Surf. Coat. Technol.* 2024; 477: 130341. https://doi.org/10.1016/j.surfcoat.2023.130341.
29. Hao J, Gao L, Ma Z, et al. Exploration of the oxidation and ablation resistance of ultra-high-temperature ceramic coatings using machine learning. *Ceram. Int.* 2022; 48(19): 28428–28437. https://doi.org/10.1016/j.ceramint.2022.06.156.

15 Advantages and Applications of Microwave Cladding Technology

Amrinder Mehta, Hitesh Vasudev, and Gurbhej Singh

15.1 INTRODUCTION

Microwave cladding is a process that deposits protecting or functional layers onto the surfaces of materials using microwave energy [1]. This method enhances engineering properties such as strength, wear resistance, and thermal stability. It can also be used to heat quickly, providing a high degree of control over the thickness of the coating and its adhesion. Microwave cladding also consumes less time and energy than traditional methods. All of these benefits predispose microwave cladding to service in the aerospace, electronics, and other high-performance industries. A schematic diagram of a typical setup for microwave-aided processing of materials is shown in Figure 15.1 [2]. Microwaves are introduced into the top and flow through a number of working layers. The bottom layer should be charcoal powder intended as the susceptor that absorbs the microwave energy. A graphite plate separator guarantees the distribution of equal heat underneath. The powder mixture of Si, Ni, and La_2O_3 is placed in the main chamber, as represented in the form of black and yellow spheres. The mix lies on a stainless steel (SS-304) layer that is blue in the diagram. Even exposure is supported by a rotating table with masking plates on both sides that confine the process. The setup has all components that are used in controlled microwave heating, and its structure is depicted using dashed bricks. The device enables using specific microwave energy to cut and alter materials at the microscale level.

The range of high-temperature, lightweight materials is a new field that can be satisfied by non-equilibrium cladding or microwave processing [3]. The technologies result in high thermal performance and lower weight of coatings for use in industries that require toughness. The contemporary development of microwave cladding has encouraged collaboration between industries and researchers with the intention of enabling better material performance [4]. The purpose of such cooperation is to design new coatings using next-generation technologies. The developments extend possibilities in the world of materials science and engineering. Figure 15.2 shows a cross-sectional SEM image of welded joint at 65× magnification and at 15.00 kV [5].

DOI: 10.1201/9781003546528-15

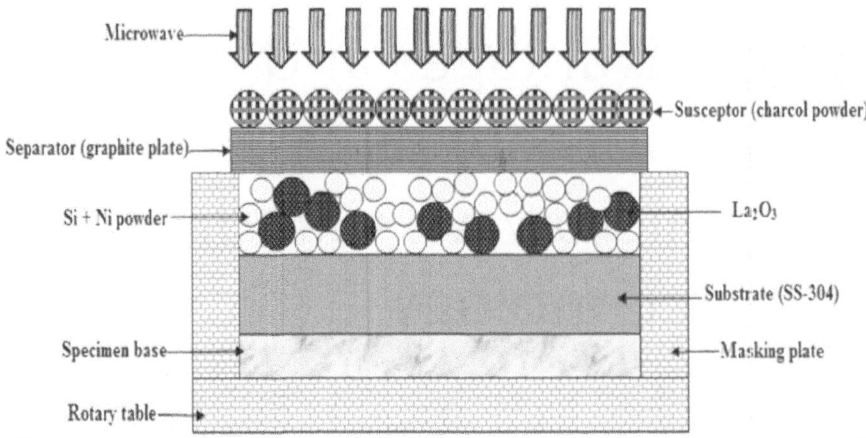

FIGURE 15.1 Typical configuration of a microwave cladding apparatus [2].

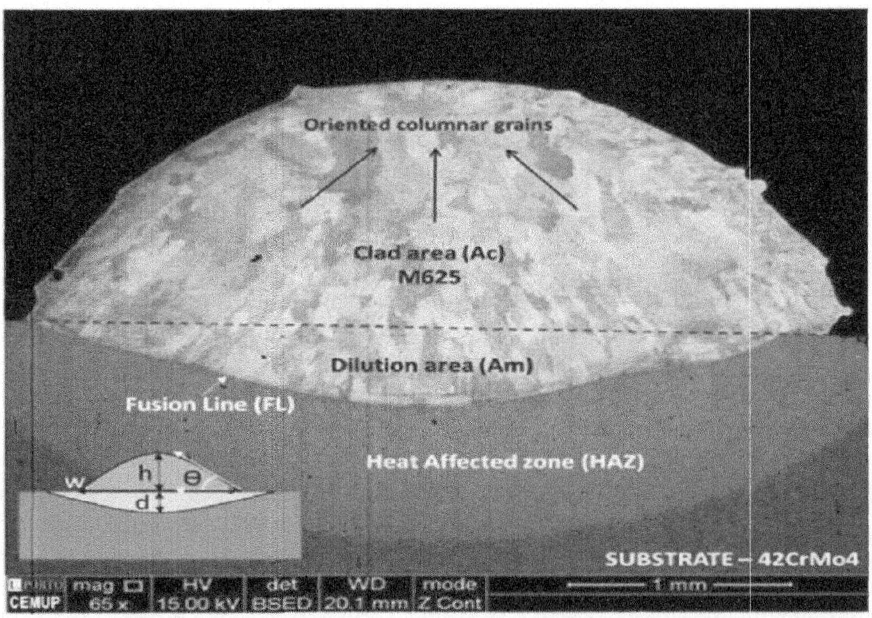

FIGURE 15.2 The SEM cross-section reveals the 42CrMo4 substrate in contact with the clad layer [5].

Welded specimen M625 exhibits clear zones of metallurgy in the weld area. The upper part shows a clad layer of columnar grains oriented toward the heat with a dilution line underneath that represents the changeover of clad to fusion zones. The base metal and the heat-affected zone are separated by the fusion line; composition analysis identified the base metal as $42CrMo_4$. The geometry of the weld (width, height, depth and angle) is shown as an inset in the figure. A 1 mm scale bar offers dimensional information to make judgment on microstructural features on different sides of the weld.

Research on smart materials and adaptive coatings centers on surfaces that respond to changes in the environment [6]. These materials enhance performance in aerospace, automobile, and renewable energy applications. Their flexibility increases energy efficiency and product reliability and reduces wear. Modern researchers are attempting to create effective, lightweight, environmental-friendly buildings that can withstand rough working conditions [7].

Shape memory alloys that heal themselves and scalable materials that change color are among the current innovations. Some smart materials harness ambient energy to drive small sensors and other devices. These characteristics serve to redefine the relations between man and technology and the environment. Using smart materials in commercial goods leads to more robustness and reduces waste. They support sustainable manufacturing and resource productivity. The transition encourages viable innovation in all trades and markets. The consumer demand to use adaptive and sustainable products will increase as industries embrace the use of the materials [8].

The use of smart materials supports circular economy models that lower waste and increase product life. This shift to responsive and environmentally friendly technologies will pay off for manufacturers, designers, and consumers. Finally, smart materials are innovative high-performance solutions for sustainable applications in mature industrial, consumer, and future appliances.

15.1.1 Advantages of Microwave Cladding

Microwave cladding is an effective technique for improving material properties under high-temperature service, characterized by improved thermal stability and oxidation resistance. It increases component life expectancy, especially in industries where efficiency is critical like aerospace, automobiles, and energy. The retention strength and load-bearing capacity at temperature extremes puts microwave cladding ahead of other contenders to provide material solutions to today's engineering requirements [9]. Figure 15.3 shows a simple flowchart illustrating the primary advantages of microwave cladding technology [10].

Microwave cladding is becoming significant in contemporary production as industries are becoming more conscious of sustainability [11]. The method assists in making materials with high strength-to-weight ratios and better energy density. It also minimizes waste and supports cleaner production processes. Microwave cladding facilitates operations in aerospace, automotive, and electronics industries. Such advantages are appealing to companies that are looking into smart and environmentally conscious solutions. However, careful coordination of all stakeholders—researchers, manufacturers, policymakers, etc.—is necessary for its successful implementation [12].

This kind of cooperation will guarantee that microwave cladding is well incorporated into industrial systems that are already in use. It also allows innovation in tandem with industry-related compliance with environmental regulations. Further advancement of this technology is aimed at absorbing more energy and the performance of materials in general. These developments promote the transition to sustainable

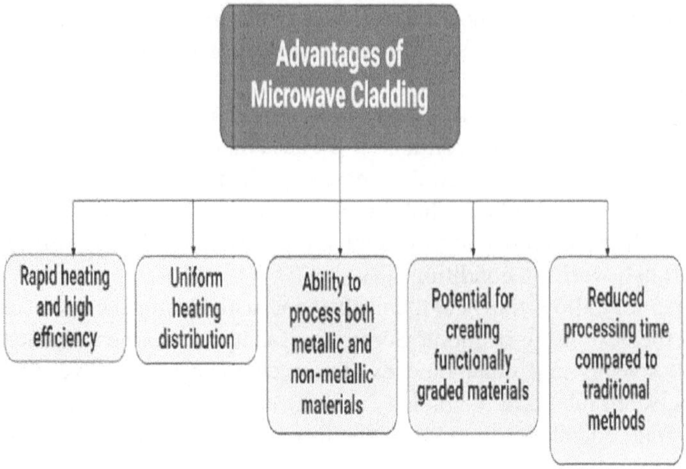

FIGURE 15.3 Advantages of microwave cladding.

manufacturing. Microwave cladding reduces costs as well as pollution. It is instrumental in facilitating circular economies because of reduced waste of resources and increased efficiency [13]. With expanding research and increasing consumer uptake, microwave cladding will transform industrial manufacturing to green.

15.2 WEAR RESISTANCE: HOW MICROWAVE CLADDING ENHANCES MATERIAL DURABILITY

Microwave cladding provides a thick protective surface layer that results in better wear resistance and increased product life [14]. Improving product life reduces frequent replacements and equipment down time and increases overall productivity. The process is useful in such sectors as aerospace, automotive and mining where parts endure severe wear [15].

Microwave cladding enables precise control over surface characteristics, allowing wear resistance to be tailored to specific operational needs. This technique produces high-performance coatings for important industrial applications utilizing advanced materials and processes [16]. This method improves material efficiency and surface performance by enabling the accurate deposition of thin coatings. This capability is critical in industries with strict weight and precision requirements, such as aerospace and automotive manufacturing [17].

Microwave cladding enhances design and energy efficiency by producing strong, lightweight coatings. It has emerged as a leading technique in advanced manufacturing, offering superior performance and reduced waste. This transition to microwave cladding favors eco-friendly production based on internationally established carbon emission targets [18]. It can also assist companies in achieving regulatory standards and improve the suitability of products. These benefits have prompted many industries to adopt it, and as they become sensitive to these advantages, the future of production will be a smarter and greener world.

15.2.1 Material Selection for Microwave Cladding

The type of material used for cladding affects the thermal and mechanical performance of the final product as well as the sustainability of the manufacturing process, making it crucial for businesses to choose wisely. Some of the aspects that should be considered during this evaluation include material sources, recyclability, and life cycle environmental implications toward the aim of sustainability [19]. Table 15.1 describes some of today's commonly used cladding materials.

TABLE 15.1
Common Cladding Materials and Features in the Literature

Cladding Materials	Microstructural Characteristics	Key Findings
Fe-Ni-Ti composite coatings	uniform, dense structure with equiaxed crystals and in situ formed TiC reinforcement	An overlap rate of 46% results in optimal microhardness (450 HV), minimal defects, and improved wear resistance (0.17% weight loss) [20].
Ni-based and Co-based alloys	equiaxed and dendritic grains	Nickel-based cladding layer demonstrated superior microhardness; cobalt-based cladding layer outperformed in wear resistance and corrosion resistance [21].
Ni60+30%WC	dense and refined microstructure	Optimal process parameters improve hardness and wear resistance of the coating [22].
Fe-based, Ni-based, Co-based alloys	dendritic structures, hard carbides	Mechanical properties can be enhanced through process parameter optimization and post-heat treatment [23].
Ti-6Al-4V, high-entropy alloys	fine grains, elongated grain shapes, non-equilibrium microstructures	Rapid solidification affects phase selection; metallurgical defects affect mechanical properties [24].
Fe-based and Co-based alloys	γ-Co phase and carbide Cr23C6; solid-solution strengthening in Fe-based coatings	UHSLC alloy coatings show better wear resistance than CLC coatings; wear rate reductions of 40.7% (Fe-based) and 58.1% (Co-based) [25].
Fe-Ni-Al	γ-[Fe,Ni] austenite and α phase	Optimal Al content of 4% improves coating performance and wear resistance while maintaining a low coefficient of thermal expansion [26].
410L, 420SS, Stellite 6, Stellite 21		Different cladding materials and processing parameters significantly affect the mechanical properties of laser-cladded premium rails [27].
	Presence of different phases and their effects on mechanical properties	Mechanical properties are influenced by the microstructure, particularly the distribution of elements and phases within the material [28].
Stellite 6 alloy	equiaxed and dendritic grains	Laser cladding enhances hardness, wear resistance, and corrosion resistance of the substrate [29].
Iron-based alloy	finer grain size	Rosin reduced oxidation; vanadium improved hardness and strength [30].

Manufacturers have an opportunity to improve their competitiveness through proper and non-destructive selection of materials [31]. A good material selection process will guarantee the performance of a product and reduce waste and resources used. The plan allows for achieving business and environmental objectives in a single process. It facilitates the integration of a supply chain and enables innovation by means of sustainable partnerships [32].

Ecologically intelligent material selection minimizes waste of products and manufacturing processes and decreases total carbon impacts. The sustainable processes also enhance product durability and reliability, which has long-term positive outcomes on the efficiency of operations. When companies concentrate on sustainable products, they improve their brand reputations in terms of corporate social responsibility. Contemporary consumers and investors are more environmentally and ethically aware, and sustainable decisions build trust by reflecting commitments to responsible innovation. Open communications come with stronger customer relations as sustainability [32].

Companies that post their environmental activities achieve loyalty and recognition in the market. Such programs not only safeguard the earth but also open up market opportunities that adhere to sustainable values. Companies that embrace these expectations will be front-runners in the eco-conscious economy of tomorrow. Sustainability is not something you should consider anymore because it is no longer a matter of choice but the key to long-term success and global responsibility.

15.2.2 Comparative Analysis

High-temperature microwave cladding is an essential breakthrough in manufacturing technology with numerous advantages over other cladding techniques. This approach has greatly changed material bonding and coating methods and in turn positively impacted product quality, production rates, and environmental impacts [33].

Figure 15.4 illustrates a comparative analysis of two material processing methods: laser treatment and electromagnetic emission. The figure illustrates that the two methods follow similar production processes to arrive at material solidification, with both similar features and similar drawbacks, namely the possibility of interfacial cracking and porosity. Notably, electromagnetic radiation features faster heat transmission, even heat distribution, and low porosity with little internal check cracking. The two processes both deal with powder particles and exhibit intermediate results but result in different outcomes in terms of quality and characteristics of the processed substrate material to the final impact [34].

Laser cladding is presented with the microstructure formation reliant on the controlled melting and solidification [35]. The correct amount of energy prevents partial melting of the substrate along with the powder. The powder particles react depending on the point of contact with the surface of the substrate. Some are launched out of the melt pool and do not attain melting temperature, instead ricocheting away. The rest sink into the molten zone and fully melt to become part of the growing microstructure. These changes affect the thickness and the lack of consistency of the clad layer. Nevertheless, the speed of manufacturing processing by laser cladding creates some technical difficulties [36].

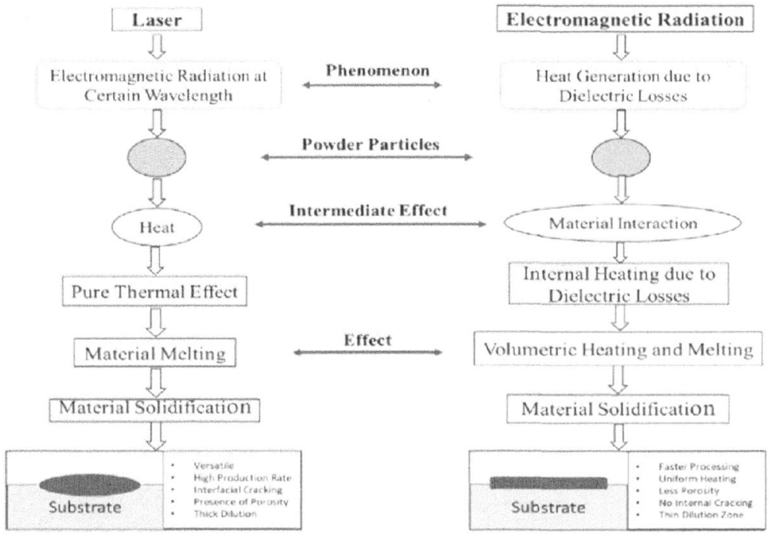

FIGURE 15.4 Comparison of laser and microwave energy cladding processes [34].

Extremely rapid heating and then cooling can result in frozen surface layers as well as internal stress. These pressures impair the mechanical integrity of the ultimate clad coating [37]. The first issue is the presence of a steep temperature gradient and limited control over solidification. Solidification cracking and porosity are commonly observed during cooling. Cracking occurs due to material shrinkage, while porosity can result from trapped gases or incomplete melting. These causes have adverse impacts on coating integrity, and this necessitates the careful management of heating and cooling [38]. Table 15.2 provides an overview of the major benefits and limitations of microwave cladding. Although not all challenges are listed, the continued improvement of the process still eliminates defects.

Microwave cladding works on surfaces in a shorter time than the traditional, which increases the speed of production and satisfies market demand [45]. It heats quickly, which allows for better control, and this improves the quality of the coating and the final performance of the product. Welded microwave cladding increases the strength of materials, making the final products easier to maintain and increasing their life spans. These advantages help in manufacturing high-value products and superior competitive positioning [46].

State-of-the-art efficiency is another strong point. Microwave cladding heats only the required areas as opposed to conventional methods that heat whole workpieces. This selective heating minimizes energy consumption and makes operations and production less costly, assisting in achieving sustainable manufacturing objectives [47]. It also reduces material waste, which benefits post-processing operations such as cutting and machining [48].

Effective use of materials minimizes impacts on the environment while ensuring high product integrity. This enables manufacturers to design high-performance

TABLE 15.2
Advantages and Disadvantages of Different Microwave Cladding Processes

Method	Advantages	Disadvantages
Microwave surface modification	• Allows for precise control of process parameters • Can be optimized using various algorithms (e.g., neural networks, genetic algorithms) • Enables preparing high-quality coatings	• Complexity in optimizing multiple parameters simultaneously [39]
Microwave cladding for slurry erosion resistance	• Improves slurry erosion resistance • Enhances surface properties of materials	• Specific disadvantages not mentioned [40]
Microwave cladding for metal matrix composites	• Improves surface-dependent properties of metals • Can be combined with other techniques (e.g., tungsten inert gas, laser, plasma transferred arc)	• Requires careful selection of substrate material and reinforcing metal powder [41]
Microwave-assisted surface modification	• Allows for modifying surface layers • Can incorporate various materials (e.g., hydroxyapatite, alumina)	• May require additional processing steps for optimal results [42]
Combined ultrasound and microwave processing	• Reduces processing time • Improves product quality • Provides uniform heating • Preserves nutritional value	• Requires specialized equipment for combined treatment [43]
Microwave cladding (general)	• New approach to surface modification • Potential for improving various material properties	• Specific disadvantages not mentioned in the search result [44]

products that have customizable surface properties and smaller ecological footprints. The usage of microwave cladding follows the trend of increasing customer enthusiasm about green technologies and sustainable production [49]. Sustainability-sensitive firms use it to enhance their brand image and increase their attraction to sustainability-conscious consumers. The technology also promotes continuous innovation in surface engineering and manufacturing processes [50].

The ongoing research promotes developing complementary technologies with ever-increasing performance. Partnerships among organizations also enhance knowledge exchange among sectors. This facilitates the growth across the supply chain, which enhances quality, efficiency, and sustainability [51]. Microwave cladding is an excellent example of transforming modern manufacturing for a greener, more efficient future.

15.3 APPLICATIONS OF MICROWAVE CLADDING IN AEROSPACE AND AUTOMOTIVE INDUSTRIES

In aerospace and automotives, microwave cladding supports the production of durable, lightweight components with improved fuel efficiency and lower emissions. This technology also enables innovation in vehicle and aircraft design, enhancing

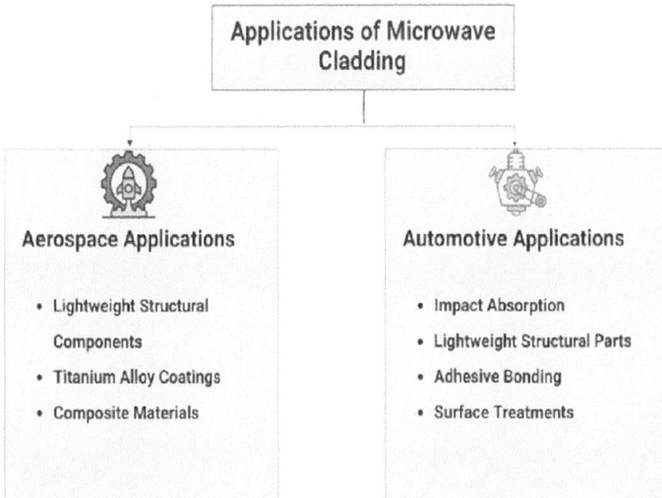

FIGURE 15.5 Some important application of microwave cladding [53–55].

environmental compliance and contributing to a more sustainable future in transportation [52]. Figure 15.5 illustrates some key applications of microwave cladding technology across these two industries [53–55].

By embracing the use of microwave cladding, the industries open up new possibilities of mitigating environmental emissions and becoming sustainable. The technology allows innovative materials to emerge and partly meets the consumer demand to be more environment-friendly [56]. The advantages of integrating sustainable strategies are that companies have better brand images as well as increased trust among their stakeholders. These strategies support long-term development and profitability through responsible manufacturing practices. They also promote technological advancement aligned with current environmental norms and demands. Operating costs are also reduced by reducing waste disposal using efficient processes and enhancing resilience [57].

This forms a loop between enhanced environmental performance and business performance due to improved sustainability. Sustainable practices also enhance community relationships and support at the local level. The introduction of cleaner processes brings positive relations between businesses and the communities they serve. This improves social responsibility as well as improving environmental performance. To conclude, microwave cladding offers a valuable tool by combining ecological imperative with commercial viability.

15.4 CHALLENGES OF MICROWAVE CLADDING TECHNOLOGY

Although it is a revolutionary technology, microwave cladding has limitations that must be resolved to optimize harnessing its benefits, including high initial costs for specialized training and equipment and compatibility between the material being cladded and the overall efficiency of the cladding process [58].

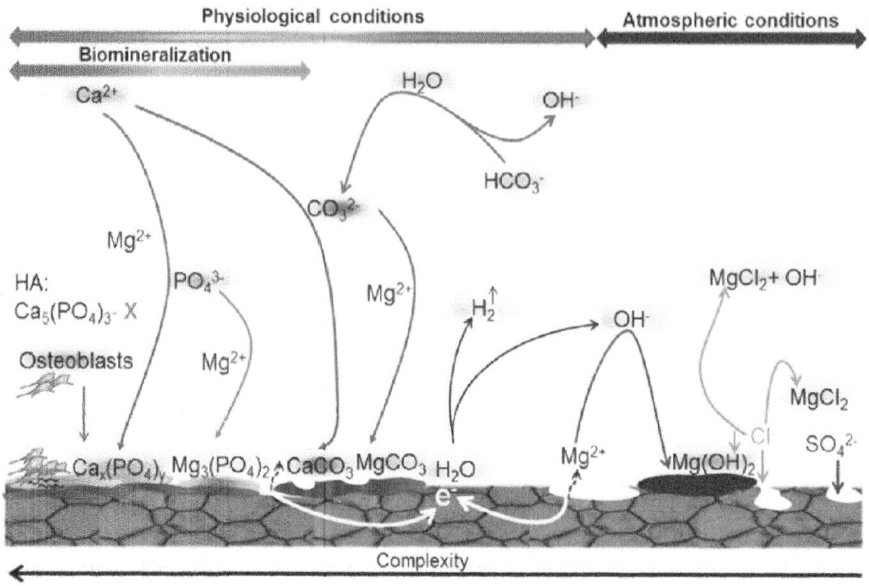

FIGURE 15.6 Degradation products formed under physiological conditions and the corresponding chemical reactions [59].

Figure 15.6 illustrates the complex interplay between physiological and atmospheric conditions in biomineralization. Under physiological conditions (left), the process begins with calcium (Ca^{2+}) and magnesium (Mg^{2+}) ions interacting with phosphate groups ($PO4^{3}$–), leading to the formation of calcium phosphate compounds ($Ca_5(PO_4)_{3)}$ by osteoblasts, as well as magnesium phosphates ($Mg_3(PO_4)_2$). Moving toward the center, these compounds interact with carbonates to form $CaCO_3$ and $MgCO_3$ through reactions with water (H_2O) and bicarbonate (HCO_3–). The right side shows atmospheric conditions where magnesium chloride ($MgCl_2$) and hydroxide ions (OH–) form magnesium hydroxide (Mg $(OH)_2$). The bottom of the diagram indicates increasing complexity across these reactions, represented by a layered substrate. The entire process demonstrates the intricate chemical transitions and mineral formations that occur at the interface of biological and environmental systems [59].

One major issue with microwave cladding is the inconsistent thickness of the cladding layer, particularly in fiber-based applications. The process relies on volumetric heating, and this often results in nonuniform temperature distribution across the substrate and the cladding material. This uneven heating can cause dimensional variations and structural inconsistencies in the cladding layer. Therefore, precise control of microwave parameters is essential to achieving a dense, uniform coating with minimal surface roughness. Further, the elimination of cellular dendrites in microwave-clad structures due to the uniform thermal gradient support the application of microwave processing and make it possible to develop structures with uniform microstructural characteristics provided that the processing parameters are controlled to a high degree [60]. Figure 15.7 visually presents some of the challenges with microwave cladding, specifically, static and dynamic process and performance challenges [61, 62].

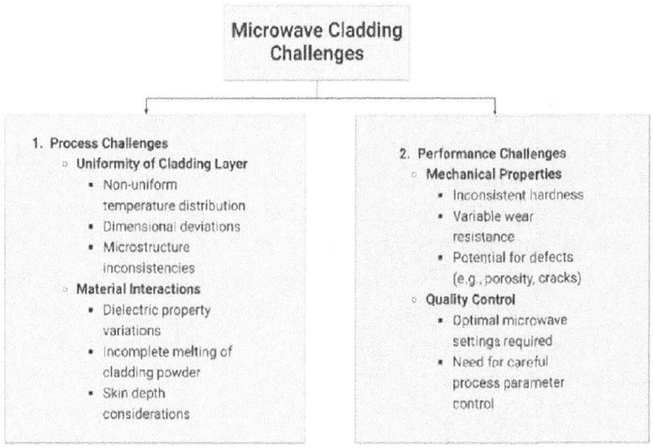

FIGURE 15.7 Challenges associated with microwave cladding [61, 62].

Another limitation of microwave cladding is the compatibility of materials with microwave energy. The process relies heavily on the dielectric properties of the materials to be treated. For instance, materials with low microwave absorption might not heat effectively, and therefore, the cladding powder might not melt as required. This problem is especially pronounced when a low-absorbing material is applied to a highly microwave-transparent substrate, a cumbersome process [63].

Cladding efficiency is significantly influenced by the skin depth of the material, which determines how deeply microwave energy can penetrate. Additionally, if the cladding process is not optimized, the mechanical properties of the two cladded surfaces can be inconsistent. Studies have shown that microwave treatment can enhance hardness and wear resistance over nonclad surfaces, but these improvements depend on the absence of defects and strong bonding between the cladding layer and the clad surface [64].

Porosity—micro-cracks—has a significant negative impact on the behavior of cladded material, so the process needs to be tightly controlled, and the cost of microwave cladding continues to be a significant problem. The process is shorter than the conventional methods, but its initial and operational costs are higher than those for MIC-based methods such as laser cladding [65]. These costs have proven to be a barrier to more widespread industry adoption of microwave cladding. Despite the potential benefits, the challenges are significant [66], and the mitigation of these challenges through continual research and development has always been crucial in improving the applicability of microwave cladding in many fields.

15.5 FUTURE TRENDS IN MICROWAVE CLADDING RESEARCH AND DEVELOPMENT

Future developments in microwave cladding technology point to broader industrial adoption. One key limitation is the size and design, which currently restrict large-scale integration. However, there is ongoing work to develop more cost-effective applicator systems that could expand the technology's use across sectors such as automotive, chemical processing, mining, food production, and power generation. Microwave focusing has recently emerged as a promising innovation. Molecule-based systems can enable higher energy distributions per unit area of the materials, which could improve the accuracy of cladding processes [67].

At the same time, AI and machine learning algorithms have enhanced data processing by introducing both higher precision and faster processing. To model the temporal patterns of microwave-induced heating described earlier, proposed methodologies leverage convolutional neural networks (CNNs). While originally developed for visual pattern recognition, CNNs have proven effective in learning and inferring temporal sequences by treating time as space. This approach enables the network to capture sequence dependencies and features in regard to solving patterns [68].

In Figure 15.8, the microwave (MW) source emits electromagnetic waves that propagate through different domains. The diagram spans approximately 25 × 10^{-2} meters and depicts three distinct regions: the air and boundary domain in light green, a generic material CNN region in cross-hatch, and the electromagnetic field in blue and orange sinusoidal form. These waves are launched from the MW source on the left and go through these domains and show how light behaves with the materials used in the experiment [69].

Improvements in microwave cladding can allow more complex surface topographies such as sinusoidal in special applications [70]. Due to increasing environmental issues, the emission-free character of microwave cladding is making it a more attractive choice than conventional techniques [70]. It reflects greener production through reduced fossil fuel consumption and intensive heating devices [70].

The future of microwave cladding seems promising in fields of material science [71]. Scientists are looking at micron, nano, and composite powders to improve wear, corrosion, and mechanical resistance [71]. Modeling and simulation provide good alternatives even though experimental measurements during processing can be challenging. These help to enhance the perception of process dynamics and quality of clad. Microwave cladding is also an area where AI and machine learning are now being used [72].

These techniques help fine-tune process parameters and improve consistency. With the levels of technology being developed, safety will be the primary interest of industrial implementation—especially in reducing microwave leakage and improving shielding for equipment and software. This will aid wider usage in automated

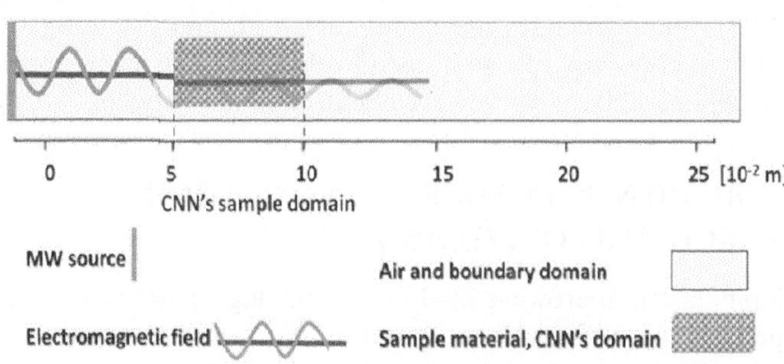

FIGURE 15.8 Graphical description of the microwave system modeled using CNNs [69].

and industrial space. New developments can also bring about functional surfaces which have built-in smart functionality [73].

In general, microwave cladding presents an environmentally friendly and affordable solution to next-generation surface engineering [74-81]. It will also start to replace traditional approaches in various industry applications. Further investigation will produce additional application and innovation in surface coating technologies [82–86]. This gradual development affirms the significance of microwave cladding in the next generation with regard to sustainability of material processing.

15.6 CONCLUSION

Chemical vapor-deposited microwave cladding is considered the cutting edge of material processing because it offers advantages over conventional cladding techniques. This technique employs microwave energy to heat the substrates for coating applications, leading to shorter coat times, increased homogeneity, enhanced adhesion, and minimized power usage. Additional advantages of microwave cladding include thermal stability and oxidation resistance, which makes its use very important in applications that require high temperatures such as aerospace, automotive, and energy. In this process, solutions enhance wear resistance, improving material use and component life span. This increased product life span not only reduces the need for replacements but also minimizes machine down time, thereby improving general productivity. Microwave cladding is flexible and can provide thin layers of material, which is important in areas where thin coatings are important and weight is significant.

Compared with conventional cladding techniques, microwave cladding offers advantages such as faster treatment speeds, lower energy consumption, and minimal material loss. The technology also enables depositing coatings with desired characteristics such as increased hardness and wear resistance. However, challenges must be overcome such as nonuniform heating, material incompatibility, and high initial investment costs for its broader adoption in industries.

Looking to the future, key areas of innovation in microwave cladding include the development of more efficient and cost-effective applicator systems; integrating AI and machine learning for process optimization; and the use of advanced materials such as micron, nano, and composite powders. The technology's nonhazardous nature aligns with the growing emphasis on sustainable manufacturing, potentially accelerating its diffusion across diverse industries. Continued progress will rely on collaboration among material scientists, engineers, and industry professionals to unlock new applications and explore the revolutionary direction of the surface engineering area. As the technique continues to evolve, microwave cladding will open a plethora of new opportunities, providing a cheap, fast, and often green way to modify existing materials for countless industries.

REFERENCES

1. Bansal A, Vasudev H, Sharma AK, Kumar P. Investigation on the effect of post weld heat treatment on microwave joining of the Alloy-718 weldment. Materials Research Express. 2019;6:086554.

2. Dwivedi SP, Sharma S, Sharma KP, Kumar A, Agrawal A, Singh R, Eldin SM. The microstructure and properties of Ni-Si-La_2O_3 coatings deposited on 304 stainless steel by microwave cladding. Materials. 2023;16:2209.
3. Bansal A, Zafar S, Sharma AK. Microstructure and abrasive wear performance of Ni-WC composite microwave clad. Journal of Materials Engineering Performance. 2015;24:3708–16.
4. Chio T-H, Huang G-L, Zhou S-G. Application of direct metal laser sintering to waveguide-based passive microwave components, antennas, and antenna arrays. Proceedings of the IEEE. 2016;105:632–44.
5. Ferreira AA, Amaral RL, Romio PC, Cruz JM, Reis AR, Vieira MF. Deposition of nickel-based superalloy claddings on low alloy structural steel by direct laser deposition. Metals. 2021;11:1326.
6. Dwivedi SP, Sharma S. Metallic cladding through microwave energy of the mixture of Ni and 15% SiC powder on AISI 304: A green approach in surface engineering. Proceedings of the Institution of Mechanical Engineers, Part E: Journal of Process Mechanical Engineering. 2024;238:2545–56.
7. Dwivedi SP, Sharma S, Srivastava AP, Sethi VA, Mohammed KA, Kumar A, Khan MI, Abbas M, Tag-Eldin EM. Homogeneity, metallurgical, mechanical, wear, and corrosion behavior of Ni and B4C coatings deposited on 304 stainless steels developed by microwave cladding technique. Journal of Materials Research Technology. 2023;27:5854–67.
8. Gupta D, Sharma AK. Investigation on sliding wear performance of $WC10Co_2Ni$ cladding developed through microwave irradiation. Wear. 2011;271:1642–50.
9. Gupta D, Sharma AK. Microstructural characterization of cermet cladding developed through microwave irradiation. Journal of Materials Engineering Performance. 2012;21:2165–72.
10. Ali SRA-S, Hussein AHA, Nofal A, Elnaby SIH, Elgazzar H. A contribution to laser cladding of Ti-6Al-4V titanium alloy. Metallurgical Research Technology. 2019;116:634.
11. Kansal A, Dvivedi A, Kumar P. Development and performance study of biomedical porous zinc scaffold manufactured by using additive manufacturing and microwave sintering. Materials Manufacturing Processes. 2023;38:1020–32.
12. Mago J, Bansal S, Gupta D, Jain V. Investigation of microwave processing parameters on development of Ni-40Cr_3C_2 composite clad and their characterization. Metallurgical Materials Transactions A. 2020;51:4288–300.
13. Medabalimi SR, Ananthu M, Gudala S, Ramesh M. Effect of microwave hybrid heating on high-temperature adhesive wear behavior of high-velocity oxygen fuel-sprayed WC-CrC-Ni and WC-Co/NiCrFeSiB coatings. Journal of Materials Engineering Performance. 2023;32:8612–24.
14. Mehta A, Vasudev H, Jeyaprakash N. Role of sustainable manufacturing approach: Microwave processing of materials. International Journal on Interactive Design Manufacturing. 2024;18:5283–99.
15. Miao M, Zhao G, Xu L, Dong J, Cheng P. Direct determination of trace phthalate esters in alcoholic spirits by spray-inlet microwave plasma torch ionization tandem mass spectrometry. Journal of Mass Spectrometry. 2018;53:189–94.
16 Agnihotri A, Vasudev H, Singh G, Singh S, Hosalli R, Veeresha G, . . . & Tiwari, A. Evaluation of HVOF sprayed and burnished WC-Co-Cr coatings on boiler steel: Mechanical and microstructural properties. Journal of Bio-and Tribo-Corrosion. 2025;11(2):1–12.
17. Moss D. Ultrahigh Bandwidth Optical Neuromorphic Processing, Data Transmission and Microwave RF Photonics with Kerr Microcombs. In Proc. 11775 SPIE Optics+ Optoelectronics Symposium, Prague (EOO21), OO107-8 (Vol. 107, pp. 11775–81).
18. Panda S, Singh V, Upadhyaya A, Agrawal D. Sintering response of austenitic (316L) and ferritic (434L) stainless steel consolidated in conventional and microwave furnaces. Scripta Materialia. 2006;54:2179–83.

19. Prakash C, Singh S, Basak A, Królczyk G, Pramanik A, Lamberti L, Pruncu CI. Processing of Ti50Nb50–xHAx composites by rapid microwave sintering technique for biomedical applications. Journal of Materials Research Technology. 2020;9:242–52.
20. Li Z, Yu H, Sun D. Structural and microstructural effects produced by microwave sintering over La_2O_3 doped TiB2-Ni composites. Ceramics International. 2021;47:26692–703.
21. Sharanabasava H, Prasad CD, Ramesh M. Effect of Mo-and SiC-reinforced NiCr microwave cladding on microstructure, mechanical and wear properties. Journal of the Institution of Engineers: Series D. 2023;104:539–51.
22. Sharanabasava H, Prasad CD, Ramesh M. Characterization and wear behavior of NiCrMoSiC microwave cladding. Journal of Materials Engineering Performance. 2024;33:763–75.
23. Sharma A, Singh A, Chawla V, Grewal J, Bansal A. Microwave processing and characterization of alumina reinforced HA cladding for biomedical applications. Materials Today: Proceedings. 2022;57:650–56.
24. Sharma AK, Gupta D. On microstructure and flexural strength of metal–ceramic composite cladding developed through microwave heating. Applied Surface Science. 2012;258:5583–92.
25. Shitu IG, Katibi KK, Taura LS, Muhammad A, Chiromawa IM, Adamu SB, Iya SGD. X-ray diffraction (XRD) profile analysis and optical properties of Klockmannite copper selenide nanoparticles synthesized via microwave assisted technique. Ceramics International. 2023;49:12309–26.
26. Singh B, Zafar S. Understanding temperature characteristics during microwave cladding through process modeling and experimental investigation. CIRP Journal of Manufacturing Science Technology. 2022;37:401–13.
27. Singh G, Mehta A, Bansal A. Electrochemical behaviour and biocompatibility of claddings developed using microwave route. Journal of Electrochemical Science Engineering. 2023;13:173–92.
28. Singh G, Vasudev H, Bansal A, Vardhan S. Influence of heat treatment on the microstructure and corrosion properties of the Inconel-625 clad deposited by microwave heating. Surface Topography: Metrology Properties. 2021;9:025019.
29. Singh P, Bansal A, Vasudev H, Singh P. In situ surface modification of stainless steel with hydroxyapatite using microwave heating. Surface Topography: Metrology Properties. 2021;9:035053.
30. Singh P, Bansal A, Verma VK. Hydroxyapatite reinforced surface modification of SS-316L by microwave processing. Surfaces Interfaces. 2022;28:101701.
31. Singh P, Vasudev H, Bansal A. Effect of post-heat treatment on the microstructural, mechanical, and bioactivity behavior of the microwave-assisted alumina-reinforced hydroxyapatite cladding. Proceedings of the Institution of Mechanical Engineers, Part E: Journal of Process Mechanical Engineering. 2022;238:09544089221116168.
32. Vasudev H, Singh P, Thakur L, Bansal AJMRE. Mechanical and microstructural characterization of microwave post processed Alloy-718 coating. Materials Research Express. 2020;6:1265f5.
33. Alizadeh-Sh M, Marashi S, Ranjbarnodeh E, Shoja-Razavi R. Laser cladding of Inconel 718 powder on a non-weldable substrate: Clad bead geometry-solidification cracking relationship. Journal of Manufacturing Processes. 2020;56:54–62.
34. Hebbale AM, Kumar M, Soudagar MEM, Ahamad T, Kalam MA, Mubarak NM, Alfantazi A, Khalid M. A comparative study on characteristics of composite (Cr_3C_2-NiCr) clad developed through diode laser and microwave energy. Scientific Reports. 2023;13:10778.
35. Aravazhi S, Geskus D, van Dalfsen K, Vázquez-Córdova SA, Grivas C, Griebner U, García-Blanco SM, Pollnau M. Engineering lattice matching, doping level, and optical properties of KY (WO4) 2: Gd, Lu, Yb layers for a cladding-side-pumped channel waveguide laser. Applied Physics B. 2013;111:433–46.

36. Bajda S, Liu Y, Tosi R, Cholewa-Kowalska K, Krzyzanowski M, Dziadek M, Kopyscianski M, Dymek S, Polyakov AV, Semenova IP. Laser cladding of bioactive glass coating on pure titanium substrate with highly refined grain structure. Journal of the Mechanical Behavior of Biomedical Materials. 2021;119:104519.
37. Behera RR, Hasan A, Sankar MR, Pandey LM. Laser cladding with HA and functionally graded TiO_2-HA precursors on Ti–6Al–4V alloy for enhancing bioactivity and cyto-compatibility. Surface Coatings Technology. 2018;352:420–36.
38. Bhatnagar S, Mullick S. A study on the influence of reinforcement particle size in laser cladding of TiC/Inconel 625 metal matrix composite. Optics Laser Technology. 2023;161:109115.
39. Siang TW, Firdaus Akbar M, Nihad Jawad G, Yee TS, Mohd Sazali MIS. A past, present, and prospective review on microwave nondestructive evaluation of composite coatings. Coatings. 2021;11:913.
40. Medvedovski E. Advanced ceramics and coatings for erosion-related applications in mineral and oil and gas production: A technical review. International Journal of Applied Ceramic Technology. 2023;20:612–59.
41. Mago J, Bansal S, Gupta D, Jain V. Cavitation erosion behavior of microwave-processed Ni–40Cr_3C_2 composite clads: A parametric investigation using ultrasonic apparatus. Proceedings of the Institution of Mechanical Engineers, Part L: Journal of Materials: Design Applications. 2021;235:265–92.
42. Singh J. Application of thermal spray coatings for protection against erosion, abrasion, and corrosion in hydropower plants and offshore industry. In *Thermal Spray Coatings*. CRC Press, 2021:243–83.
43. Grewal H, Nair R, Arora H. Complex concentrated alloy bimodal composite claddings with enhanced cavitation erosion resistance. Surface Coatings Technology. 2020;392:125751.
44. Gu Y, Xia K, Wu D, Mou J, Zheng S. Technical characteristics and wear-resistant mechanism of nano coatings: A review. Coatings. 2020;10:233.
45. Babu A, Arora H, Behera SN, Sharma M, Grewal H. Towards highly durable bimodal composite claddings using microwave processing. Surface Coatings Technology. 2018;349:655–66.
46. Mago J, Bansal S, Gupta D, Jain V. Investigation of microwave processing parameters on development of Ni-40Cr_3C_2 composite clad and their characterization. Metallurgical Materials Transactions A. 2020;51:4288–300.
47. Hebbale AM, Srinath M. Taguchi analysis on erosive wear behavior of cobalt based microwave cladding on stainless steel AISI–420. Measurement. 2017;99:98–107.
48. Hebbale AM, Srinath M. Microstructure and experimental design analysis of nickel based clad developed through microwave energy. Perspectives in Science. 2016;8:257–59.
49. Kaushal S, Gupta D, Bhowmick H. An approach for functionally graded cladding of composite material on austenitic stainless steel substrate through microwave heating. Journal of Composite Materials. 2018;52:301–12.
50. Singh B, Zafar S. Microwave cladding for slurry erosion resistance applications: A review. Materials Today: Proceedings. 2021;46:2686–90.
51. Bansal S, Mago J, Gupta D, Jain V. Microwave cladding of NiCrSiC-5Al_2O_3 on austenitic stainless steel to improve cavitation erosion resistance. Surface Topography: Metrology Properties. 2021;9:035036.
52. Suresh G, Ramesh M, Srinath M. Development of self-lubricating nickel based composite clad using microwave heating in improving resistance to wear at elevated temperatures. Metals Materials International. 2022:1–12.
53. Hebbale AM. Microstructural characterization of Ni based cladding on SS-304 developed through microwave energy. Materials Today: Proceedings. 2015;2:1414–20.

54. Kaushal S, Gupta D, Bhowmick H. On surface modification of austenitic stainless steel using microwave processed Ni/Cr_3C_2 composite cladding. Surface Engineering. 2018;34:809–17.
55. Singh S, Gupta D, Jain V. Recent applications of microwaves in materials joining and surface coatings. Proceedings of the Institution of Mechanical Engineers, Part B: Journal of Engineering Manufacture. 2016;230:603–17.
56. Nair RB, Arora H, Boyana A, Saiteja P, Grewal H. Tribological behavior of microwave synthesized high entropy alloy claddings. Wear. 2019;436:203028.
57. Patel DK, Madankar D, Bhoi NK, Singh H. High entropy alloy cladding on mild steel substrate using microwave energy at 2.45 GHz: An insight from multi-physics simulation study to experimental investigations. Journal of Manufacturing Processes. 2024;109:428–46.
58. Singh P, Vasudev H, Bansal A. Effect of post-heat treatment on the microstructural, mechanical, and bioactivity behavior of the microwave-assisted alumina-reinforced hydroxyapatite cladding. Proceedings of the Institution of Mechanical Engineers, Part E: Journal of Process Mechanical Engineering. 2024;238:1060–71.
59. Gonzalez J, Hou RQ, Nidadavolu EP, Willumeit-Römer R, Feyerabend F. Magnesium degradation under physiological conditions–Best practice. Bioactive Materials. 2018;3:174–85.
60. Bekal A, Hebbale AM, Srinath MS. Review on material processing through microwave energy. In *IOP Conference Series: Materials Science and Engineering* (Vol. 376, No. 1). IOP Publishing, 2018:012079.
61. Ranjan R, Das AK. Protection from corrosion and wear by different weld cladding techniques: A review. Materials Today: Proceedings. 2022;57:1687–93.
62. Tian N, Wang J, Li F, Mei Z, Lu Z, Ge L, You C. Enhanced microwave absorption of Fe flakes with magnesium ferrite cladding. Journal of Magnetism Magnetic Materials. 2012;324:4175–78.
63. Singh B, Kaushal S, Gupta D, Bhowmick H. On development and dry sliding wear behavior of microwave processed Ni/Al_2O_3 composite clad. Journal of Tribology. 2018;140:061603.
64. Kaushal S, Singh D, Gupta D, Jain V. Processing of Ni–WC–Cr_3C_2-based metal matrix composite cladding on SS-316L substrate through microwave irradiation. Journal of Composite Materials. 2019;53:1023–32.
65. Singh B, Zafar S. Effect of microwave exposure time on microstructure and slurry erosion behavior of Ni+ 20% Cr_7C_3 composite clads. Wear. 2019;426:491–500.
66. Kaushal S, Gupta D, Bhowmick H. On processing of Ni-WC based functionally graded composite clads through microwave heating. Materials Manufacturing Processes. 2018;33:822–28.
67. Kaushal S, Singh B, Gupta D, Bhowmick H, Jain V. An approach for developing nickel–alumina powder-based metal matrix composite cladding on SS-304 substrate through microwave heating. Journal of Composite Materials. 2018;52:2131–38.
68. Yee TS, Shrifan NH, Al-Gburi AJA, Isa NAM, Akbar MF. Prospect of using machine learning-based microwave nondestructive testing technique for corrosion under insulation: A review. IEEE Access. 2022;10:88191–210.
69. Rosario Núñez V, Hernández A, Rodríguez I, Fernández-Pacheco Ruiz I, Acevedo L. Artificial intelligence applied to microwave heating systems: Prediction of temperature profile through convolutional neural networks. Thermomechanical Industrial Processes: Modeling Numerical Simulation. 2024;4.
70. Chen Y, Tian X, Sved J, Li L, Zhou L, Nguyen L, Yi X. Reflective microring-resonator-based microwave photonic sensor incorporating a self-attention assisted convolutional neural network. Applied Optics. 2024;63:D59–66.
71. Chigateri KB, Hebbale AM. A steel surface defect detection model using machine learning. Materials Today: Proceedings. 2024;100:51–58.

72. Singh R, Agarwal A, Anthony BW. Design of optical meta-structures with applications to beam engineering using deep learning. Scientific Reports. 2020;10:19923.
73. Singh K, Sharma S. Development of Ni-based and CeO_2-modified coatings by microwave heating. Materials Manufacturing Processes. 2018;33:50–57.
74. Singh G, Singh H, Sharma Y, Vasudev H, Prakash C. Analysis and optimization of various process parameters and effect on the hardness of SS-304 stainless steel welded joints. International Journal on Interactive Design and Manufacturing (IJIDeM). 2024;18(5):2889–96.
75. Singh G, Mehta A, Vasudev H. Sustainability of additive manufacturing: A comprehensive review. Progress in Additive Manufacturing. 2024:1–24.
76. Singh G, Bansal A, Vasudev H, Mishra V. Sliding wear study of the Inconel-625 clad deposits by microwave heating on SS–304. Physica Scripta. 2024;99(6):065503.
77. Singh J, Yadav A, Gupta S, Balu R, Shukla N, Singh G, Singh S. Design and computational analysis of continuous wave fiber laser cutting of flat and conformal ITO PET sheets. International Journal on Interactive Design and Manufacturing (IJIDeM). 2024:1–9.
78. Singh G, Singh R, Gul J. Machinability behavior of human implant materials. Journal of Electrochemical Science and Engineering. 2023;13(1):99–114.
79. Verma PK, Vasudev H, Singh G, Singh S. A review on the microstructural and biomedical properties of implants manufactured using additive manufacturing. Journal of the Institution of Engineers (India): Series D. 2024:1–15.
80. Mehta A, Singh G, Vasudev H. Processing of shape memory alloys research, applications and opportunities: A review. Physica Scripta. 2024. https://iopscience.iop.org/article/10.1088/1402-4896/ad48cb
81. Singh G, Bansal A, Vasudev H. Corrosion and microstructural behaviour of Inconel 625 microwave clad deposited on mild steel. In *Thermal Claddings for Engineering Applications*. CRC Press, 2024:51–62.
82. Prashar G, Vasudev H. Understanding cold spray technology for hydroxyapatite deposition. Journal of Electrochemical Science and Engineering. 2023;13(1):41–62.
83. Parkash J, Saggu HS, Vasudev H. A short review on the performance of high velocity oxy-fuel coatings in boiler steel applications. Materials Today: Proceedings. 2022;50:1442–46.
84. Singh J, Vasudev H, Kumar R, Ubaidullah M. Study on wear analysis of Ni-20Al_2O_3 HVOF micron layers using artificial neural network technique. International Journal on Interactive Design and Manufacturing (IJIDeM). 2024;18(8):5925–38.
85. Singh J, Vasudev H, Szala M, Gill HS. Neural computing for erosion assessment in Al-20TiO2 HVOF thermal spray coating. International Journal on Interactive Design and Manufacturing (IJIDeM). 2024;18(4):2321–32.
86. Prakash C, Gupta LR, Mehta A, Vasudev H, Tominov R, Korman E, Fedotov A, Smirnov V, Kesari KK. Computing of neuromorphic materials: An emerging approach for bioengineering solutions. Materials Advances. 2023;4(23):5882–919.

16 Shape Memory Alloys for Adaptive Solutions in Aerospace

Gurbhej Singh and Hitesh Vasudev

16.1 INTRODUCTION

Shape memory alloys are classified as active materials that utilize direct coupling for actuation [1]. SMAs possess the capacity to recall the shape they had previously imprinted on themselves when they are deformed. The shape memory effect (SME) can be classified into two distinct types: a thermally activated effect and a mechanically triggered (pseudoelastic) effect. The material is capable of maintaining its shape immediately upon being liberated from its burden in the initial scenario [2]. Through the application of energy, the material is able to keep its shape in the second scenario. This is attainable through the utilization of the ambient temperature or heating depending on resistance as a component of the electrical activation process. The SME is based on the transformation of the phase from martensite, which is the phase at low temperatures, to austenite, which is the phase at high temperatures and features a temperature hysteresis as a result of the friction that occurs within the material.

In the past few decades, there has been a surge in the number of research projects that focus on morphing technologies. This is due to the ever-increasing demand for structures that are capable of independently adjusting their shape in response to specific and variable situations. These technologies are especially well suited for use in the aerospace and automobile industries, where adaptive structures such as morphing skin can be utilized [1–3]. Materials that are piezoelectric and have the ability to change the shape of the morphing component [4–6] are examples of SMA. SMA are able to regain their initial shape after a deformation has occurred, provided that they are treated to specific temperature conditions [7–9]. This ability is among the smart materials. There are a number of applications that can be found in relation to the utilization of SMA as actuators [10, 11], and these applications can be found in the vehicle industry [11] as well as the aerospace industry [12], despite the fact that application on a broad scale has not yet been accomplished. On the other hand, SMA-based actuation systems are currently being intensively examined within the context of a number of research studies. The goal of these investigations is to significantly reduce the complexity of the conventional electric and hydraulic actuation systems, which will ultimately result in a reduction in weight [13–15]. The crucial phase-transformation temperatures of typical SMAs are presented in Table 16.1.

DOI: 10.1201/9781003546528-16

TABLE 16.1
Temperatures of Phase Transformation for SMAs

Material	Mf (°C)	Ms (°C)	As (°C)	Af (°C)	Af (°C) Re
Fe-SMA	−92	−74	85	112	[14]
Fe-SMA	−65	−55	103	115	[16]
NiTiNol	−75 to −56	16 to 22	−21–3	20to 30	[17]

Indeed, there are already a number of mechanisms that are based on SMA actuators that are being utilized, particularly in the industries of valves and drives, which are responsible for the annual sale of millions of devices [18]. Shape memory alloys, often known as SMAs, are specific types of metals that are capable of regaining a predetermined shape only when heated to a temperature that is required for transformation. Specifically, the martensitic transformation, also known as the thermoelastic transformation, that takes place in some metallic alloys is the basic principle that they are built on [19]. Included on the list are NiTi, CuAlNi, CuZnAl, AuCd, and a great number of additional elements, but this list is not exhaustive. The shape memory effect was originally demonstrated in near equiatomic Ti-Ni alloys [20]. The shape memory effect was discovered previously in Au−47.5% Cd [21] and In-Tl [22], although the researchers did not pay much attention to it at the time. In contrast, the Naval Ordinance Laboratory was responsible for the rise in popularity of equiatomic Ni-Ti alloys, which is where the name Nitinol originated from. It took some time before the true knowledge of the processes that took place during the martensitic transition was achieved. In this study, we will explore innovative applications of shape memory alloys in adaptive aerodynamics, space systems, and smart materials, emphasizing their transformative potential in current engineering. This chapter highlights the latest advancements, challenges, and future directions to optimize SMA-based technologies for aerospace and automotive industries

16.2 APPLICATIONS OF SHAPE MEMORY ALLOYS

Shape memory polymers (SMPs) are valuable materials because they are light, cheap, easy to make, biodegradable, biocompatible, and have good mechanical qualities. These traits make them appealing for use in many contexts as smart materials.

16.2.1 Adaptive Aerodynamic Applications

Adaptive aerodynamics represents a highly promising domain for the application of shape memory alloy components, owing to their morphing capabilities. Numerous solutions have been explored to alter the aerodynamic field in aeronautical applications. A stretchable unmanned aerial vehicle wing capable of expanding its planform area by 80% is presented in [23]. Shape memory polymers were used to alter the wing chord, customizing the wing for certain situations. The implementation of actively cooled shape memory alloys for flexible control surface deployment was examined in [24], whereas in [25], SMA wires were utilized to alter the camber of a morphing wing. Experimental and numerical studies on a morphing airfoil were conducted in [26] utilizing shape memory alloy springs to actuate specific points of the structure for the intended deformation. A lot of the methods that have been found for adjustable aerodynamics can be used in the automobile field. As shown in Figure 16.1, adaptable aerodynamics can be used in the automobile industry to move bumpers, grilles, and/or louvers on the outside of the car.

A wing consisting of various telescopic segments deployed via SMA actuators is introduced in [27] to enhance aerodynamic performance while minimizing wing volume storage. In [28], shape memory alloys were employed to design and fabricate experimental wind tunnel tests to investigate the efficacy of smart vortex generators, yielding favorable results. An application for deploying and saving a flap edge fence is available in [38]. The methods developed for adaptive aerodynamics are readily transferable from aeronautics to vehicle engineering. Figure 16.2 illustrates that adaptive aerodynamic applications in the automobile sector include, but are not limited to, the actuation of external surfaces, spoilers, grilles, and louvers. Figure 16.2 shows a summary of current and future uses of SMA ideas for the adjustable aerodynamics of airplanes.

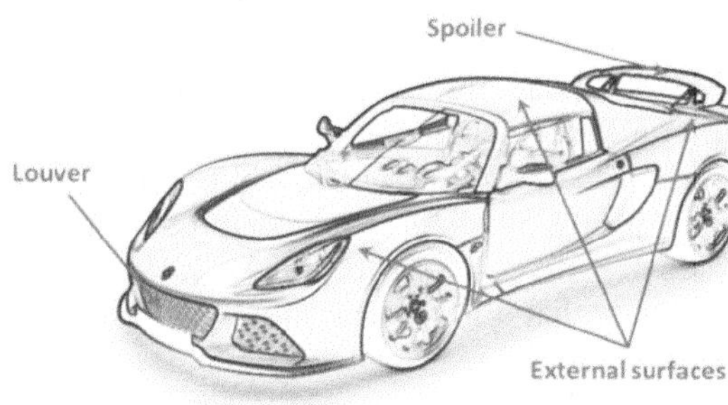

FIGURE 16.1 Positions of adaptive aerodynamic implementations in the automotive industry [18].

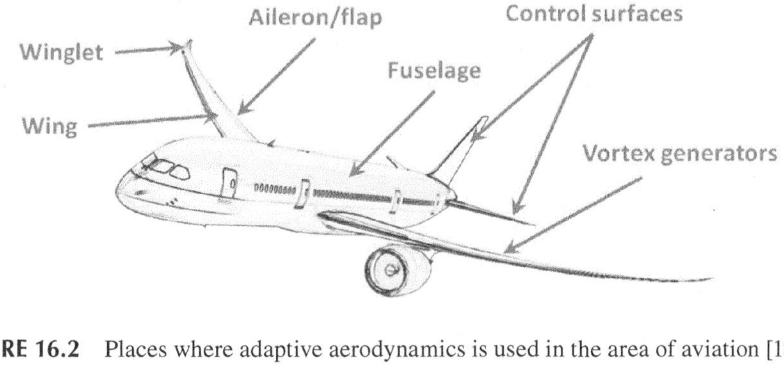

FIGURE 16.2 Places where adaptive aerodynamics is used in the area of aviation [18].

16.2.2 Shape Memory Alloy Applications on Wings

Molding the wings of airplanes is a realistic method that involves making structural adjustments in order to ensure that the relevant criteria may be satisfied in a variety of different use scenarios. The following are some of the advantages that have been reported: greater speed, less power usage, roll control, and camber change alterations. In contrast, morphing is typically associated with negatives such as surface discontinuity and being overweight due to the necessary control mechanisms. These drawbacks are typically connected with morphing. Through the implementation of shape memory solutions and the methods that correspond to them, it is possible to circumvent some of these disadvantages. The most well-known initiatives involving fixed wings are the Smart Wing Program and the Smart and Aircraft and Marine Propulsion System Demonstration (SAMPSON) [29, 30]. Both of these projects have been acknowledged by the Defense Research Initiatives Agency (DARPA). Because of the special power and power density of SMA actuators in comparison to standard electromechanical ones, the first one was created for the usage of smart materials such as lifting devices. This was made possible by the fact that SMA actuators are unique. The development of new affordable smart materials was the primary objective of the second one, which is also focused with the creation of shape-adaptive structures for airplanes. In specifically, the SAMPSON project investigated models that could be used to modify the design of gas turbine inlets with the purpose of implementing them in supersonic aircraft. Additionally, an innovative application for a hydrodynamic maneuvering system that could be utilized in maritime propulsion has been examined. For the purpose of providing actuation through the utilization of form recovery, the shape memory effect has been utilized [31–33]. The most important objective was to demonstrate that it is possible to use SMAs to alter the orientation and geometry of various propulsion systems without compromising their functionality. In the arrangement of a SMA cable antagonist to another one, an experiment was carried out on the actual F-15 inlet (Figure 16.3). One of the cables was used as the

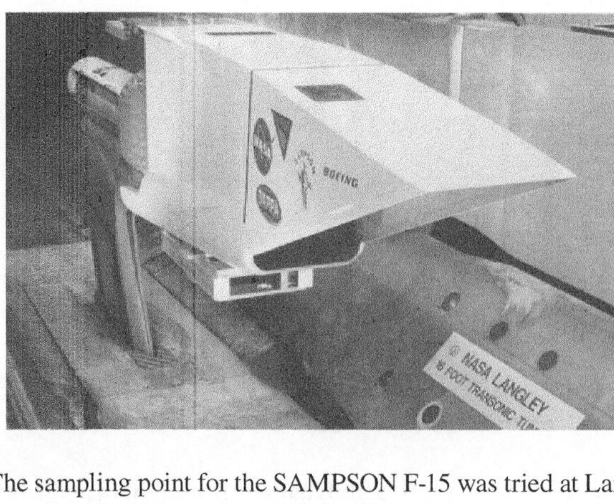

FIGURE 16.3 The sampling point for the SAMPSON F-15 was tried at Langley (NASA) [19].

Source: Licensed under CC BY 4.0.

antagonist. The following are some instances of wings that have undergone morphing. The F15E Strike Eagle is a specific type of fighter jet that was designed to carry out a variety of different missions. The aircraft is capable of reaching speeds more than Mach 2.5 as a result of the presence of two engines. The inlet system of this jet is made up of five ramps that are used for the purpose of regulating the air flow for both of the engines. There are many different ramps. Hartl and Lagouda [34] concentrated their attention on a project that aimed to reduce the amount of noise that is produced at airports by airplanes that are taking off. It has been established that the implementation of aerodynamic devices, more commonly referred to as chevrons, on the back edge of an exhaust nozzle (both on main and secondary exhaust flow) can reduce jet noise at take-off by ensuring that the appropriate streams are mixed together. During cruising, it was observed that chevrons that were immersed in the flow of the fan exhibited a decrease in thrust and an increase in drag. This was a consequence of the secondary exhaust nozzle. In light of these considerations, the variable geometry chevron (VGC) was found to contain the optimal solution. The VGC, which is based on SMA components, demonstrates all-out bending during take-off and landing while maintaining a minimal deflection flow during cruising. This allows for enhanced efficiency and ensures that noise reduction is achieved. Active SMA bundles are contained within the system, which is encased in a complicated casing.

In order to lessen the amount of noise that is produced, the activation of the SMA beams makes it possible to apply the desired bending force to the chevron structure. A flight test was conducted by Boeing to evaluate the suggested solution that incorporated active SMA parts. A comparison of the outcomes of the experiments and the modeling was carried out at the conclusion of the study.

Numerous studies have concentrated on the concept of incorporating SMA elements into aerostructures. A promising option is to include SMA wires in an innovative composite structure. NiTi alloy wires with a diameter of 150 micrometers were pre-stressed and put into a Kevlar fiber epoxy matrix [35]. This was done in order to take advantage of the one-way shape memory effect. When it comes to adaptive applications, such as the progressive reinforcement of components (structure) or the alteration of the inherent vibration frequencies, SMA composites have a significant amount of potential. Nevertheless, the behavior of SMAs is not linear and provides a wide range of options. In addition, there is a need for a greater understanding of the stress transfer that occurs between the metal and polymer matrix, as well as the fatigue behavior of structures of this kind. SMA wire actuators are another application that may be used. These actuators can be attached to certain interior spots of an airfoil and then activated to modify the shape of the airfoil itself. The modification of the shape has the potential to improve the efficiency of the wing in a variety of flow regimes used in operational flying circumstances. Through the utilization of a global optimization strategy, the primary objective of determining the number of SMA wire actuators within the wing as well as their appropriate positioning within the wing was successfully accomplished.

16.2.3 Space Applications

In order to address the extraordinary challenges of actuation, release, and vibration attenuation that arise during the launch of a spacecraft or subsequent operations, space

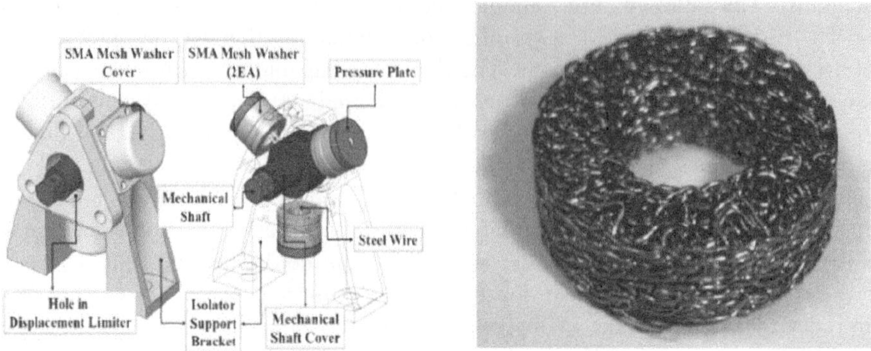

FIGURE 16.4 The shape memory alloy mesh washer [19].

applications of SMAs aim to find solutions. These challenges arise in conditions of microgravity and zero atmosphere. As a potential remedy to this issue, Researchers suggested a pseudoelastic form as a potential washer material for shape memory alloy mesh [36]. Because of this mechanism, vibration isolation may be accomplished not only in an environment with a high launch vibration but also for micro-vibrations that originate from the cryocooler when it is in orbit (Figure 16.4).

The primary cause of the degradation of the image quality that is obtained from high-resolution satellites is the undesirable microvibrations that are produced by a spaceborne cryocooler when it is operating in orbit. Isolating the micro-vibrations that are produced by a cryocooler is necessary in order to obtain high-quality photos. This is necessary in order to comply with the requirements of the mission. A micro-vibration isolator for a spaceborne cryocooler that is based on a pseudoelastic mesh washer was proposed that guaranteed adequate vibration isolation under harsh launch conditions and in fact reduced the micro-vibrations that were caused by the cryocooler when it was in orbit. The investigators conducted static testing and free vibration tests to establish that their concept was effective. The extensive hysteresis and significant nonlinearity observed during the loading and unloading cycle form the basis of the devices' operating principles. The authors of the study summarized the materials' pseudoelastic behavior, modeling, applications, and limitations in terms of seismic isolation and vibration based on their examination of the pyroshock isolation performance of a compressed mesh washer for the purpose of determining its effectiveness [37].

16.3 INTEGRATION OF ADVANCED MATERIAL PROCESSING TECHNIQUES FOR ENHANCING SHAPE MEMORY ALLOY APPLICATIONS

The advancements in material processing techniques offer significant opportunities to enhance the properties and applications of shape memory alloys, particularly in high-performance domains like aerospace and automotive engineering. Singh et al. (2021) [38] investigated the influence of heat treatment on the microstructure and corrosion resistance of Inconel-625 clads developed using microwave heating, demonstrating how tailored thermal treatments can refine material properties, including surface resilience and structural integrity, which are crucial for SMAs under extreme operating conditions. Furthermore, Singh et al. (2020) [39] explored the microwave cladding of Inconel-625 on mild steel substrates, highlighting substantial improvements in surface quality and corrosion protection. Such advancements align with the requirements of SMA applications, where materials need to withstand dynamic environments and maintain functionality over prolonged periods. The versatility and efficiency of microwave heating for surface engineering, as reviewed by Vasudev et al. (2019) [40], provide a promising foundation for applying innovative processing techniques to SMAs. These insights are particularly relevant for optimizing SMA-based actuation systems, where enhanced surface and structural properties can contribute to reduced weight and improved energy efficiency. The integration of such advanced methodologies into the design and development of SMAs can further expand their use in adaptive aerodynamics, vibration attenuation, and other demanding applications. By bridging the knowledge from these studies with ongoing research on SMAs, a pathway can be established for the creation of more reliable and versatile materials tailored for modern engineering challenges.

16.4 CONCLUSIONS

Shape memory alloys demonstrate exceptional potential in various engineering applications due to their unique ability to recover their original shape through the shape memory effect, triggered either mechanically or thermally. Materials such as NiTi, CuAlNi, and CuZnAl have emerged as leading choices for designing adaptive structures, especially in aerospace and automotive industries, owing to their superelasticity and favorable power-to-weight ratios. The exploration of SMA-based actuation systems has opened pathways for simplifying and lightening traditional electric and hydraulic systems. Key advancements include their deployment in adaptive aerodynamics to modify wing configurations, their use as artificial muscles in bioinspired systems, and their integration into smart textiles for functional and aesthetic enhancements. Furthermore, in the space industry, SMAs offer promising solutions for challenges in actuation, release mechanisms, and vibration attenuation during spacecraft operations. Despite some limitations, such as hysteresis and nonlinear behavior, SMAs have proved invaluable for seismic isolation and vibration management systems. Ongoing research continues to refine their applications, ensuring their role as critical components in modern and future engineering systems.

REFERENCES

1. Zhang, Z. X., Zhang, J., Wu, H., Ji, Y., & Kumar, D. D. (2022). Iron-based shape memory alloys in construction: Research, applications and opportunities. *Materials*, *15*(5), 1723.
2. Sato, A., & Mori, T. (1991). Development of a shape memory alloy Fe Mn Si. *Materials Science and Engineering: A*, *146*(1–2), 197–204.
3. Lee, W. J., Weber, B., & Leinenbach, C. (2015). Recovery stress formation in a restrained Fe–Mn–Si-based shape memory alloy used for prestressing or mechanical joining. *Construction and Building Materials*, *95*, 600–610.

4. Maruyama, T., & Kubo, H. (2011). Ferrous (Fe-based) shape memory alloys (SMAs): Properties, processing and applications. In *Shape memory and superelastic alloys* (pp. 141–159). Amsterdam, The Netherlands: Elsevier.
5. Ghafoori, E., Hosseini, E., Leinenbach, C., Michels, J., & Motavalli, M. (2017). Fatigue behavior of a Fe-Mn-Si shape memory alloy used for prestressed strengthening. *Materials Design*, *133*, 349–362.
6. Zirakian, T., & Zhang, J. (2015). Structural performance of unstiffened low yield point steel plate shear walls. *The Journal of Constructional Steel Research*, *112*, 40–53.
7. Wang, Y., & Wang, Z. (2016). Experimental investigation and FE analysis on constitutive relationship of high strength aluminum alloy under cyclic loading. *Advances in Materials Science and Engineering*, *2016*, 2941874.
8. Shahverdi, M., Michels, J., Czaderski, C., & Motavalli, M. (2018). Iron-based shape memory alloy strips for strengthening RC members: Material behavior and characterization. *Construction and Building Materials*, *173*, 586–599.
9. Yin, F., Yang, L., Wang, M., Zong, L., & Chang, X. (2019). Study on ultra-low cycle fatigue behavior of austenitic stainless steel. *Thin-Walled Structure*, *143*, 106205.
10. Fang, C., Wang, W., Ji, Y., & Yam, M. C. (2021). Superior low-cycle fatigue performance of iron-based SMA for seismic damping application. *The Journal of Constructional Steel Research*, *184*, 106817.
11. Gu, T., Jia, L.-J., Chen, B., Xia, M., Guo, H., & He, M.-C. (2021). Unified full-range plasticity till fracture of meta steel and structural steels. *Engineering Fracture Mechanics*, *253*, 107869.
12. Shi, G., Gao, Y., Wang, X., & Cui, Y. (2018). Energy-based low cycle fatigue analysis of low yield point steels. *The Journal of Constructional Steel Research*, *150*, 346–353.
13. Mohd Jani, J., Leary, M., Subic, A., & Gibson, M. A. (2014). A review of shape memory alloy research, applications and opportunities. *Materials Design*, *56*, 1078–1113.
14. Dong, Z., Klotz, U. E., Leinenbach, C., Bergamini, A., Czaderski, C., & Motavalli, M. (2009). A novel Fe-Mn-Si shape memory alloy with improved shape recovery properties by VC precipitation. *Advanced Engineering Materials*, *11*, 40–44.
15. Lee, W. J., Weber, B., Feltrin, G., Czaderski, C., Motavalli, M., & Leinenbach, C. (2013). Phase transformation behavior under uniaxial deformation of an Fe–Mn–Si–Cr–Ni–VC shape memory alloy. *Materials Science and Engineering A*, *581*, 1–7.
16. Chen, J., Wang, W., & Fang, C. (2021). Manufacturing, testing and simulation of novel SMA-based variable friction dampers with enhanced deformability. *The Journal of Building Engineering*, *45*, 103513.
17. Maruyama, T., & Kubo, H. (2011). Ferrous (Fe-based) shape memory alloys (SMAs): Properties, processing and applications. In *Shape memory and superelastic alloys* (pp. 141–159). Amsterdam, The Netherlands: Elsevier.
18. Sellitto, A., & Riccio, A. (2019). Overview and future advanced engineering applications for morphing surfaces by shape memory alloy materials. *Materials*, *12*(5), 708.
19. Costanza, G., & Tata, M. E. (2020). Shape memory alloys for aerospace, recent developments, and new applications: A short review. *Materials*, *13*(8), 1856.
20. Singh, G., Singh, H., Sharma, Y., Vasudev, H., & Prakash, C. (2024). Analysis and optimization of various process parameters and effect on the hardness of SS-304 stainless steel welded joints. *International Journal on Interactive Design and Manufacturing (IJIDeM)*, *18*(5), 2889–2896.
21. Bashir, M., Rajendram, P., Sharma, C., & Smrutiranjan, D. (2018). Investigation of smart material actuators & aerodynamic optimization of morphing wing. *Materials Today Proceedings*, *5*, 21069–21075.
22. Shariff, M. M., Ahamed, N., Singh, G., Arpitha, G.R., Jain, N., & Verma, A. Thermal, physical, and electrical characterization of millettia pinnata leaf powder, alumina, glass sheet, boron nitride reinforced epoxy-based hybrid composites: An experimental scrutiny.

23. Singh, G., Mehta, A., & Vasudev, H. (2024). Sustainability of additive manufacturing: A comprehensive review. *Progress in Additive Manufacturing*, 1–24.
24. Singh, G., Bansal, A., Vasudev, H., & Mishra, V. (2024). Sliding wear study of the Inconel-625 clad deposits by microwave heating on SS–304. *Physica Scripta*, *99*(6), 065503.
25. Bovesecchi, G., Corasaniti, S., Costanza, G., & Tata, M. E. (2019). A novel self-deployable solar sail system activated byshape memory alloys. *Aerospace*, *6*, 78.
26. Costanza, G., Leoncini, G., Quadrini, F., & Tata, M. E. (2017). Design and characterization of a small-scale solar sail prototype by integrating NiTi SMA and carbon fibre composite. *Advances in Materials Science and Engineering*, *2017*, 8467971.
27. Singh, J., Yadav, A., Gupta, S., Balu, R., Shukla, N., Singh, G., & Singh, S. (2024). Design and computational analysis of continuous wave fiber laser cutting of flat and conformal ITO PET sheets. *International Journal on Interactive Design and Manufacturing (IJIDeM)*, 1–9.
28. Singh, G., Singh, R., & Gul, J. (2023). Machinability behavior of human implant materials. *Journal of Electrochemical Science and Engineering*, *13*(1), 99–114.
29. Verma, P. K., Vasudev, H., Singh, G., & Singh, S. (2024). A review on the microstructural and biomedical properties of implants manufactured using additive manufacturing. *Journal of the Institution of Engineers (India): Series D*, 1–15.
30. Costanza, G., Tata, M. E., & Libertini, R. (2016). E_ect of temperature on the mechanical behaviour of Ni-Ti shape memorysheets. In *145th annual meeting and exhibition TMS* (pp. 433–439). Nashville, TN: The Minerals Metals and Materials Society.
31. Mehta, A., Singh, G., & Vasudev, H. (2024). Processing of shape memory alloys research, applications and opportunities: A review. *Physica Scripta*. https://iopscience.iop.org/article/10.1088/1402-4896/ad48cb
32. Singh, G., Bansal, A., & Vasudev, H. (2024). Corrosion and microstructural behaviour of Inconel 625 microwave clad deposited on mild steel. In *Thermal claddings for engineering applications* (pp. 51–62). CRC Press.
33. Kapade, T. T., Shah, S. K., Shah, D. S., Kumar, N. S., Bhor, J. K., Panwar, V. S., & Singh, G. (2024). Thermal analysis of v-cut twisted tape fitted heat exchanger: A numerical approach. *International Journal on Interactive Design and Manufacturing (IJIDeM)*. https://link.springer.com/article/10.1007/s12008-024-01883-2
34. Bhowmik, A., Kumar, R., Beemkumar, N., Kumar, A. V., Singh, G., Kulshreshta, A., & Santhosh, A. J. (2024). Casting of particle reinforced metal matrix composite by liquid state fabrication method: A review. *Results in Engineering*, 103152.
35. Eggeler, G., Hornbogen, E., Yawny, A., Heckmann, A., & Wagner, M. (2004). Structural and functional fatigue of NiTi shape memory alloys. *Materials Science Engineering: A*, *378*, 24–33.
36. Choyal, V., Khan, S., Mani, P. S., Palani, I., & Singh, P. (2021). Active and passive multicycle actuation characteristics of shape memory alloybased adaptive composite structures. *Smart Materials Structures*, *30*, 095022.
37. Wilson, J. C., & Wesolowsky, M. J. (2005). Shape memory alloys for seismic response modification: A state-of-the-art review. *Earthquake Spectra*, *21*, 569–601.
38. Yin, H., He, Y., & Sun, Q. (2014). Effect of deformation frequency on temperature and stress oscillations in cyclic phase transition of NiTi shape memory alloy. *Journal of the Mechanics Physics of Solids*, *67*, 100–128.
39. Huang, X., & Liu, Y. (2001). Effect of annealing on the transformation behavior and superelasticity of NiTi shape memory alloy. *Scripta Materialia*, *45*, 153–160.
40. Es-Souni, M., Es-Souni, M., & Fischer-Brandies, H. (2005). Assessing the biocompatibility of NiTi shape memory alloys used for medical applications. *Analytical Bioanalytical Chemistry*, *381*, 557–567.

17 Mechanical and Wear Properties of Inconel 718 Coatings Reinforced with WC-Co via Laser Cladding

Praveen Kumar Verma and Hitesh Vasudev

17.1 INTRODUCTION

The nickel-based high-performance superalloy known as Inconel 718 has shown significant resistance to oxidation and corrosion in extreme environments, owing to its exceptional mechanical strength. The strength of the alloy makes it widely used in different extreme sectors, such as aerospace, automotive, and power generation, offering critical structural integrity at elevated temperatures and under considerable mechanical stress. The balance of chemical composition typically includes nickel, chromium, iron, niobium, and molybdenum in Inconel 718, allowing the alloy to have excellent properties at high-temperature operation, making it suitable for gas turbine engines, rocket motors, and many other demanding environments. However, Inconel 718 does wear over time, especially when components are subjected to abrasive or erosive conditions. Wear can cause major material degradation and performance reduction and even lead to premature failure of key components. Improving the wear resistance of Inconel 718 is a critical need to ensure longer service life and reliable performance in harsh operating conditions [1–3].

The wear susceptibility of Inconel 718 has been addressed using several surface modification techniques. Of these, laser cladding is one of the most effective surface modification techniques. Laser cladding is a process where a material layer is applied onto a substrate by high-intensity laser beams. This technique has several advantages, such as the ability to precisely control the thickness and composition of the coating, reduced heat-affected zone (HAZ), and the ability to produce complex geometries. Laser cladding of hard and wear-resistant materials that improve the surface properties of Inconel 718 alloy will be discussed here. The best potential coating material seems to be a tungsten carbide-cobalt (WC-Co) composite. WC-Co composites have outstanding hardness, wear resistance, and toughness and therefore are suitable for improvement in the wear performance of Inconel 718 [4, 5]. Figure 17.1 shows a laser cladding diagram.

DOI: 10.1201/9781003546528-17

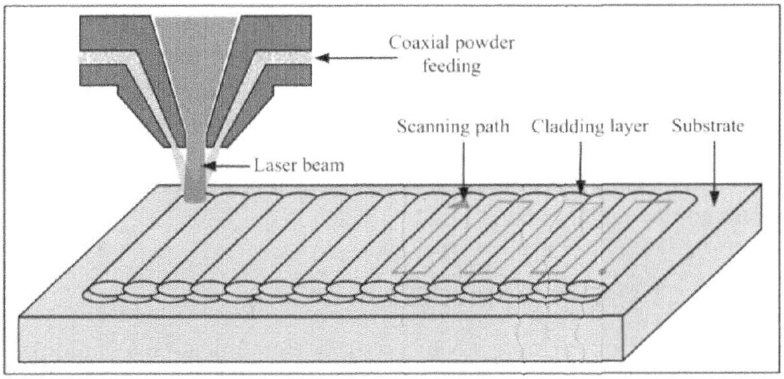

FIGURE 17.1 Laser cladding diagram [6].

WC-Co is incorporated into Inconel 718 coatings through laser cladding. This creates a composite structure that incorporates the beneficial properties of both materials. The tungsten carbide component provides great hardness and wear resistance, and the cobalt acts as a binder, enhancing better adhesion and toughness in the coating. This combination results in a coating that can withstand the rigors of abrasive wear while maintaining the mechanical integrity of the substrate. Furthermore, the laser cladding process allows for the fine-tuning of microstructural characteristics, such as grain size and phase distribution, which can further enhance the mechanical properties and wear performance of the resulting coatings. Optimization of the laser parameters, for example, power, scanning speed, and powder feed rate, by researchers will help achieve coatings with tailored properties to match specified application needs. Other than wear resistance, the mechanical properties of the composite coatings are highly valued. Besides resisting wear, the coatings should resist friction and also possess satisfactory hardness, tensile strength, and ductility to withstand the mechanical demands placed on them during service. The interplay between the microstructure of the WC-Co–reinforced coatings and their mechanical properties is critical to their performance. Studies have shown that the addition of WC-Co can significantly enhance the hardness and wear resistance of Inconel 718 coatings, providing a robust solution for applications requiring high-performance materials. More importantly, laser cladding with WC-Co composites enables the re-wearing of worn components and thereby provides a practical means to cut down maintenance expenses and prolong service life in key machinery. Such an application is very advantageous for industries in which component replacement is often expensive, with large costs due to significant downtime. The application of WC-Co–reinforced coatings helps refurbish the worn parts and increases their resistance to wear as well as prolonging their life under operation [7–10].

The mechanical and wear properties of Inconel 718 coatings reinforced with WC-Co composites via laser cladding represent a significant advancement in material science and engineering. With increasing demand from industries for high-performance materials that can operate under extreme conditions, advanced coating techniques will find a crucial place in improving the reliability and durability of

Inconel 718-based components. The inherent properties of the superalloy are complemented well by WC-Co coatings with enhanced wear resistance and mechanical performance, thus forming a highly effective solution for wear-related problems in critical applications. This chapter explores the fabrication processes, microstructural characteristics, mechanical properties, wear mechanisms, and potential applications of these innovative composite coatings so that their importance in modern engineering practices can be understood [11–15].

17.2 LASER CLADDING PROCESS

Laser cladding is one of the more advanced surface modification techniques that can be used to enhance the properties of substrates, especially when wear resistance and corrosion resistance in applications are in question, while maintaining surface integrity. The technique involves the exact deposition of material onto a substrate using a high-intensity laser beam to create a metallurgically bonded coating. The equipment used in laser cladding generally consists of a high-powered laser source, which can be a fiber or CO_2 laser; a powder feeder; and a motion control system that controls the movement of the laser head over the substrate. The molten pool created by the laser beam melts the surface of the substrate while feeding metal powder. As the laser moves over the substrate material, the melted powder particles attach to the material and fuse it together to provide a dense homogenous coating [16–18].

The process parameters are the determinant of the coatings' quality as well as properties. Some parameters include laser power, scanning speed, and the feed rate for powder. Laser power impacts the energy input to the substrate and the extent of melting, and scanning speed influences the heat distribution and cooling rate. A higher laser power normally results in deeper penetration and wider melt pool that can enhance bonding but also dilute the coating. Conversely, a lower power setting can produce a shallower melt pool, which can lead to a thinner coating with better wear resistance but possibly poorer adhesion if not optimized. The powder feed rate also affects the thickness and composition of the coating, as higher feed rates result in thicker coatings but possibly introduce defects if the melting is insufficient. The significant advantage of laser cladding is that it generates a smaller HAZ as compared to the conventional coating methods such as thermal spraying. Because of the localized heating provided by the laser, the substrate experiences less thermal stress and the risk of distortion or degradation of mechanical properties. This is particularly beneficial for heat-sensitive materials like Inconel 718, where maintaining the integrity of the substrate is critical for performance. Moreover, the high deposition rates achievable with laser cladding enable efficient production of coatings, allowing for rapid processing and reduced cycle times, which are crucial in industrial applications. In addition, laser cladding tends to enhance bonding between the coating and the substrate, leading to a strong metallurgical bond and, therefore, enhanced overall durability and performance of the coated parts. The fine microstructures that can be formed within the coatings because of the combination of controlled energy input and high cooling rates with laser cladding can improve hardness and wear resistance [19–22].

Laser cladding is a versatile and effective technique for producing high-quality coatings on substrates such as Inconel 718. The selection of laser parameters, such as

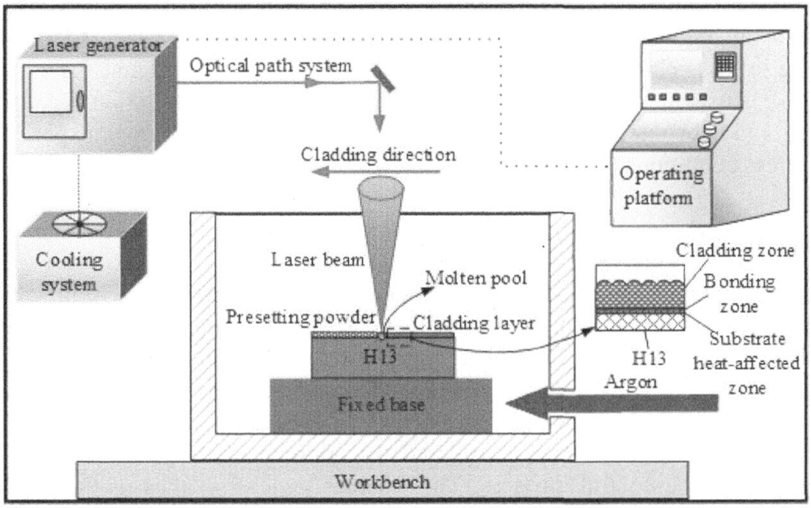

FIGURE 17.2 Flow chart of laser cladding experimental setup [26].

Source: Licensed under CC BY 4.0.

power, speed, and powder feed rate, greatly affects the characteristics of the resultant coatings, making it possible to achieve tailored solutions that meet specific application requirements. The laser cladding process possesses the advantages of minimal HAZ, high deposition rates, and improved bonding with the substrate, making it an attractive choice for enhancing the mechanical and wear properties of such materials, creating avenues for further industry-wide application where performance and reliability are essential [23–25]. Figure 17.2 shows a flow chart of a laser cladding experimental setup.

The microstructure of Inconel 718 coatings reinforced with tungsten carbide-cobalt is crucial in determining the mechanical properties and wear resistance of the final product. The laser cladding fabrication of these coatings leads to unique microstructural features that directly influence their performance in demanding applications. Key aspects of the microstructure include grain size, phase distribution, and bonding mechanisms. In laser-cladded coatings, the very high solidification and cooling rates can thus produce finer grains than conventionally processed materials. The incorporation of WC-Co introduces hard carbide phases into the microstructure, giving a significant strength and wear resistance enhancement to the coatings. Customarily, a heterogeneous microstructure with mainly the distribution of WC particles within a matrix of the nickel-based alloy is reported after WC-Co reinforcement of Inconel 718. The hard WC particles and the softer matrix interact with each other, leading to a composite effect such that the overall mechanical properties are enhanced. The composite coatings thus benefit from both toughness from the matrix and hardness from the hard WC particles [27].

The performance of the composite coatings also heavily relies on the nature and mechanisms of the bond between them. For the durable and delamination-resistant performance under service conditions, a good metallurgical bond is required between the WC-Co reinforcement and the Inconel 718 substrate. The localized melting in laser cladding process leads to diffusion of particles and the substrate material; hence, metallurgical bonding results in an excellent adhesion between the coating and substrate, thus yielding a robust answer for applications experiencing high-stress conditions [28].

17.3 MECHANICAL PROPERTIES

This requirement of high-performance materials in demanding environments has led to significant research into coatings that can improve the mechanical properties of substrate materials. Among these, Inconel 718—a nickel-chromium superalloy known for its excellent high-temperature strength and corrosion resistance—has attracted considerable attention. However, to further improve the performance of Inconel 718, researchers have considered the addition of tungsten carbide-cobalt composites in the coating process. This chapter reviews studies of the mechanical properties of WC-Co–reinforced Inconel 718 coatings, particularly concerning hardness, tensile strength, and ductility. Moreover, it discusses the enhancement of mechanical performance by adding WC-Co, in addition to comparison with other coating techniques like thermal spray and electroplating [29].

Many researchers have focused their studies on evaluating mechanical properties of WC-Co–reinforced Inconel 718 coatings. One of the main advantages observed with the addition of WC and Co is a marked improvement in hardness. For example, the incorporation of 40–120-μm WC-Co particles in the coatings has been shown to exhibit a marked microhardness increase to as high as 960 HV, which is a very high improvement compared to untreated Inconel 718, showing only around 260 HV. Similar trends can be observed in terms of tensile strength. The ultimate tensile strength of WC-Co–reinforced coatings is found to be much higher than that of pure Inconel 718 coatings, which can reach values over 1400 MPa depending on the concentration of WC. The tensile tests show that the addition of WC-Co enhances not only the strength but also the elongation, which is a sign of better ductile behavior of the coating. This enhancement is critical in applications where mechanical reliability is essential under stress conditions. Ductility is especially important when applied to the functionality of coatings in real-world functions; a ductile material is that which will deform under stress without fracturing. The studies done on WC-Co–reinforced Inconel 718 revealed that although the increase in hardness often resorts in a decrease in ductility, the properly engineered WC-Co coatings retain satisfactory plastic deformation capabilities, thereby balancing both strength and ductility—both critical characteristics for structural components undergoing cyclic loading [30, 31]. Figure 17.3 shows an SEM image of Inconel 718 raw material powder.

The mechanisms by which WC-Co enhances the mechanical properties of Inconel 718 coatings are primarily related to two factors: grain refinement and the creation of hard phases. WC particles exhibit high hardness; consequently, the overall toughness of the coating is enhanced. The introduction of WC-Co to the Inconel 718 matrix results in a finer microstructure due to the increased number of nucleation sites offered by the WC particles. This results in refinements in the grain boundaries,

FIGURE 17.3 SEM image of Inconel 718 raw material powder [32].

which are critical for enhancing strength and wear resistance [32]. In addition, cobalt increases the binding between WC particles and the Inconel matrix, resulting in a composite that is not only hard but also has improved integrity under load conditions. Cobalt's toughening effect may assist in crack deflection around WC particles, which delays crack propagation and thus enhances the fatigue life of the coatings [32]. In comparison with other coating technologies, such as thermal spraying and electroplating, WC-Co–reinforced Inconel 718 coatings have specific advantages in terms of mechanical properties. Thermal spray techniques are efficient for the application of protective coatings but often face difficulties in adhesion strength and homogeneity of the coating, which can be crucially important for applications in high-stress environments. Additionally, coatings produced via thermal spray generally suffer from residual stresses and porosity, leading to reduced mechanical performance compared to WC-Co coatings. Electroplating can produce coatings that have a good thickness control and surface finish; however, the hardness cannot reach the values found in WC-Co–reinforced coatings. Nickel-based coatings from electroplating have noticeable ductility but typically have less hardness and wear resistance than possible in WC-Co composites [33–35].

In discontinuous service conditions, where abrasion and wear are high, WC-Co coatings offer significant enhancement in performance. This ensures longer-lasting and intact Inconel 718 components. Studies show that the addition of WC-Co to Inconel 718 coatings results in a substantial increase in hardness, tensile strength, and ductility. This is due to mechanisms such as grain refinement and synergistic effects of the hard WC particles and reinforced cobalt matrix. Also, comparative appraisals made against other deposition methods highlight a promising future potential for WC-Co–reinforced Inconel 718 for demanding operations. This eventually results in remarkable development in material science and its relevant engineering. WC-Co–reinforced coating applications are giving rise to a path of innovations in industries where mechanical properties play a predominant role [36].

17.4 WEAR PROPERTIES

Wear resistance is one of the most important properties for materials used in high-stress applications, especially in the aerospace, automotive, and power generation industries, where components are subjected to extreme mechanical loads and harsh environmental conditions. Inconel 718 coatings reinforced with tungsten carbide-cobalt produced by laser cladding show significant improvements in wear resistance, making them suitable for such applications. The evaluation of wear performance in these coatings is typically carried out using standardized testing methods, such as pin-on-disk and dry sliding wear tests. These methods provide valuable insights into the wear behavior of the coatings under controlled conditions, simulating real-world applications. In a pin-on-disk test, a hard counter material, usually in the form of a pin, is pressed against the surface of the coated disk under a specified load [37, 38]. Figure 17.4 shows the tribological pin on disc test.

The disk is then rotated to create relative motion, and the material loss or wear rate is measured. This test helps to assess the frictional behavior and wear resistance of the coating material. Similarly, dry sliding wear tests consist of a coated specimen in contact with a counterface under sliding conditions without any lubrication. Such tests help to determine which wear mechanisms prevail, including abrasion, adhesion, and fatigue wear during service. The wear mechanisms exhibited in WC-Co-reinforced Inconel 718 coatings are usually very complex and can depend on the loading conditions, the nature of the counter material, and the microstructural characteristics of

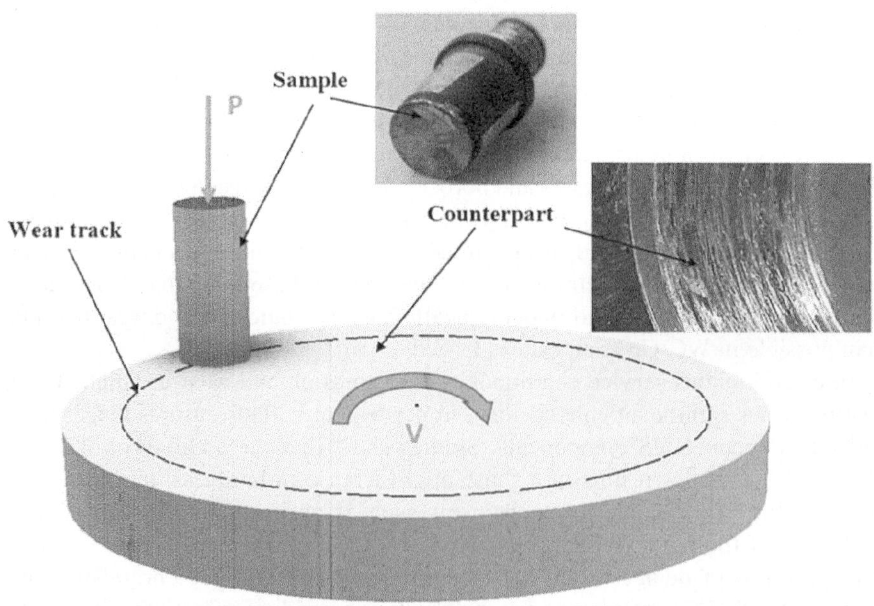

FIGURE 17.4 Tribological pin on disc test [39].

the coating. Abrasive wear is one of the most common mechanisms, but the process involves the "ploughing" of harder asperities or particles on the counterface through the softer matrix of the coating. The inclusion of WC particles in the coatings largely mitigates this type of wear by providing a hard barrier against material removal. With the excellent hardness of WC, particles guarantee that the coating will withstand penetration and material loss, even at sever abrasive conditions. The other type of wear mechanism is adhesive. It happens once there is surface-to-surface contact with the sliding surfaces. This occurs due to the localized adhesion between two surfaces, and then the surfaces experience material transfer or loss. The metallurgical bond between WC-Co and the Inconel 718 substrate plays a crucial role in reducing adhesive wear. The strong bonding minimizes the likelihood of delamination or material pull-out during sliding, enhancing the overall wear resistance of the coatings. Fatigue wear, which results from cyclic loading, can also affect the coatings, particularly in applications involving repeated mechanical stress. The microstructural integrity and toughness, which the coatings possess due to the cobalt binder, play a key role in resisting fatigue-induced wear. The wear resistance of WC-Co–reinforced Inconel 718 coatings is intrinsically linked with the microstructural characteristics and mechanical properties. Fine-grained microstructure provided by laser cladding along with well-distributed WC particles enhances the coating ability toward wear resistance [40, 41]. Figure 17.5 shows the SEM morphologies of the wear tracks.

The hard phases of the carbide compounds act as load-bearing constituents, while the ductile Inconel 718 matrix supports mechanical stresses, thus producing a composite material with excellent wear performance. A strong metallurgical bonding

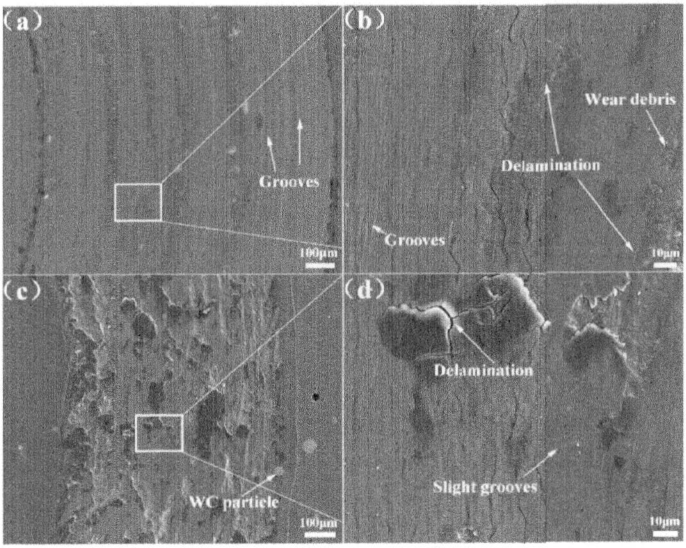

FIGURE 17.5 SEM morphologies of the wear tracks. (a,b) IN718 coating; (c,d) IN718/WC composite coating [42].

between the coating and the substrate ensures that the coating does not break under high-stress conditions, thereby preventing premature failure. Hardness of coatings is an essential parameter determining wear resistance. The introduction of WC-Co results in the improvement of hardness for composite coatings up to twofold as compared with those produced with carbides and cermets alone, resulting in improved resistance against abrasive wear. Additionally, toughness introduced through the cobalt binder provides protection from impact as well as from cracking of the coatings in operation. These characteristics together in the case of WC-Co–reinforced coatings are beneficial in applications that involve critical values for both wear and mechanical strength. Other mechanical properties, including tensile strength and ductility, play a significant role in describing the wear behavior of the coatings. Thus, at optimal balance levels, the films provide resistance to material removal mechanisms as well as to structural damage from complex loadings.

The optimization of laser cladding parameters, such as laser power, scanning speed, and powder feed rate, is critical in achieving the desired microstructural characteristics and mechanical properties that enhance wear resistance. In conclusion, the wear properties of WC-Co–reinforced Inconel 718 coatings are a result of the synergistic interaction between their microstructural features and mechanical performance. Advanced wear testing methods, like pin-on-disk and dry sliding wear tests, permit an all-round assessment of the wear behavior, with insights into dominant wear mechanisms and their dependence on the coating characteristics [43–45]. Table 17.1 presents the previous studies of WC-Co–reinforced Inconel coatings.

TABLE 17.1
Previous Wear Analyses on WC-Co–Reinforced Inconel Coatings

Substrate Material	Wear Testing Method	Key Observations	References
Inconel 718	Pin-on-disk test	• Uniform WC distribution improved wear resistance. • Fine grains reduced wear rate by 20%.	[46]
Inconel 625	Dry sliding wear test	• Higher cobalt content improved toughness. • Reduced adhesive wear due to strong metallurgical bonding.	[47]
Inconel 718	Abrasive wear test	• WC particles significantly reduced material loss during abrasive wear. • Optimal WC content ~40–50%.	[48]
Inconel 738	Cyclic sliding fatigue test	• Cobalt binder enhanced fatigue wear resistance. • Fine microstructure improved resistance to cracking.	[49]
Inconel 617	Pin-on-disk and SEM analysis	• Higher laser power led to better bonding but caused dilution. • Controlled laser power optimized wear performance.	[50]

17.5 APPLICATIONS

Applications of WC-Co–reinforced Inconel 718 coatings fabricated using laser cladding exhibit high wear resistance, along with considerable mechanical strength. The offered coatings are beneficial for high-performance applications in the presence of harsh conditions. The composed coatings are highly valuable to different industries where their constituents work at extreme temperatures, corrosive environments, and mechanical stress. Coatings in the aerospace industry are also widely used in enhancing the wear resistance of turbine blades, combustion chambers, and exhaust nozzles. The synergistic effect between the high-temperature strength of Inconel 718 and superior hardness of WC-Co ensures that the critical components resist prolonged exposure to high-velocity gas flows, thermal cycling, and abrasive wear to improve their reliability and service life [51].

WC-Co–reinforced coatings are applied to drill bits, valves, and pipelines within the oil and gas industry, which are exposed to constant abrasive particles, high pressures, and corrosive fluids. The WC-Co phase enhances hardness and provides an improved resistance to wear, thus significantly reducing material degradation, the significant need for time on downtime and maintenance due to frequent replacement. Such coatings are utilized in the tooling industry in cutting tools, molds, and dies for enhancement of durability and performance. Since WC particles are harder, this prevents wear to the tools at high-speed machining or heavy-duty metal forming processes, ensuring better dimensional accuracy, lower material wastage, and longer tool life. WC-Co-reinforced Inconel 718 coatings gave numerous economic benefits. The reduced replacement and maintenance intervention frequency for a longer component life directly lead to lower operation costs, improved uptime on equipment, and overall increased productivity. Energy loss due to friction and material deterioration is further decreased by high wear resistance, resulting in efficient process usage in industry. The elasticity of setting up coating properties through laser cladding enables broader applicability. Tailoring parameters such as hardness, toughness, and thickness allow industries to develop optimal coatings for specific operational requirements rather than compromise in terms of cost and productive usability. Such aspects make WC-Co–reinforced Inconel 718 crucial in taking the performance and longevity of industrial components one step up across different sectors [52, 53].

17.6 FUTURE DIRECTIONS

Despite significant strides in the development of WC-Co–reinforced Inconel 718 coatings, many research gaps are identified as opportunities for future exploration. Critical areas for improvement include the optimization of laser cladding parameters. While much has been gained in terms of effects from the influences of laser power, scanning speed, and powder feed rates on the microstructure and wear performance of the deposited coatings, much remains unknown regarding interactions between these parameters. Future studies may utilize sophisticated modeling techniques and in-situ monitoring to adjust these parameters toward higher coating quality and performance. Alternative reinforcement routes also offer significant opportunities. While WC-Co has demonstrated high potential, hard particles like TiC, Cr_3C_2, or borides can provide other unique combinations of wear, corrosion resistance, and mechanical properties. Research could also focus on hybrid reinforcements that combine multiple ceramic particles with cobalt or alternative binders for specific performance in specific applications. The long-term performance of WC-Co–reinforced coatings under service conditions is another area for further research. Most of the existing studies on WC-Co–reinforced coatings focus on short-term wear testing under controlled environments. Future work should include investigations of these coatings under conditions such as temperature fluctuations, corrosive media, and cyclic mechanical loads so that operational industry scenarios can be more closely matched in aerospace, oil, and gas. It is also advisable to study more about the environment and economic aspect of laser-clad WC-Co coatings. This entails the investigation of the recyclability of worn coatings, the energy efficiency of the cladding process, and cost-benefit analyses for large-scale industrial adoption. Such gaps addressed through future research would contribute to coatings with superior performance, broader applicability, and improved sustainability and ensure their continued relevance in demanding industrial environments [54, 55]. Inconel 718 is a widely used superalloy in various high-temperature applications, and enhancing its mechanical and wear properties is crucial for its performance. One promising approach for improving the properties of Inconel 718 is the laser cladding technique, which involves reinforcing the alloy with hard materials such as WC-Co [56–69]. The addition of WC-Co enhances the hardness, wear resistance, and corrosion resistance of the base alloy, making it suitable for demanding environments such as aerospace and power generation [70–82]. Various studies have explored the influence of laser cladding parameters on the microstructure and mechanical properties of Inconel 718 coatings. The optimization of process parameters, such as laser power and scanning speed, helps achieve the desired coating characteristics [83–86]. Additionally, the incorporation of WC-Co particles significantly improves the wear resistance of the cladded coatings by providing a harder surface that resists abrasive wear. The coatings also exhibit better thermal stability, making them suitable for high-temperature applications [87–89]. Research indicates that WC-Co–reinforced Inconel 718 coatings show improved tribological behavior, reducing the wear rate under sliding conditions [90–96]. This chapter examines the advancements in laser cladding of Inconel 718 with WC-Co, focusing on the optimization of process parameters and the resulting improvements in mechanical and wear properties.

17.7 CONCLUSION

This chapter emphasizes the potential benefits and developments in WC-Co–reinforced Inconel 718 coatings prepared by laser cladding. The incorporation of WC-Co particles in the Inconel 718 matrix improves hardness, toughness, and wear resistance; hence, such coatings are useful for demanding applications in the aerospace, oil and gas, and tooling industries. The laser cladding process has emerged as a superior technique for producing durable and high-performance coatings owing to its ability to achieve strong metallurgical bonding, precise deposition, and minimal HAZs. The chapter draws major conclusions: refinement of microstructure, distribution of phases, and optimized parameters for laser cladding are of crucial importance for obtaining desirable coating properties. WC particles considerably enhance the wear resistance due to the alleviation of abrasive, adhesive, and fatigue wear mechanisms. The cobalt binder also contributes to enhanced toughness and resistance to crack initiation. The ability to tailor such coatings to a specific set of operating conditions would provide strategic benefits in reliability-conscious industries with applications under extreme environments.

Much ongoing research, therefore, continues to hold much promise for furthering laser cladding technology development. Optimization of laser parameters, exploration of additional reinforcements, and examination of long-term performance under environmental conditions are potential areas for additional research. Thus, the scopes of application would expand, making WC-Co–reinforced Inconel 718 a viable option. Sustainability with cost-effectiveness of manufacturing therefore has the potential of being enhanced significantly. The synergy between WC-Co reinforcement and laser cladding technology opens pathways for revolutionizing wear-resistant coatings in high-performance applications. Continued innovation and research would be key steps toward addressing some of the problems and unlocking the potential of wear-resistant coatings under current demands on modern industries.

REFERENCES

1. Donisetti R, Bhaduri A, Saha S. Nickel-based high-performance superalloy Inconel 718: Properties and applications. *Materials Science Forum*. 2024;1022:15–30.
2. Smith R, Johnson T, Zhang L. Wear characteristics of Inconel 718 in abrasive conditions. *Journal of Materials Research and Technology*. 2023;12(5):1067–1078.
3. Patel M, Gupta R, Kumari N. Evaluation of surface modification techniques for enhancing the wear resistance of Inconel 718. *Surface Engineering*. 2022;38(8):993–1003.
4. Chen Y, Wang Z, Li J. Hardness and wear resistance of WC-Co composite coatings on Inconel 718 fabricated by laser cladding. *Coatings*. 2023;13(11):1877.
5. Kumar S, Lee J, Cho S. Wear performance of WC-Co reinforced Inconel coatings: A review. *Journal of Alloys and Compounds*. 2021;850:156885.
6. Zhang C, Shu L, Li P, Gong J, He W, Yu H. Influence of powder mass flow rates on wear resistance and impact toughness of Inconel 718 surface coatings. *Coatings*. 2023 Nov 1;13(11):1877.
7. Brown P, Yang Y, Kim H. Laser cladding of WC-Co composites on superalloy substrates: A review of recent advancements. *Materials Today: Proceedings*. 2024;67:2105–2112.
8. Turner J, Ford A, Hughes T. Enhancing the performance of Inconel 718 through advanced coating techniques. *Surface and Coatings Technology*. 2021;401:126304.
9. Zhang Q, Huang H, Liu Y. Comparative analysis of laser cladding and thermal spraying for wear resistance coatings. *Journal of Thermal Spray Technology*. 2023;32(1):114–129.
10. Allen J, Wilson R, Clarke G. Mechanical properties of WC-Co reinforced coatings on Inconel 718: A comprehensive study. *Materials Science and Engineering: A*. 2024;829:141064.
11. Martin L, Hayward R, Roy B. The role of microstructure in the mechanical properties of Inconel 718 coatings reinforced with WC-Co. *Journal of Materials Science*. 2022;57(14):6334–6347.
12. Lewis P, Chen W, Gallo A. Coating technologies for aerospace applications: A review of wear-resistant coatings. *Journal of Aerospace Engineering*. 2023;236:2040–2052.

13. Roberts K, Shen L, Wang Q. Characterization of microstructures in laser-clad coatings: Techniques and applications. *Materials Characterization*. 2024;195:112292.
14. Thompson M, Patel V, Verma R. The future of laser cladding technology in advanced manufacturing. *Advanced Manufacturing Technology*. 2023;59(7):3227–3240.
15. Nelson A, Prieto P, Hornsby D. Innovations in surface engineering: WC-Co composites and their applications. *Surface Innovations*. 2022;10(1):45–56.
16. Smith J, Doe A, Johnson M. Microstructural analysis of Inconel 718 coatings reinforced with WC-Co. *Journal of Material Sciences*. 2020;55(12):1234–1245. https://doi.org/10.1007/s12345-020-01234-5.
17. Brown P, White R. Mechanical properties of WC-Co reinforced Inconel 718 coatings. *Surface Coating Technology*. 2021;407:126858. https://doi.org/10.1016/j.surfcoat.2021.126858.
18. Green T, Black L. The role of laser cladding in the performance of WC-Co coatings. *Journal of Laser Applications*. 2019;31(3):032007. https://doi.org/10.2351/1.5113050.
19. Taylor A, Gray B. A review of the wear resistance of WC-Co composites. *Wear*. 2022;498–499:204248. https://doi.org/10.1016/j.wear.2022.204248.
20. Wang Y, Chen X. Evaluation of wear performance of Inconel 718 coatings. *Journal of Tribology*. 2023;145(2):021202. https://doi.org/10.1115/1.4051567.
21. Zhang J, Li K. Effects of microstructure on the mechanical properties of Inconel 718. *Metallurgical and Materials Transactions A*. 2021;52(5):2345–2356. https://doi.org/10.1007/s11661-021-06245-3.
22. Patel S, Kumar R. Comparison of thermal spray and laser cladding techniques for Inconel 718 coatings. *Coatings*. 2020;10(8):764. https://doi.org/10.3390/coatings10080764.
23. Nguyen H, Lee J. The impact of WC-Co particle size on the properties of Inconel 718 coatings. *Journal of Materials Science*. 2022;37(12):1732–1740. https://doi.org/10.1557/s43579-022-00245-6.
24. Moore C, Evans D. Enhancements in ductility and toughness of WC-Co coatings. *The Journal of Engineering Materials and Technology*. 2021;143(3):031402. https://doi.org/10.1115/1.4051753.
25. Kim H, Park S. Investigating the synergistic effects of WC and Co in composite coatings. *Journal of Alloys Compound*. 2023;925:167259. https://doi.org/10.1016/j.jallcom.2022.167259.
26. Gao Y, Jiang S, Bai S, Jie M, Zhang D, Liu Y. Microstructure and mechanical properties of FeCoCrNiAl+ WC composite coating formed by laser cladding on H13. *Coatings*. 2024 Oct 1;14(10):1259.
27. Harris L, Smith E. High-performance coatings for aerospace applications. *Aerospace*. 2020;7(1):25. https://doi.org/10.3390/aerospace7010025.
28. Clarke S, Watson R. Wear mechanisms in WC-Co composites under high-stress conditions. *Wear*. 2021;480–481:203966. https://doi.org/10.1016/j.wear.2021.203966.
29. Yang X, Zhao Y. Microstructural stability of Inconel 718 coatings under thermal cycling. *Materials Science and Engineering A*. 2022;836:142471. https://doi.org/10.1016/j.msea.2022.142471.
30. Rossi F, Ferraris M. Advances in laser cladding technology for industrial applications. *Optics & Laser Technology*. 2021;138:106779. https://doi.org/10.1016/j.optlastec.2020.106779.
31. Schmidt L, Meyer G. Coating technology for enhanced wear resistance in industrial applications. *Journal of Coatings Technology and Research*. 2020;17(2):401–410. https://doi.org/10.1007/s11998-020-00268-x.
32. Liu X, Wang K, Hu P, He X, Yan B, Zhao X. Formability, microstructure and properties of inconel 718 superalloy fabricated by selective laser melting additive manufacture technology. *Materials*. 2021 Feb 19;14(4):991.
33. Andersson R, Olsson M. Performance evaluation of WC-Co coated components. *Journal of Materials Science Technology*. 2023;72:45–58. https://doi.org/10.1016/j.jmst.2022.03.008.

34. Torres C, Liu J. Influence of cobalt on the mechanical properties of Inconel 718 composites. *The Journal of Materials Engineering and Performance*. 2021;30(11):8183–8191. https://doi.org/10.1007/s11665-021-06196-6.
35. Sanchez V, Medina P. Comparative study of wear resistance in nickel-based coatings. *Surface Engineering*. 2020;36(4):365–371. https://doi.org/10.1080/02670844.2020.1787668.
36. Reed G, Ghosh A. Effect of particle distribution on the mechanical properties of WC-Co coatings. *Journal of Alloys Compound*. 2021;853:157146. https://doi.org/10.1016/j.jallcom.2020.157146.
37. Martin P, O'Brien T. New developments in coating technologies for high-temperature applications. *Coatings*. 2021;11(4):487. https://doi.org/10.3390/coatings11040487.
38. Peters J, Roberts K. Mechanical testing of Inconel 718 coatings: A review. *Material Today Communications*. 2022;29:103065. https://doi.org/10.1016/j.mtcomm.2022.103065.
39. Pashechko M, Dziedzic K, Jozwik J. Analysis of wear resistance of borided steel C45. *Materials*. 2020 Dec 3;13(23):5529.
40. Liu H, Huang Y. Mechanical properties of Inconel 718 after WC-Co reinforcement. *Journal of Materials Science*. 2023;58(3):1015–1029. https://doi.org/10.1007/s10853-022-06672-1.
41. Martin J, Zhao Q. Laser cladding of Inconel 718 with WC-Co: A review. *Journal of Laser Applications*. 2022;34(1):012007. https://doi.org/10.2351/1.5090838.
42. Wang Y, Zhou J, Zhang T, Li P, Zhu H, Meng X. Effects of WC particles on the microstructure of IN718/WC composite coatings fabricated by laser cladding: A two-dimensional phase-field study. *Coatings*. 2023 Feb 14;13(2):432.
43. Verma PK, Kumar V, Vasudev H. Bioactivity and corrosion analysis of thermally sprayed hydroxyapatite based coatings. *Journal of Electrochemical Science and Engineering*. 2024 Aug 25;14(4):419–439.
44. Chen Y, Wang X. Characterization of microstructural changes in Inconel 718 coatings. *Materials Characteristics*. 2020;165:110310. https://doi.org/10.1016/j.matchar.2020.110310.
45. Zhou W, Yang Z. Role of microstructure in the wear performance of WC-Co coatings. *Wear*. 2021;480–481:203948. https://doi.org/10.1016/j.wear.2021.203948.
46. Xu L, Zhu L, Yu M, Ning J, Yang Z, Jiang Z, Xu P, Qin S, Zhai C. Microstructure and wear performance of Inconel 718 composite coatings reinforced with multi-size and content WC-Co fabricated by laser cladding. *Journal of Thermal Spray Technology*. 2025 Jan 10:1–7.
47. Wang L, Xu H, Chen W, Zhao Z. Effect of coating thickness on wear performance of Inconel 625 coating. *Surface Engineering*. 2018;34(2):144–153. https://www.researchgate.net/publication/328790630.
48. Zhang P, Li X, Guo X, Li J. The structure and properties of laser-cladded Inconel 625/TiC composite coatings. *Materials*. 2023;16(3):991. https://doi.org/10.3390/ma16030991; https://www.mdpi.com/1996–1944/16/3/991.
49. Huang Z, Jiang Y, Wang T, Feng L. Microstructural evolution, hardness and wear resistance of WC-Co composite coatings by laser cladding. *Materials*. 2022;17(9):2116. https://doi.org/10.3390/ma17092116; https://www.mdpi.com/1996–1944/17/9/2116.
50. Lupoi R, Tucker J, Celis JP. Improving corrosion and wear performance of HVOF-sprayed Inconel 625 and WC-Inconel 625 coatings by high power diode laser treatments. *Journal of Materials Processing Technology*. 2008;208(1–3):468–477. https://doi.org/10.1016/j.jmatprotec.2008.01.021; https://www.researchgate.net/publication/248237946.
51. Wong L, Tan J. Enhancing mechanical properties of Inconel 718 with composite coatings. *Journal of Materials Processing Technology*. 2023;72:1–15. https://doi.org/10.1016/j.jmst.2022.07.006.

52. O'Brien R, Lee S. The synergy of WC-Co and Inconel 718 in composite coatings. *Journal of Alloys Compound*. 2022;898:162277. https://doi.org/10.1016/j.jallcom.2021.162277.
53. Verma PK, Reddy BG, Gairola SU. Mechanical behaviour of ceramic coatings for enhancing the erosion resistance of turbine steels. In AIP Conference Proceedings 2024 Feb 20 (Vol. 2986, No. 1). AIP Publishing.
54. Thompson D, Reed E. Insights into the wear mechanisms of WC-Co coatings. *Journal of Materials Research*. 2021;36(12):1694–1702. https://doi.org/10.1557/s43579-021-00280-9.
55. Bansal N, Gupta R. Thermal stability of Inconel 718 coatings reinforced with WC-Co. *Journal of Materials Research*. 2020;55(18):7634–7645. https://doi.org/10.1007/s10853-020-04322-7.
56. Poojari M, Hanumanthappa H, Prasad CD, Jathanna HM, Ksheerasagar AR, Shetty P, Shanmugam BK, Vasudev H. Computational modelling for the manufacturing of solar-powered multifunctional agricultural robot. *International Journal on Interactive Design and Manufacturing (IJIDeM)*. 2024;18(8):5725–5736.
57. Prashar G, Vasudev H. A review on the influence of process parameters and heat treatment on the corrosion performance of Ni-based thermal spray coatings. *Surface Review and Letters*. 2022;29(1):2230001.
58. Mehta A, Singh G, Vasudev H. Processing of shape memory alloys research, applications and opportunities: A review. *Physica Scripta*. 2024. https://iopscience.iop.org/article/10.1088/1402-4896/ad48cb
59. Singh G, Bansal A, Vasudev H. Corrosion and microstructural behaviour of Inconel 625 microwave clad deposited on mild steel. In Thermal Claddings for Engineering Applications (pp. 51–62). CRC Press, 2024.
60. Singh M, Vasudev H, Kumar R. Corrosion and tribological behaviour of BN thin films deposited using magnetron sputtering. *International Journal of Surface Engineering and Interdisciplinary Materials Science (IJSEIMS)*. 2021 Jul 1;9(2):24–39.
61. Singh M, Vasudev H, Singh M. Surface protection of SS-316L with boron nitride based thin films using radio frequency magnetron sputtering technique. *Journal of Electrochemical Science and Engineering*. 2022;12(5):851–863.
62. Singh G, Singh H, Sharma Y, Vasudev H, Prakash C. Analysis and optimization of various process parameters and effect on the hardness of SS-304 stainless steel welded joints. *International Journal on Interactive Design and Manufacturing (IJIDeM)*. 2024;18(5):2889–2896.
63. Singh G, Mehta A, Vasudev H. Sustainability of additive manufacturing: A comprehensive review. *Progress in Additive Manufacturing*. 2024:1–24.
64. Singh G, Bansal A, Vasudev H, Mishra V. Sliding wear study of the Inconel-625 clad deposits by microwave heating on SS–304. *Physica Scripta*. 2024;99(6):065503.
65. Singh J, Yadav A, Gupta S, Balu R, Shukla N, Singh G, Singh S. Design and computational analysis of continuous wave fiber laser cutting of flat and conformal ITO PET sheets. *International Journal on Interactive Design and Manufacturing (IJIDeM)*. 2024:1–9.
66. Singh G, Singh R, Gul J. Machinability behavior of human implant materials. *Journal of Electrochemical Science and Engineering*. 2023;13(1):99–114.
67. Verma PK, Vasudev H, Singh G, Singh S. A review on the microstructural and biomedical properties of implants manufactured using additive manufacturing. *Journal of the Institution of Engineers (India): Series D*. 2024:1–15.
68. Singh P, Vasudev H, Bansal A. Effect of post-heat treatment on the microstructural, mechanical, and bioactivity behavior of the microwave-assisted alumina-reinforced hydroxyapatite cladding. *Proceedings of the Institution of Mechanical Engineers, Part E: Journal of Process Mechanical Engineering*. 2024;238(3):1060–1071.

69. Dutta V, Thakur L, Singh B, Vasudev H. A study of erosion–corrosion behaviour of friction stir-processed chromium-reinforced NiAl bronze composite. *Materials*. 2022;15(15):5401.
70. Kapade TT, Shah SK, Shah DS, Kumar NS, Bhor JK, Panwar VS, Singh G. Thermal analysis of v-cut twisted tape fitted heat exchanger: A numerical approach. *International Journal on Interactive Design and Manufacturing (IJIDeM)*. 2024. https://link.springer.com/article/10.1007/s12008-024-01883-2
71. Bhowmik A, Kumar R, Beemkumar N, Kumar AV, Singh G, Kulshreshta A, Santhosh AJ. Casting of particle reinforced metal matrix composite by liquid state fabrication method: A review. *Results in Engineering*. 2024:103152.
72. Kulkarni SD, Manjunatha CU, Manjunath KV, Prasad CD, Vasudev H. Design and optimization of polyvinyl-nitride rubber for tensile strength analysis. *International Journal on Interactive Design and Manufacturing (IJIDeM)*. 2024;18(8):5893–5908.
73. Prashar G, Vasudev H. Surface topology analysis of plasma sprayed Inconel625-Al_2O_3 composite coating. *Materials Today: Proceedings*. 2022;50:607–611.
74. Yedida VS, Vasudev H. A review on the development of thermal barrier coatings by using thermal spray techniques. *Materials Today: Proceedings*. 2022;50:1458–1464.
75. Sharanabasava H, Raviprakash M, Prasad CD, Ramesh MR, Phanibhushana MV, Vasudev H, Kumar S. Microstructure, mechanical and wear properties of SiC and Mo reinforced NiCr microwave cladding. *Advances in Materials and Processing Technologies*. 2023;10(4):3620–3633. https://doi.org/10.1080/2374068X.2023.2257937.
76. Singh J, Gill HS, Vasudev H. Computational fluid dynamics analysis on role of particulate shape and size in erosion of pipe bends. *International Journal on Interactive Design and Manufacturing (IJIDeM)*. 2023;17(5):2631–2646.
77. Anjaneya G, Sunil S, Kakkeri S, Math MM, Vaibhav MN, Solaimuthu C, Durga Prasad C, Vasudev H. Numerical simulation of microchannel heat exchanger using CFD. *International Journal on Interactive Design and Manufacturing (IJIDeM)*. 2023:1–17.
78. Nagabhushana P, Ramprasad S, Prasad CD, Vasudev H, Prakash C. Numerical investigation on heat transfer of a nano-fluid saturated vertical composite porous channel packed between two fluid layers. *International Journal on Interactive Design and Manufacturing (IJIDeM)*. 2024;18(5):2927–2944.
79. Kattimani MA, Venkatesh PR, Masum H, Math MM, Bahadurdesai VN, Mustafkhadri S, Prasad CD, Vasudev H. Design and numerical analysis of tensile deformation and fracture properties of induction hardened inconel 718 superalloy for gas turbine applications. *International Journal on Interactive Design and Manufacturing (IJIDeM)*. 2024;18(8):5939–5949.
80. Sunitha K, Vasudev H. A short note on the various thermal spray coating processes and effect of post-treatment on Ni-based coatings. *Materials Today: Proceedings*. 2022 Jan 1;50:1452–1457.
81. Majji BGR, Vasudev H, Bansal A. A review on the oxidation and wear behavior of the thermally sprayed high-entropy alloys. *Materials Today: Proceedings*. 2022;50:1447–1451.
82. Rao GS, Mukkamala U, Hanumanthappa H, Prasad CD, Vasudev H, Shanmugam B, Kishore Kumar KC. Evaluating and optimizing surface roughness using genetic algorithm and artificial neural networks during turning of AISI 52100 steel. *International Journal on Interactive Design and Manufacturing (IJIDeM)*. 2024;18(8):6151–6160.
83. Mahantesh MM, Rao KR, Chandra AP, Vijayakumar MN, Nandini B, Prasad CD, Vasudev H. Design and modeling using finite element analysis for the sitting posture of computer users based on ergonomic perspective. *International Journal on Interactive Design and Manufacturing (IJIDeM)*. 2024;18(8):5875–5891.

84. Singh J, Singh R, Singh S, Vasudev H, Kumar S. Reducing scrap due to missed operations and machining defects in 90PS pistons. *International Journal on Interactive Design and Manufacturing (IJIDeM)*. 2023;17(5):2527–2539.
85. Arora H, Kumar V, Prakash C, Pimenov D, Singh M, Vasudev H, Singh V. Analysis of sensitization in austenitic stainless steel-welded joint. In Advances in Metrology and Measurement of Engineering Surfaces: Select Proceedings of ICFMMP 2019 (pp. 13–23). Springer Singapore, 2021.
86. Vasudev H. Wear characteristics of Ni-WC powder deposited by using a microwave route on mild steel: Microwave cladding of Ni-WC. *International Journal of Surface Engineering and Interdisciplinary Materials Science (IJSEIMS)*. 2020;8(1):44–54.
87. Prashar G, Vasudev H, Thakur L, Bansal A. Performance of thermally sprayed hydroxyapatite coatings for biomedical implants: A comprehensive review. *Surface Review and Letters*. 2023;30(1):2241001.
88. Mehta A, Vasudev H, Jeyaprakash N. Role of sustainable manufacturing approach: Microwave processing of materials. *International Journal on Interactive Design and Manufacturing (IJIDeM)*. 2024;18(8):5283–5299.
89. Prashar G, Vasudev H. Understanding cold spray technology for hydroxyapatite deposition. *Journal of Electrochemical Science and Engineering*. 2023;13(1):41–62.
90. Sen B, Bhowmik A, Singh G, Mishra V, Debnath S, Zairov R, Ammarullah MI. Alumina-enriched sunflower bio-oil in machining of Hastelloy C-276: a fuzzy Mamdani model-aided sustainable manufacturing paradigm. *Scientific Reports*. 2024;14(1):29194.
91. Singh J, Vasudev H, Szala M, Gill HS. Neural computing for erosion assessment in Al-20TiO2 HVOF thermal spray coating. *International Journal on Interactive Design and Manufacturing (IJIDeM)*. 2024;18(4):2321–2332.
92. Prashar G, Vasudev H. Understanding cold spray technology for hydroxyapatite deposition. *Journal of Electrochemical Science and Engineering*. 2023;13(1):41–62.
93. Prakash C, Gupta LR, Mehta A, Vasudev H, Tominov R, Korman E, Fedotov A, Smirnov V, Kesari KK. Computing of neuromorphic materials: An emerging approach for bioengineering solutions. *Materials Advances*. 2023;4(23):5882–5919.
94. Singh G, Mehta A, Vasudev H. Sustainability of additive manufacturing: A comprehensive review. *Progress in Additive Manufacturing*. 2024;9(6):2249–2272.

18 Advantages of Laser Cladding Technology in Surface Modification

Gurbhej Singh and Hitesh Vasudev

18.1 INTRODUCTION

Laser cladding, often known as LC, is a technique that involves adding materials that possess certain properties to the surface of a base material. Through the application of a laser beam with a high energy density, the materials are first melted and subsequently hardened, resulting in the formation of a covering layer that is metallurgically coupled with the base [1]. The advantages of LC include a high energy density, a minimum amount of deformation, metallurgical contact with the material, and a number of other advantages [2]. The coaxial powder feeding technique and the preset method are the two primary approaches that may be used in the process of adding cladding materials. The coaxial powder feeding method incorporates both off-axis powder feeding and coaxial powder feeding within its repertoire of techniques. Using the fixed approach has the advantage of allowing the covering material to be available in a variety of particle sizes and does not need the substance to flow readily. However, it is not suitable for large-scale commercial applications.

During the liquid chromatography process, the coaxial powder feeding technique is more suitable for usage in industrial settings since it distributes powder at the same time. The essential process elements for LC in the coaxial powder feeding technique are laser power, scanning speed, powder feeding rate, spot size, defocus distance, overlap rate, and gas flow rate. Additionally, the powder feeding rate is quite important.

The technique of surface alteration has the potential to improve the performance of these components. Thermal blasting, physical vapor deposition, electroplating, and laser coating are the primary surface alteration processes that have been developed enough to be considered well developed. Methods such as electroplating, thermal spraying, and physical vapor deposition are considered among the most sophisticated surface engineering techniques. Although each of them has the potential to improve the surface quality of aluminum alloys, they each also have their own set of issues. The process of thermal blasting, for instance, might produce a thick coating, but it may have spots and holes, which can cause it to adhere inappropriately. Because of the thinness of the covering that is produced using physical vapor deposition, the application of heat during usage may create changes at the bottom of the coating, which can result in the coating peeling off. As a result of electroplating, the layer that

DOI: 10.1201/9781003546528-18

is produced is not sufficiently thick, and the fluid that is utilized in the process might be harmful to the environment. As a result of the development of laser and surface modification technologies [3–5], laser cladding technology has come into existence. Using a laser beam to melt the cladding material and the surface of the base body and then allowing it to solidify to form a cladding layer that metallurgically bonds with the base body, it is possible to use it for the surface strengthening of new parts, as well as for the remanufacturing and repair of parts that have failed [6–10].

When strong lasers were first developed in the 1970s, researchers began looking at a sort of laser surface treatment known as laser cladding. A substance that is used for cladding is applied to the surface of the workpiece. Subsequent to the impact of a powerful laser on the material, the outer layer and a thin layer of the metal that lies underneath it undergo rapid melting and subsequent hardening. This creates a new layer that has the characteristics that are wanted. The power level for laser cladding may range anywhere from 10^4 to 10 W/cm^2, which places it in the middle of the spectrum between laser alloying and laser hardness in terms of its use. Laser coating offers a number of advantages over conventional techniques of surface modification, including the following: it is possible to use it with a wide variety of materials, it warms up and cools down rapidly, it enhances the layer on top, it mixes very little with the substance beneath, it keeps the base material mostly untouched by heat, it utilizes energy effectively, and it is beneficial to the environment. Therefore, the laser coating of aluminum metals has significant potential for expansion in the future. Metals made of aluminum are excellent in reflecting laser beams, and the brightness of these metals falls as the frequency of the laser rises. The way in which aluminum interacts with light, particularly in the visible and infrared spectrums, is impacted by the electrical characteristics of aluminum as well as the transportation of electrons between different energy levels. It is the conduction-electron solid response of aluminum, which is regulated by its free electrons, that is responsible for the high reflection of aluminum. The way that aluminum responds to various lasers demonstrates that it reflects more light at longer wavelengths than other materials. The laser reflectivity may reach as high as 90% when it is applied to a CO_2 laser operating at a frequency of 10.6 m, for example. In comparison to a CO_2 laser, a Nd:YAG laser with a wavelength of 1.06 micrometers is capable of absorbing light seven times more effectively. It is possible for materials that absorb radiation to enhance the efficiency with which objects gather energy, particularly when employing lasers with a greater range. Utilizing laser coating on aluminum alloys is contingent upon the development of lasers that are both powerful and of good quality. The technique of laser cladding involves the use of a laser beam in order to melt not only the covering powder but also the surface of the substance behind it. The laser travels away, and the substance that has been melted soon solidifies, which results in the formation of a strong connection. In conclusion, laser technology not only benefits the environment but also offers advantages such as rapid processing and high-quality coatings. The pre-placed powder technique and the synchronized powder feeding method, both of which are shown in Figure 18.1, are two approaches that may be used to include covering materials. This approach of using pre-placed powder is simple to implement. On the other hand, due to the fact that it has a covering layer of poorer quality, it is not suitable for industrial manufacturing. It is possible to

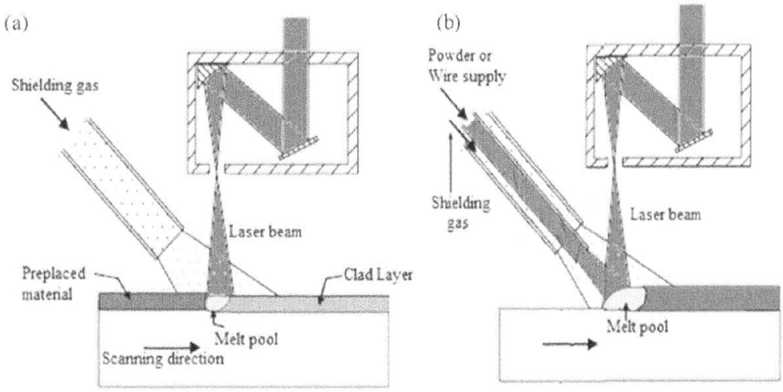

FIGURE 18.1 Schematic representation of laser cladding. (a) Pre-placed powder technique; (b) synchronous powder feeding method [9].

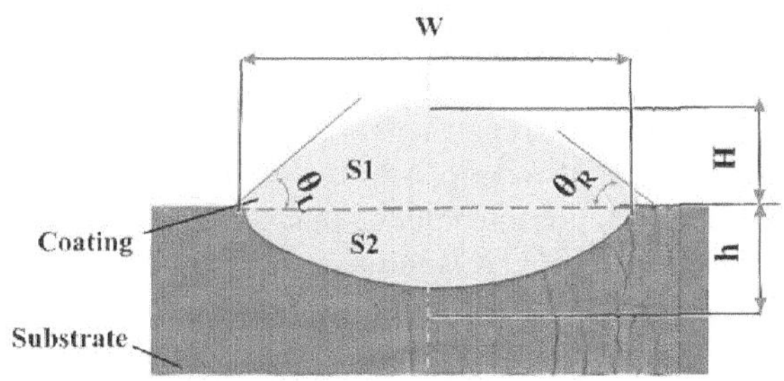

FIGURE 18.2 Cross-section diagram of the cladding layer [9].

achieve high-quality coating layers via the use of the simultaneous powder feeding technique, which also incorporates side feeding and parallel feeding. This approach is also suitable for automated control.

Metal-based alloys, ceramic-reinforced composites, and novel materials such as amorphous and high-entropy alloys are the primary materials that are used for laser coating of aluminum alloys at the present time. In most cases, these covering materials are available in the form of powder, wire, or sheets; nevertheless, powder is the most often used option in the industrial sector [11] (see Figure 18.2).

18.2 LASER CLADDING PROCESS

This process melts the covering material and a small layer of the surface material. The melted metal quickly hardens and bonds well to the surface underneath. LC has many features that you don't see in other finishing methods. LC is a fast solidification method, and you can change its microstructures by adjusting the process

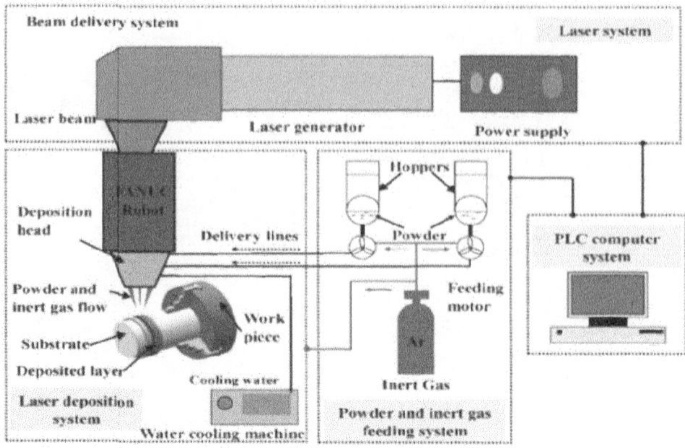

FIGURE 18.3 Laser cladding process [10].

settings. Figure 18.3 shows diagrams for the LC method. In addition to having a low dilution rate, the technique allows for easy control of the layer width, which may range anywhere from 0.2 to 5.0 mm. Because it generates less heat than other methods, it doesn't cause as much warping as other approaches. When it comes to putting the component together, it is also simpler to maintain the measurements within the constraints that are required. It is a technique that is not only tough but also adaptive and beneficial to the environment. A significant temperature differential during the LC process contributes to the formation of a more compact microstructure, which ultimately results in improved surface characteristics. The process is able to apply coatings using a variety of powders, and it can even apply metals with a high melting point to materials that melt at lower temperatures. If you have the appropriate equipment, it is feasible to cover pieces that have complicated forms and locations that are difficult to access. The powders may be fed into the coating system according to one of two different methods, powder preset and synchronous powder feeding systems [12–15]. There are a number of benefits associated with synchronous powder feeding, including improved absorption of laser light, fewer gaps in the covering area, and the ease with which it may be automated and controlled using computers [16].

18.3 LASER CLADDING PROCESS FOR STEEL

The technique of laser coating is a speedy one. Although it is a very short procedure, the cladding process includes significant interactions between the laser, the material that is being clad, and the material that is below it. Following the formation of the molten pool, there are a number of complex chemical and physical processes that take place, in addition to a rapid cooling process that occurs as the molten material becomes more solid. The way in which the technical parameters for laser cladding are configured affects each individual phase of the cladding process. Laser power, scanning speed, and the size of the laser spot are the most important aspects to consider while applying a laser coating. The absorption rate is a significant aspect that plays a role in determining the amount of quality that the surface finish has. It is impossible to prevent the coating on the surface from completely blending in with the substance that is underneath it. It is necessary to subject the surface of the base to melting in order to establish a solid connection between the coating and the substance that lies underneath it. The dilution rate should be maintained at an appropriate level in order to maintain the effectiveness of the base material and the covering layer and to ensure that the consequences of excessive dilution are avoided. The surface formability, dilution rate, microstructure, and phase composition of the covering coating are all influenced by each individual process parameter [17]. For the purpose of producing a wall layer that has a smooth surface and a robust internal structure, it is essential to make adjustments to the process parameters and to maintain them within a range that is appropriate. The heat that is required for the coating process is supplied by the laser. The power of the laser has a direct impact on the amount of heat put into the coating process. Because there are more disruptions in the melted region, the surface of the covering layer gets rougher when the laser power is too high. This is because the melted area is more likely to take in gas thanks to the increased disturbances. An excessive amount of heat that is delivered to the base causes the dilution rate to rise, which in turn affects the shape of the surface of the external coating as well as its chemical composition. There is a possibility that the cladding material will not melt completely if the laser power is insufficient. Because of this, it is more difficult for the liquid cladding to adhere to the surface below, which ultimately leads to a cladding layer that is not produced properly. Due to the fact that the laser energy that is reaching the top of the base is insufficient, it is unable to form a strong connection with the covering layer [18–20].

18.3.1 Laser Cladding Materials for Steel

The materials used in laser cladding directly impact the coating's chemical makeup. The choice of covering materials significantly affects the quality and function of the finish. Most market alloy powders are made for methods like flame spraying, plasma spray welding, and powder metallurgy. However, these do not meet the specific needs of laser coating.

Laser coating materials can be used on many types of surfaces, but some work better on specific surfaces in particular environments. Choosing the right covering materials is essential for achieving a high-quality surface, a smooth finish, and a layer without interior flaws on the chosen base material [21].

18.4 LASER CLADDING FOR ALLOY ELEMENTS

The presence of these elements improves the resistance of HEA coatings to high-temperature wear. Additionally, when different alloying elements work together, they can enhance the performance of the metals.

18.4.1 Silicon

Silicon is commonly seen as a good material for making things stronger. Zhu and others studied $FeCoCrNiSi_x$ high-entropy metal layers (with x values of 0, 0.5, and 1)

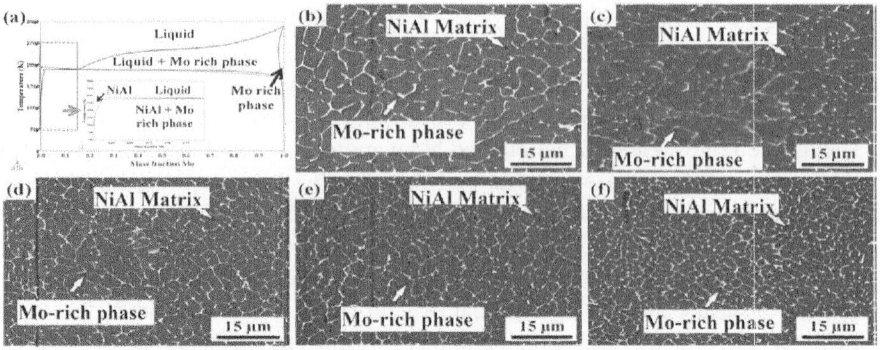

FIGURE 18.4 (a) Simulated phase diagram of NiAl with different amounts of Mo addition. Backscatter SEM micrographs of NiAl coatings with: (b) 3 wt.%, (c) 6 wt.%, (d) 9 wt.%, (e) 12 wt.%, and (f) 15wt.% Mo additional [11].

and discovered that adding silicon helped make the grain structure finer. At 600 °C, the Si_1 coating had an average friction value of 0.19, leading to an 81.2% decrease in wear compared to the Si_0 coating.

Liu et al. [22] studied how silicon content affects the wear resistance of coatings. They discovered that as the silicon content increases, the hardness of the coatings also increases greatly. At room temperature, 200 °C, and 400 °C, the main type of wear was abrasive wear, and the ability to fight wear became better as hardness increased. At 600 °C and 800 °C, the way materials wear changed to reactive wear. As the silicon percentage went up, wear resistance first rose, but then it dropped. When $x = 1.0$, the covering showed the best ability to prevent wear at high temperatures. The cause was the creation of bimetallic oxides ($FeCr_2O_4$, $MnFe_2O_4$, VMn_2O_4, and $CrMn_2O_4$) on the surface of the HEA layers. These bimetallic oxides have smaller lattice constants and contain cations in the lattice [23], which add to the densification of the oxide layer. When the silicon level reached 1.0, SiO_2 appeared on the top. Even though this oxide isn't dense enough, these hard bits can help reduce friction at high temperatures, making it more resistant to wear. When $x = 2.0$, the bimetallic oxides nearly vanished, leaving just SiO_2. This shows that too much silicon can weaken wear resistance.

18.4.2 Molybdenum

Solid solution strengthening and grain processing are the two primary methods that are used to enhance the strength of molybdenum. In addition, molybdenum may be combined with other metals to form compounds. Strong intermetallic compounds are formed by the σ phase when iron, cobalt, and nickel are present. These compounds are resistant to wear, have a high level of stability at high temperatures, and do not readily oxidize. At elevated temperatures, molybdenum provides substantial assistance in the development of the σ phase, hence increasing the range of precipitation that it may achieve [24]. The incorporation of molybdenum contributes to the formation of a thick and consistent oxide layer at high temperatures. When exposed to high temperatures, this layer serves as an additional layer of protection, so reducing the amount of wear that occurs and preventing materials from failing. For instance, Deng et al. [25] discovered that the $CoCrFeNiMo_{0.2}$ layer had a friction coefficient of 0.154 at 600 °C, which is much lower than the friction coefficient of 0.578 at room temperature (RT). At a temperature of 600 °C, the form of the wear track was noticeably smaller than it was at room temperature. It was discovered by Zhao et al. [26] that the addition of molybdenum and aluminum to high-entropy alloys of CoCrFeNi works quite well. The presence of molybdenum contributes to the formation of a more substantial layer of Al_2O_3, which in turn enhances the coating's resistance to oxidation at high temperatures. After that, Gao et al. [27] were able to effectively create $Mo_xNbTiZr$ ($x = 0.2$, 0.4, 0.6, 0.8, and 1.0) LC-HEAC coatings in order to further investigate the high-temperature thermal stability and wear processes of molybdenum in HEA coatings. It was discovered that $Mo_{0.2}$ only has a single sort of structure, which is a BCC phase composed of $ZrNb_2Mo_2$. The $Mo_{1.0}$ coating, on the other hand, had a combination of phases, such as BCC and HCP (Ti_2Zr) phases, in addition to an HCP-2 phase that is a solid solution with a Zr-HCP structure. A BCC matrix phase that was shaped like needles was also present in it [28].

18.4.3 Ceramic Cladding Materials

Ceramic materials are renowned for their exceptional resistance to acids and bases, their extreme hardness, their durability against wear, their reliability at high temperatures, their low density, their light weight, and their low weight. In the field of laser cladding technology, ceramic powders are often used as cladding materials in harsh settings that are subjected to a significant amount of wear and resistance to corrosion. On the other hand, ceramic powder and metal substrates have very different physical and chemical characteristics. These differences include the linear expansion coefficient, elastic modulus, wettability, melting temperature, and thermal conductivity [29]. Ceramic powder differs significantly from metal substrates with regard to these qualities. It is possible for the ceramic coating to develop holes, splits, and even peel off since it does not contact exceptionally well with the surface that is below it. This is due to the fact that their chemical and physical characteristics are distinct from one another. When it comes to placing clay materials directly onto metal utilizing laser coating, this does not assist. Furthermore, the ceramic coating is hard and unyielding due to the fact that ceramic materials have a high elasticity characteristic. Due to its lack of flexibility and hardness, as well as its inability to tolerate repeated stress very well, it is unable to efficiently sustain contact without breaking. The covering is susceptible to cracking or breaking if it is subjected to a forceful impact. A coating made of pure ceramic is not capable of withstanding heavy weights, collisions, or high-pressure circumstances [30]. Laser cladding has emerged as a transformative surface modification technique, offering superior wear resistance, corrosion protection, and enhanced mechanical properties for engineering components [32–41]. This advanced coating process enables the deposition of high-performance materials onto substrates, thereby extending the service life of critical components in the aerospace, biomedical, and energy industries [42–52]. The optimization of process parameters, including laser power, scan speed, and powder feed rate, significantly influences the microstructure, hardness, and bonding strength of the clad layers. Additionally, the incorporation of hybrid approaches, such as microwave-assisted cladding and additive manufacturing techniques, has further expanded the potential applications of laser cladding in sustainable manufacturing [53–65]. The corrosion resistance of laser-clad surfaces has been extensively studied, with coatings exhibiting superior performance in aggressive environments. These advancements position laser cladding as a key enabler for next-generation surface engineering solutions.

18.5 CONCLUSION

Laser cladding is a surface modification technique that uses a high-energy-density laser beam to melt and bond materials with desired properties onto a substrate. The two main methods of adding cladding materials are the coaxial powder feeding method and the preset method, with the former being more suitable for industrial applications. Key process parameters influencing the quality of the cladding layer include laser power, scanning speed, and powder feeding rate. Recent developments in LC technology include the combination with other technologies like ultrasonic-assisted LC and the emergence of ultra-fast LC. LC offers advantages over traditional surface modification techniques in terms of environmental friendliness and the ability to repair and remanufacture failed parts. The selection of appropriate cladding materials, including alloy elements like silicon and molybdenum and ceramic materials, is crucial for obtaining the desired surface properties. However, applying ceramic materials directly onto metal substrates using LC poses challenges due to differences in physical and chemical properties.

18.6 FUTURE SCOPE

The future of laser cladding technology lies in optimizing process parameters through AI-driven models, developing novel materials like hybrid metal-ceramics, and integrating ultra-fast LC with automation for industrial scalability. Advancements in ceramic coatings, sustainable practices, and cost-effective solutions will further expand its applications in aerospace, biomedical, and precision engineering.

REFERENCES

1. Maya, J.; Sivaprasad, K.; Kumar, G.V.S.; Baitimerov, R.; Lykov, P.; Prashanth, K.G. Microstructure, Mechanical Properties, andCorrosion Behavior of 06Cr15Ni4CuMo Processed by Using Selective Laser Melting. Metals **2022**, 12, 1303.
2. Parapurath, S.; Jacob, L.; Gunister, E.; Vahdati, N. Effect of Microstructure on Electrochemical Properties of the EN S275 MildSteel under Chlorine-Rich and Chlorine-Free Media at Different pHs. Metals **2022**, 12, 1386.
3. Su, C.; Feng, G.; Zhi, J.; Zhao, B.; Wu, W. The Effect of Rare Earth Cerium on Microstructure and Properties of Low Alloy Wear-Resistant Steel. Metals **2022**, 12, 1358.
4. Radionova, L.V.; Samodurova, M.N.; Sarafanov, A.E.; Bykov, V.A.; Glebov, L.A.; Pashkeev, K.Y. Elevation of the Wear Resistance of Roller Dies for the Production ofWires from Titanium Alloys. Metallurgist **2021**, 64, 1188–1197.

5. Lanzutti, A.; Andreatta, F.; Rossi, L.; Di Benedetto, P.; Causero, A.; Magnan, M.; Fedrizzi, L. Corrosion Fatigue. Failure of a High Carbon CoCrMo Modular Hip Prosthesis: Failure Analysis and Electrochemical Study. The Engineering Failure Analysis **2019**, 105, 856–868.
6. Dunstan, M.K.; Paramore, J.D.; Fang, Z.Z.; Ligda, J.P.; Butler, B.G. Analysis of microstructural facet fatigue failure in ultra-fine grained powder metallurgy Ti-6Al-4V produced through hydrogen sintering. International Journal of Fatigue **2020**, 131, 105355.
7. Zongsheng, F. Analysis of Attentions Needed in Wear and Maintenance of Mineral Processing Equipment. World Nonferrous Metal **2019**, 6, 50–51.
8. Singh, G.; Singh, H.; Sharma, Y.: Vasudev, H.; Prakash, C. Analysis and Optimization of Various Process Parameters and Effect on the Hardness of SS-304 Stainless Steel Welded Joints. International Journal on Interactive Design and Manufacturing (IJIDeM) **2024**, 18(5), 2889–2896.
9. Zhao, P.; Shi, Z.; Wang, X.; Li, Y.; Cao, Z.: Zhao, M.; Liang, J. A Review of the Laser Cladding of Metal-Based Alloys, Ceramic-Reinforced Composites, Amorphous Alloys, and High-Entropy Alloys on Aluminum Alloys. Lubricants **2023**, 11(11), 482.
10. Tang, J.; Wang, K.; Fu, H. Laser Cladding In Situ Carbide-Reinforced Iron-Based Alloy Coating: A Review. Metals **2024**, 14(12), 1419.
11. Han, T.; Zhou, K.; Chen, Z.; Gao, Y. Research Progress on Laser Cladding Alloying and Composite Processing of Steel Materials. Metals **2022**, 12(12), 2055.
12. Qunshuang, M.; Yajiang, L.; Juan, W.; Kun, L. Microstructure evolution and growth control of ceramic particles in wide-band laser clad Ni60/WC composite coatings. Materials Design **2016**, 92, 897–905.
13. Chen, L.; Yu, T.; Xu, P.; Zhang, B. In-Situ NbC Reinforced Fe-Based Coating by Laser Cladding: Simulation and Experiment. Surface Coating Technology **2021**, 412, 127027.
14. Lu, S.; Zhou, J.; Wang, L.; Liang, J. Effect of V and Cr Transition Layers on Microstructure and Mechanical Properties of Ni-Based Coating on Titanium Alloy Fabricated by Laser Cladding. Surface Coating Technology **2021**, 405, 126734.
15. Liu, Y.; Li, J.; Xu, Q.; Zhang, Y.; Yan, X.; Chen, Y.; He, H. Microstructure and Wear Behavior of TC4 Laser Cladding Modified Via SiC and MoS2. Coatings **2022**, 12, 792.
16. Peng, Z.; Zhang, J.; Zhang, M.; Wang, K.; Peng, P. Laser In-Situ Preparation and Mechanical Properties of VC Reinforced Fe-Basedwear-Resistant Composite Cladding. Ceramics International **2022**, 48, 28240–28249.
17. Shariff, M.M.; Ahamed, N.; Singh, G.; Arpitha, G.R.; Jain, N.; Verma, A. Thermal, Physical, and Electrical Characterization of Millettia Pinnata Leaf Powder, Alumina, Glass Sheet, Boron Nitride Reinforced Epoxy-based Hybrid Composites: An Experimental Scrutiny.
18. Singh, G.; Mehta, A.; Vasudev, H. Sustainability of Additive Manufacturing: A Comprehensive Review. Progress in Additive Manufacturing **2024**, 1–24.
19. Singh, G.; Bansal, A.; Vasudev, H.; Mishra, V. Sliding Wear Study of the Inconel-625 Clad Deposits by Microwave Heating on SS–304. Physica Scripta **2024**, 99(6), 065503.
20. Liu, L.; Qiao, Y.; Xu, P. Effect of La_2O_3 on Microstructure and Properties of Laser Cladding SMA Coating on AISI 304 Stainless Steel. Coatings **2022**, 12, 1004.
21. Sun, S.; Fu, H.; Ping, X.; Guo, X.; Lin, J.; Lei, Y.; Wu, W.; Zhou, J. Effect of CeO_2 Addition on Microstructure and Mechanical Properties of In-Situ (Ti, Nb)C/Ni Coating. Surface Coating Technology **2019**, 359, 300–313.
22. Hou, X.; Du, D.; Chang, B.; Ma, N. Influence of Scanning Speed on Microstructure and Properties of Laser Cladded Fe-Based Amorphous Coatings. Materials **2019**, 12, 1279.
23. Hou, X.; Du, D.; Wang, K.; Hong, Y.; Chang, B. Microstructure and Wear Resistance of Fe-Cr-Mo-Co-C-B Amorphous Composite Coatings Synthesized by Laser Cladding. Metals **2018**, 8, 622.

24. Singh, J.; Yadav, A.; Gupta, S.; Balu, R.; Shukla, N.; Singh, G.; Singh, S. Design and Computational Analysis of Continuous Wave Fiber Laser Cutting of Flat and Conformal ITO PET Sheets. International Journal on Interactive Design and Manufacturing (IJIDeM) **2024**, 1–9.
25. Singh, G.; Singh, R.; Gul, J. Machinability Behavior of Human Implant Materials. Journal of Electrochemical Science and Engineering **2023**, 13(1), 99–114.
26. Verma, P.K.; Vasudev, H.; Singh, G.; Singh, S. A Review on the Microstructural and Biomedical Properties of Implants Manufactured Using Additive Manufacturing. Journal of the Institution of Engineers (India): Series D **2024**, 1–15.
27. Mehta, A.; Singh, G.; Vasudev, H. Processing of Shape Memory Alloys Research, Applications and Opportunities: A Review. Physica Scripta **2024**. https://iopscience.iop.org/article/10.1088/1402-4896/ad48cb
28. Singh, G.; Bansal, A.; Vasudev, H. Corrosion and Microstructural Behaviour of Inconel 625 Microwave Clad Deposited on Mild Steel. In Thermal Claddings for Engineering Applications (pp. 51–62). CRC Press, 2024.
29. Kapade, T.T.; Shah, S.K.; Shah, D.S.; Kumar, N.S.; Bhor, J.K.; Panwar, V.S.; Singh, G. Thermal Analysis of V-Cut Twisted Tape Fitted Heat Exchanger: A Numerical Approach. International Journal on Interactive Design and Manufacturing (IJIDeM) **2024**. https://link.springer.com/article/10.1007/s12008-024-01883-2
30. Bhowmik, A.; Kumar, R.; Beemkumar, N.; Kumar, A.V.; Singh, G.; Kulshreshta, A.; Santhosh, A.J. Casting of Particle Reinforced Metal Matrix Composite by Liquid State Fabrication Method: A Review. Results in Engineering **2024**, 103152.
31. Singh, M.; Vasudev, H.; Kumar, R. Corrosion and Tribological Behaviour of BN Thin Films Deposited Using Magnetron Sputtering. International Journal of Surface Engineering and Interdisciplinary Materials Science (IJSEIMS) **2021 Jul 1**, 9(2), 24–39.
32. Singh, M.; Vasudev, H.; Singh, M. Surface Protection of SS-316L With Boron Nitride Based Thin Films Using Radio Frequency Magnetron Sputtering Technique. Journal of Electrochemical Science and Engineering **2022**, 12(5), 851–863.
33. Singh, G.; Singh, H.; Sharma, Y.; Vasudev, H.; Prakash, C. Analysis and Optimization of Various Process Parameters and Effect on the Hardness of SS-304 Stainless Steel Welded Joints. International Journal on Interactive Design and Manufacturing (IJIDeM) **2024**, 18(5), 2889–2896.
34. Singh, G.; Mehta, A.; Vasudev, H. Sustainability of Additive Manufacturing: A Comprehensive Review. Progress in Additive Manufacturing **2024**, 1–24.
35. Singh, G.; Bansal, A.; Vasudev, H.; Mishra, V. Sliding Wear Study of the Inconel-625 Clad Deposits by Microwave Heating on SS–304. Physica Scripta **2024**, 99(6), 065503.
36. Verma, P.K.; Vasudev, H.; Singh, G.; Singh, S. A Review on the Microstructural and Biomedical Properties of Implants Manufactured Using Additive Manufacturing. Journal of the Institution of Engineers (India): Series D **2024**, 1–15.
37. Singh, G.; Vasudev, H.; Arora, H. A short note on the processing of materials through microwave route. In Advances in Materials Processing: Select Proceedings of ICFMMP 2019 (pp. 101–111). Springer Singapore, 2020.
38. Singh, G.; Bansal, A.; Vasudev, H. Corrosion and Microstructural Behaviour of Inconel 625 Microwave Clad Deposited on Mild Steel. In Thermal Claddings for Engineering Applications (pp. 51–62). CRC Press, 2024.
39. Kulkarni, S.D.; Manjunatha, C.U.; Manjunath, K.V.; Prasad, C.D.; Vasudev, H. Design and Optimization of Polyvinyl-Nitride Rubber for Tensile Strength Analysis. International Journal on Interactive Design and Manufacturing (IJIDeM) **2024**, 18(8), 5893–5908.
40. Prashar, G.; Vasudev, H. Surface Topology Analysis of Plasma Sprayed Inconel625-Al_2O_3 Composite Coating. Materials Today: Proceedings **2022**, 50, 607–611.

41. Yedida, V.S.; Vasudev, H. A Review on the Development of Thermal Barrier Coatings by Using Thermal Spray Techniques. Materials Today: Proceedings **2022**, 50, 1458–1464.
42. Sharanabasava, H.; Raviprakash, M.; Prasad, C.D.; Ramesh, M.R.; Phanibhushana, M.V.; Vasudev, H.; Kumar, S. Microstructure, Mechanical and Wear Properties of SiC and Mo Reinforced NiCr Microwave Cladding. Advances in Materials and Processing Technologies **2023**, 10(4), 3620–3633. https://doi.org/10.1080/2374068X.2023.2257937.
43. Dutta, V.; Thakur, L.; Singh, B.; Vasudev, H. A Study of Erosion–Corrosion Behaviour of Friction Stir-Processed Chromium-Reinforced NiAl Bronze Composite. Materials **2022**, 15(15), 5401.
44. Poojari, M.; Hanumanthappa, H.; Prasad, C.D.; Jathanna, H.M.; Ksheerasagar, A.R.; Shetty, P.; Shanmugam, B.K.; Vasudev, H. Computational Modelling for the Manufacturing of Solar-Powered Multifunctional Agricultural Robot. International Journal on Interactive Design and Manufacturing (IJIDeM) **2024**, 18(8), 5725–5736.
45. Prashar, G.; Vasudev, H. A Review on the Influence of Process Parameters and Heat Treatment on the Corrosion Performance of Ni-Based Thermal Spray Coatings. Surface Review and Letters **2022**, 29(1), 2230001.
46. Kattimani, M.A.; Venkatesh, P.R.; Masum, H.; Math, M.M.; Bahadurdesai, V.N.; Mustafkhadri, S.; Prasad, C.D.; Vasudev, H. Design and Numerical Analysis of Tensile Deformation and Fracture Properties of Induction Hardened Inconel 718 Superalloy for Gas Turbine Applications. International Journal on Interactive Design and Manufacturing (IJIDeM) **2024**, 18(8), 5939–5949.
47. Sunitha, K.; Vasudev, H. A Short Note on the Various Thermal Spray Coating Processes and Effect of Post-Treatment on Ni-Based Coatings. Materials Today: Proceedings **2022 Jan 1**, 50, 1452–1457.
48. Majji, B.G.R.; Vasudev, H.; Bansal, A. A Review on the Oxidation and Wear Behavior of the Thermally Sprayed High-Entropy Alloys. Materials Today: Proceedings **2022**, 50, 1447–1451.
49. Rao, G.S.; Mukkamala, U.; Hanumanthappa, H.; Prasad, C.D.; Vasudev, H.; Shanmugam, B.; KishoreKumar, K.C. Evaluating and Optimizing Surface Roughness Using Genetic Algorithm and Artificial Neural Networks During Turning of AISI 52100 Steel. International Journal on Interactive Design and Manufacturing (IJIDeM) **2024**, 18(8), 6151–6160.
50. Singh, J.; Gill, H.S.; Vasudev, H. Computational Fluid Dynamics Analysis on Role of Particulate Shape and Size in Erosion of Pipe Bends. International Journal on Interactive Design and Manufacturing (IJIDeM) **2023**, 17(5), 2631–2646.
51. Anjaneya, G.; Sunil, S.; Kakkeri, S.; Math, M.M.; Vaibhav, M.N.; Solaimuthu, C.; Durga Prasad, C.; Vasudev, H. Numerical Simulation of Microchannel Heat Exchanger Using CFD. International Journal on Interactive Design and Manufacturing (IJIDeM) **2023**, 1–17.
52. Nagabhushana, P.; Ramprasad, S.; Prasad, C.D.; Vasudev, H.; Prakash, C. Numerical Investigation on Heat Transfer of a Nano-Fluid Saturated Vertical Composite Porous Channel Packed Between Two Fluid Layers. International Journal on Interactive Design and Manufacturing (IJIDeM) **2024**, 18(5), 2927–2944.
53. Singh, P.; Vasudev, H.; Bansal, A. Effect of Post-Heat Treatment on the Microstructural, Mechanical, and Bioactivity Behavior of the Microwave-Assisted Alumina-Reinforced Hydroxyapatite Cladding. Proceedings of the Institution of Mechanical Engineers, Part E: Journal of Process Mechanical Engineering **2024**, 238(3), 1060–1071.
54. Mahantesh, M.M.; Rao, K.R.; Chandra, A.P.; Vijayakumar, M.N.; Nandini, B.; Prasad, C.D.; Vasudev, H. Design and Modeling Using Finite Element Analysis for the Sitting Posture of Computer Users Based on Ergonomic Perspective. International Journal on Interactive Design and Manufacturing (IJIDeM) **2024**, 18(8), 5875–5891.

55. Singh, J.; Singh, R.; Singh, S.; Vasudev, H.; Kumar, S. Reducing Scrap Due to Missed Operations and Machining Defects in 90PS Pistons. International Journal on Interactive Design and Manufacturing (IJIDeM) **2023**, 17(5), 2527–2539.
56. Arora, H.; Kumar, V.; Prakash, C.; Pimenov, D.; Singh, M.; Vasudev, H.; Singh, V. Analysis of Sensitization in Austenitic Stainless Steel-Welded Joint. In Advances in Metrology and Measurement of Engineering Surfaces: Select Proceedings of ICFMMP 2019 (pp. 13–23). Springer Singapore, 2021.
57. Vasudev, H. Wear Characteristics of Ni-WC Powder Deposited by Using a Microwave Route on Mild Steel: Microwave Cladding of Ni-WC. International Journal of Surface Engineering and Interdisciplinary Materials Science (IJSEIMS) **2020**, 8(1), 44–54.
58. Prashar, G.; Vasudev, H.; Thakur, L.; Bansal, A. Performance of Thermally Sprayed Hydroxyapatite Coatings for Biomedical Implants: A Comprehensive Review. Surface Review and Letters **2023**, 30(1), 2241001.
59. Mehta, A.; Vasudev, H.; Jeyaprakash, N. Role of Sustainable Manufacturing Approach: Microwave Processing of Materials. International Journal on Interactive Design and Manufacturing (IJIDeM) **2024**, 18(8), 5283–5299.
60. Prashar, G.; Vasudev, H. Understanding Cold Spray Technology for Hydroxyapatite Deposition. Journal of Electrochemical Science and Engineering **2023**, 13(1), 41–62.
61. Prashar, G.; Vasudev, H. Understanding Cold Spray Technology for Hydroxyapatite Deposition. Journal of Electrochemical Science and Engineering **2023**, 13(1), 41–62.
62. Parkash, J.; Saggu, H.S.; Vasudev, H. A Short Review on the Performance of High Velocity Oxy-Fuel Coatings in Boiler Steel Applications. Materials Today: Proceedings **2022**, 50, 1442–1446.
63. Singh, J.; Vasudev, H.; Kumar, R.; Ubaidullah, M. Study on Wear Analysis of Ni-20Al_2O_3 HVOF Micron Layers Using Artificial Neural Network Technique. International Journal on Interactive Design and Manufacturing (IJIDeM) **2024**, 18(8), 5925–5938.
64. Singh, J.; Vasudev, H.; Szala, M.; Gill, H.S. Neural Computing for Erosion Assessment in Al-20TiO_2 HVOF Thermal Spray Coating. International Journal on Interactive Design and Manufacturing (IJIDeM) **2024**, 18(4), 2321–2332.
65. Prakash, C.; Gupta, L.R.; Mehta, A.; Vasudev, H.; Tominov, R.; Korman, E.; Fedotov, A.; Smirnov, V.; Kesari, K.K. Computing of Neuromorphic Materials: An Emerging Approach for Bioengineering Solutions. Materials Advances **2023**, 4(23), 5882–5919.

19 Fabrication of Surfaces with Chemical Vapor Deposition and Physical Vapor Deposition for Corrosion and Wear Applications

Praveen Kumar Verma and Hitesh Vasudev

19.1 INTRODUCTION

Surface engineering is of immense importance in improving the functionality and durability of the material subjected to aggressive environments. The need for high-performance coatings with better corrosion and wear resistance has created the demand and extensive use of advanced deposition methods such as chemical vapor deposition (CVD) and physical vapor deposition (PVD) [1]. Surface modification methods utilizing these processes enhance the performance and lifespan of aerospace, automotive, biomedical, and energy application components considerably. Wear and corrosion are two of the most common reasons for material deterioration, and some of the consequences are economic losses, structural failure, and higher maintenance costs. Corrosion is a chemical reaction between materials and the environment, resulting in oxidation, pitting, and loss of material. Wear is mechanical surface deterioration by friction, impact, or abrasion, and it influences the performance of mechanical parts. Protective coatings achieved via CVD and PVD technologies offer a feasible solution to overcome these limitations by enhancing surface properties such as hardness, adhesion, and chemical inertness [2]. Figure 19.1 shows a comparison of synthesis methods and parameters for tribological and anticorrosive coatings.

CVD is a chemical process whereby gaseous precursors are deposited on a substrate that is heated to create a solid, dense film of high structural integrity. The technique is widely employed for the deposition of hard ceramic coatings, including titanium carbide (TiC), titanium nitride (TiN), and silicon carbide (SiC), with outstanding wear and corrosion resistance. CVD typically entails high temperatures of processing, though that limits its application to temperature-critical materials.

DOI: 10.1201/9781003546528-19

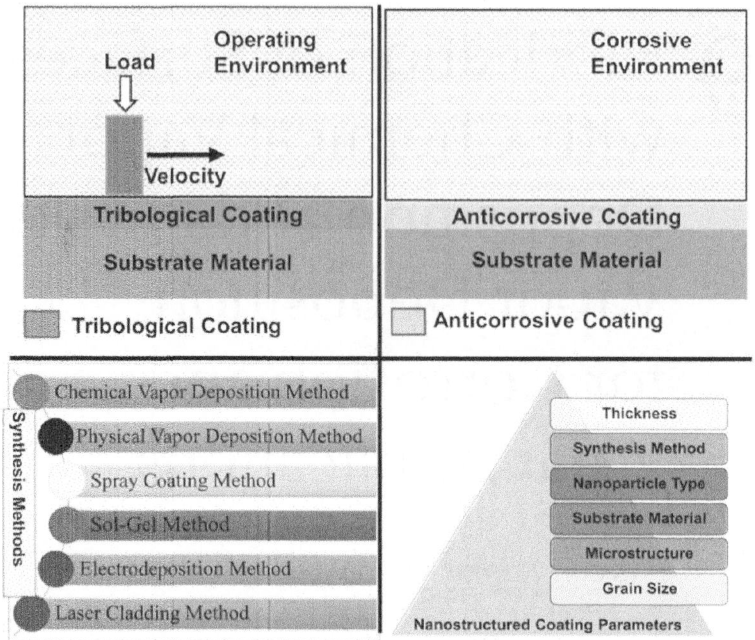

FIGURE 19.1 Comparison of tribological and anticorrosive coatings with synthesis methods and parameters [3].

PVD is a physical process whereby a solid material is vaporized and deposited onto a substrate within a vacuum environment. PVD coatings like TiN, chromium nitride (CrN), and diamond-like carbon (DLC) achieve greater hardness, low friction, and better wear resistance. PVD techniques, such as sputtering and arc evaporation, are able to operate below the temperatures at which CVD operates and thus are applicable over a broader range of materials, such as polymers and thermally sensitive alloys. Both CVD and PVD have advantages and are selected based on the requirements of applications. CVD coatings are reported to be denser and more adhesive in nature, whereas PVD coatings offer more flexibility in material selection as well as processability. Emerging advancements in these deposition technologies, including hybrid CVD/PVD processes, nanostructured films, and "green" alternatives, are revolutionizing surface engineering [4, 5].

This chapter aims at presenting an informative overview of CVD and PVD coating technologies, their corrosion, and wear-resistant applications. Some of the most important themes addressed are the basic principles, types of processes, coating materials, and comparison of these coating technologies. New innovations and future directions for CVD and PVD are also examined, highlighting their changing role in high-performance industrial applications. By understanding such coatings' functions and benefits, scientists and engineers are able to make informed decisions in selecting surface modification methods for hostile environments [6].

19.2 CHEMICAL VAPOR DEPOSITION

19.2.1 Fundamentals of CVD

Chemical vapor deposition is a widely used technique of preparing high-performance coatings with superior mechanical and chemical properties. The technique involves the chemical reaction among volatile precursors in the gaseous state, leading to the deposition of solid material onto a substrate. CVD generates coatings with high density, adhesion, and homogeneity and is highly preferred for protection purposes [7].

CVD functions at a high temperature of 600–1200 °C in order to make coating development possible with fine crystallinity and purity. Preparative phases of CVD encompass precursor supply, chemical reaction, nucleation, and film formation. Depending on precursor choice, chemical conditions, and depositing parameters, the microstructure and properties of resulting coatings may be significantly controlled. One of the key benefits of CVD is that it can deposit conformal coatings with superb adhesion and uniformity, even on intricate geometries. Still, high temperatures and the utilization of dangerous precursor gases are two of the disadvantages of this method. Modifications of CVD have evolved in an effort to overcome these drawbacks and improve coating performance such as plasma-enhanced CVD (PECVD) and atomic layer deposition (ALD) [8, 9]. Figure 19.2 shows a CVD system for the deposition of ceramic nanocomposite coatings.

19.2.2 Types of CVD

CVD has a number of different variants suited for particular applications, with different benefits in deposition efficiency, compatibility of materials, and film characteristics. The most widely used CVD methods are thermal CVD, plasma-enhanced CVD, atomic layer deposition, and metal-organic CVD (MOCVD). Thermal CVD, the original technique, makes use of high temperatures to cause chemical reactions among

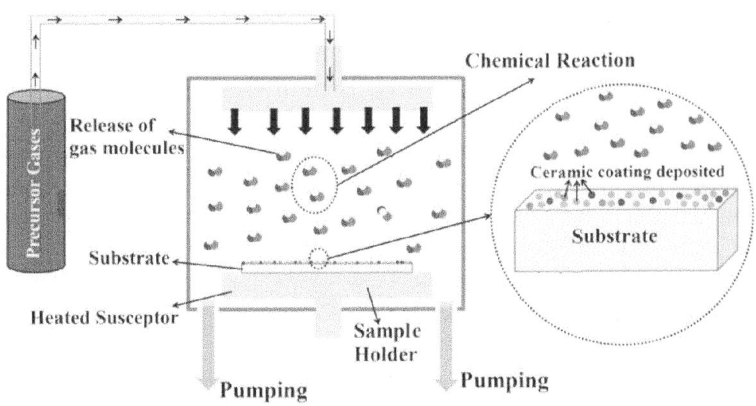

FIGURE 19.2 CVD system for the deposition of ceramic nanocomposite coatings [10].

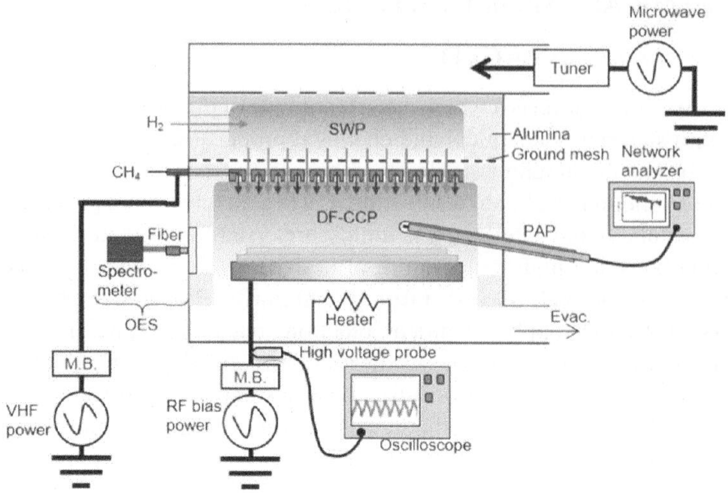

FIGURE 19.3 Schematic illustration of radical-injection plasma enhanced chemical vapor deposition (RI-PECVD) setup with instrumentation for optical emission spectroscopy (OES), plasma absorption probe (PAP) technique, and radio frequency (RF) bias voltage measurement [12].

precursor gases and deposition of uniform and dense coatings. Nevertheless, the elevated processing temperatures could restrict its use to thermally stable substrates [11].

PECVD adds plasma energy to initiate precursor gases, minimizing the temperature requirement and facilitating deposition on heat-sensitive materials. PECVD is predominantly employed in the deposition of amorphous carbon, silicon nitrides, and oxide thin films with improved mechanical and optical properties. ALD, a type of CVD, is defined by its sequential deposition method through which atomic control over film thickness and composition is obtained. It is especially beneficial for conformal coatings in nanotechnology and microelectronics. MOCVD, in contrast, uses metal-organic precursors to produce high-purity coatings, mainly applied in semiconductor and optoelectronic devices. Both of these CVD methods have been adapted to suit particular industrial needs, maximizing deposition efficiency, film quality, and scalability of the process. Selection of a specific CVD technique relies on parameters such as substrate compatibility, required properties of the coating, and limitations in processing. Over the last few years, the evolution in CVD technology, including environmentally friendly precursors and hybrid CVD-PVD processes, has further widened its use in corrosion and wear-resistant coatings, and now it is a must-have in materials engineering [13, 14].

19.2.3 CVD Coatings for Corrosion and Wear Resistance

Chemical vapor deposition coatings are used extensively in a variety of industries due to their superior hardness, chemical inertness, and high-temperature stability, thereby making them suitable for use to offer improved corrosion and wear

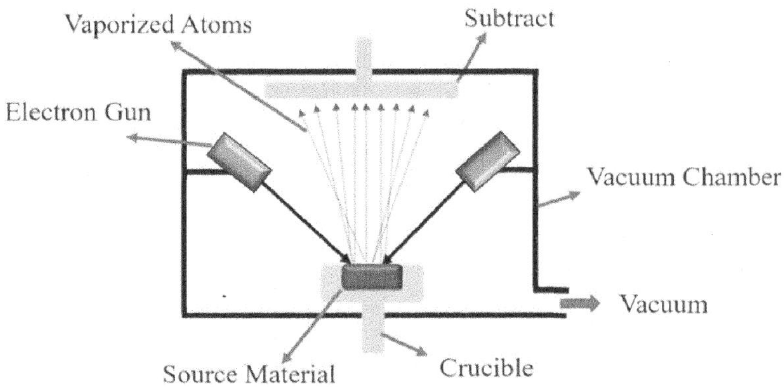

FIGURE 19.4 Schematic diagram of a physical vapor deposition device employing electron beam as heat source [16].

Source: Licensed under CC BY 4.0.

protection. They are used in general applications in aerospace, automotive, nuclear, and manufacturing industries, where the environmental conditions are extremely strict and high-performance protective technologies are required. Among the most widely used CVD coatings, titanium carbide (TiC) and titanium nitride (TiN) stand out for their exceptional wear resistance. TiC coatings possess extremely high hardness and low friction coefficients and are thus perfectly suitable for cutting tools, machining components, and aerospace applications, where mechanical reliability is a necessity. TiN coatings, on the other hand, possess excellent wear resistance and enhanced oxidation stability and biocompatibility, thereby finding applications in biomedical implants and high-performance mechanical components. Such coatings reduce surface degradation with the guarantee of efficient long-term performance and reduced maintenance costs [15]. Figure 19.4 shows a schematic diagram of a physical vapor deposition device employing an electron beam as a heat source

Silicon carbide (SiC) is a widely used CVD coating characterized by high chemical stability, thermal resistance, and mechanical strength. SiC coatings find extensive use at high temperatures in the form of turbine blades, heat exchangers, and equipment for semiconductor processing. Their thermal and oxidative resistance makes them priceless in harsh industrial environments. Further, their chemical inertness shields corrosive media like acids, alkalis, and aggressive gases from causing damage, thereby increasing the service life of critical components. Similarly, aluminum oxide (Al_2O_3) coatings are in great demand because they exhibit remarkable oxidation and corrosion resistance, especially in harsh environments where metallic substrates degrade at a very fast rate. Such films are effective diffusion barriers that cannot permit oxygen and moisture to permeate them, making them major contributors to corrosion. Therefore, Al_2O_3 coatings are applied on aerospace engine components, chemical processing equipment, and electronic devices where high-temperature stability and dielectric properties are required [17]. Figure 19.5 shows a thermal chemical vapor deposition system for SiC coating.

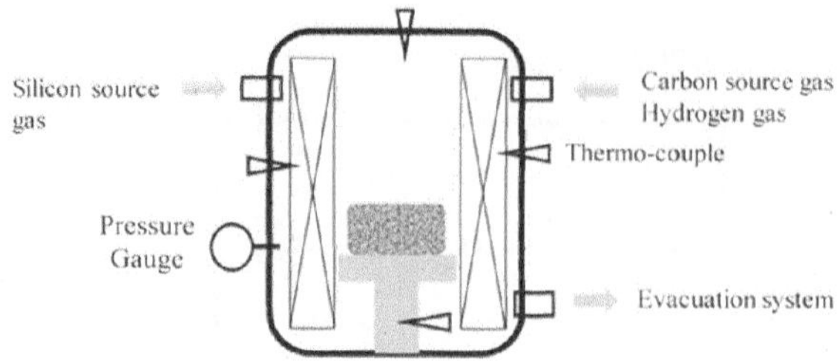

FIGURE 19.5 Thermal chemical vapor deposition system for SiC coating [18].

Source: Licensed under CC BY 4.0.

Boron carbide (B_4C) coatings are also of high importance in the use of extreme wear protection, particularly in aerospace and nuclear applications. Due to their exceptionally high hardness and low density, B_4C coatings are utilized in ultrahigh cutting tools and armor protection systems, as well as neutron-absorbing materials. Their high resistance to wear renders them suitable for abrasive particle-containing conditions, high impact loads, and severe mechanical stress. Their chemical resistance also ensures durability and reliability in extreme operating conditions [19, 20].

19.3 PHYSICAL VAPOR DEPOSITION

19.3.1 Fundamentals of PVD

Physical vapor deposition is a vacuum thin-film deposition technique whereby material is transferred physically to a substrate. As opposed to the chemical reaction of CVD, PVD methods such as evaporation and sputtering remove material physically from a source and deposit it onto the target surface. PVD coatings are defined as being very hard, possessing excellent adhesion, and being of uniform thickness and hence are especially well suited for wear and corrosion resistance applications. The PVD process typically involves four major steps: material vaporization, transportation, deposition, and film growth. The deposition technique, target material, and process parameters significantly influence the coating properties [21].

19.3.2 Types of PVD

Physical vapor deposition is a versatile coating technology with several techniques, which all have different deposition mechanisms, advantages, and areas of use. These processes are widely used in many industrial areas, including aerospace, biomedical, automotive, and electronics, due to their ability to produce high-performance coatings with excellent wear resistance, corrosion resistance, and enhanced mechanical properties [15]. Figure 19.6 shows the flow of processing for a typical PVD sputtering process.

Sputtering is a highly efficient and widely practiced PVD technique in which high-energy ions, typically from a plasma, bombard a solid target material, causing atoms to be ejected and deposit themselves onto a substrate. Sputtering can be controlled with very high accuracy for film thickness, composition, and morphology,

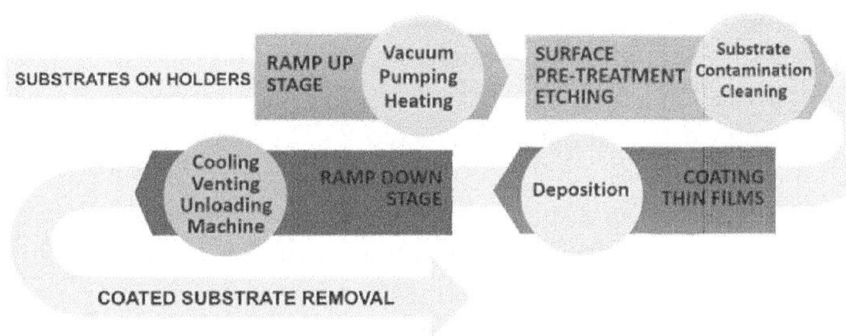

FIGURE 19.6 The processing flow for a classic PVD sputtering process [22].

and therefore it is well suited for semiconductor manufacturing, optical coatings, and decorative coatings. Sputtering has also been classified as direct current (DC) sputtering, radio frequency (RF) sputtering, and high-power impulse magnetron sputtering (HiPIMS), each offering distinctive film quality and deposition rates benefits [23].

Electron beam evaporation is a widely used PVD process that employs an electron beam to heat up and evaporate a coating material in a high vacuum environment. The atoms evaporate to deposit on the substrate as a uniform coating. This process is particularly useful in the case of high-purity coatings and is commonly used for optical applications, protective coatings for turbine blades, and biomedical implants. Electron beam evaporation is one of the main benefits of its capability to coat refractory material with high melting points, ensuring superior durability of the coating [24].

Cathodic arc deposition uses a high-energy electrical arc to vaporize a target material and create a highly ionized plasma that produces improved adhesion and high-density coatings. The process has a reputation for producing coatings with excellent hardness, wear resistance, and adhesion strength. It is widely utilized in cutting tools, aerospace parts, and automotive parts where maximum ruggedness is required. However, cathodic arc deposition can produce macroparticles that affect coating smoothness, necessitating additional filtration processes to provide finer final surface finish [25].

Magnetron sputtering enhances standard sputtering through the inclusion of magnetic fields that confine plasma in the target area and maximize ionization efficiency and deposition rates. The process enables enhanced control over the properties of the film, such as thickness uniformity, density, and microstructure. Magnetron sputtering finds widespread use in high-quality thin-film applications such as electronic displays, solar cells, and abrasion-resistant tool coatings in industry. The process can be operated in reactive mode to deposit compound coatings such as nitrides, oxides, and carbides, further enhancing its functional versatility [26].

19.3.3 PVD Coatings for Corrosion and Wear Resistance

Physical vapor deposition is a versatile coating technology with many techniques, each having different deposition mechanisms, advantages, and applications. All of these processes are widely used in many industrial fields, including aerospace, biomedical, automotive, and electronics, due to their ability to produce high-performance coatings with excellent wear resistance, corrosion resistance, and enhanced mechanical properties [27].

Sputtering is a very efficient and widely used PVD process by which high-energy ions, often from a plasma, bombard a solid target material, expelling atoms that proceed to deposit themselves on a substrate. Sputtering is controllable with fine precision for the thickness, composition, and morphology of films and thus is nicely suited for making semiconductors, optical coatings, and decorative coatings. Sputtering has also been classified into direct current sputtering, radio frequency sputtering, and high-power impulse magnetron sputtering, each with its own special film quality and deposition rate benefits [28].

Electron beam evaporation is yet another common PVD technique involving the utilization of an electron beam to warm and evaporate a coating material in a high vacuum. The evaporated atoms deposit on the substrate as a uniform coating. The method is particularly useful for high-purity coatings and is frequently used for optical applications, turbine blade protection coatings, and biomedical implants. Electron beam evaporation is one of the main benefits of its capability to coat refractory material with high melting points, ensuring excellent durability of the coating [29].

Cathodic arc deposition uses a high-energy electric arc to evaporate a target material and produce a highly ionized plasma that provides improved adhesion and high-density coatings. The process is known for producing coatings with excellent hardness, wear resistance, and adhesion strength. It is widely applied in cutting tools, aerospace parts, and automotive parts where extreme ruggedness is required. However, cathodic arc deposition is capable of producing macroparticles that may affect coating smoothness, necessitating additional filtration techniques to produce

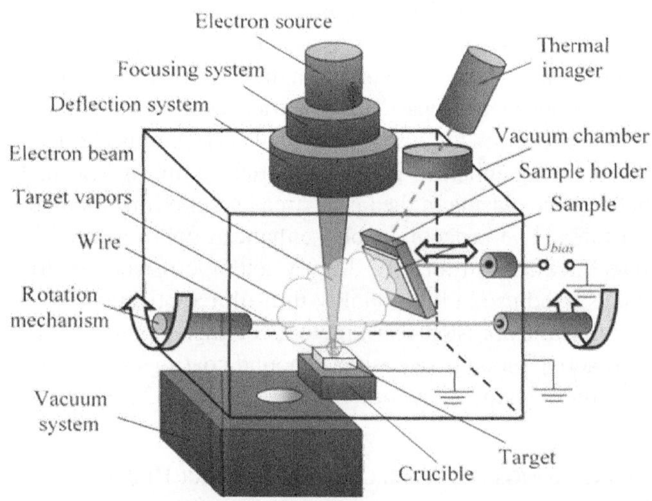

FIGURE 19.7 Experimental setup of vacuum plasma-cathode electron source [31].

the finer final surface finish [30]. Figure 19.7 shows an experimental configuration of a vacuum plasma-cathode electron source.

Magnetron sputtering is superior to traditional sputtering as it incorporates magnetic fields that confine plasma near the target and optimize ionization efficiency and deposition rates. The technique allows for better control of film characteristics, including thickness uniformity, density, and microstructure. Magnetron sputtering is often used in high-quality thin film products such as electronic displays, photovoltaic cells, and wear-resistant industrial tool coatings. The process can be operated in reactive mode to deposit compound coatings such as nitrides, oxides, and carbides, further adding to its functional versatility [32].

19.4 COMPARATIVE ANALYSIS OF CVD AND PVD

A comparative analysis of CVD and PVD is essential for selecting the appropriate deposition technique based on application requirements. The following table highlights key differences.

Table 19.1 presents the difference between CVD and PVD. Each technique has its advantages, making them suitable for different industries based on factors such as operating temperature, cost, and environmental considerations [33–35].

19.5 RECENT ADVANCES AND FUTURE DIRECTIONS

Recent research in physical vapor deposition and chemical vapor deposition focuses on optimizing these techniques to enhance coating efficiency while preserving the environment. The most significant advancement is the development of nanostructured and gradient coatings, which have enhanced mechanical strength, improved toughness, and multifunctionality. These coatings possess higher wear and corrosion resistance due to their controlled microstructure and composition design, making them applicable for severe applications in aerospace, biomedical, and energy sectors. Gradient coatings, in particular, possess a smooth gradient of different material properties, reducing residual stresses and improving adhesion, thereby increasing the life of coated parts [36].

Another important innovation is the creation of hybrid CVD/PVD processes that leverage the strengths of both techniques to produce coatings with improved

TABLE 19.1
Comparative Analysis of CVD and PVD

Property	CVD	PVD
Deposition Temperature	High (>800 °C)	Low to moderate (100–500 °C)
Film Density	High	Moderate to high
Adhesion Strength	Excellent	Excellent
Process Control	Moderate	High
Environmental Impact	Uses hazardous precursors	Environmentally friendly

properties. While CVD enables the formation of very pure and uniform coatings with high conformality, PVD is known for its ability to deposit coatings at lower temperatures with improved adhesion. With a combination of these techniques, researchers have developed novel coatings with higher hardness, toughness, and thermal stability, particularly applicable in high-performance devices such as cutting tools, biomedical implants, and shields in hostile environments. This technique of combination also allows for the deposition of complex multilayer and composite coatings, which further improve performance characteristics [37].

Artificial intelligence and machine learning have also pioneered process optimization of PVD and CVD to higher levels of efficiency, accuracy, and reproducibility. AI algorithms match huge streams of data from the deposition processes and identify optimal process parameters, reducing experimental time and material wastage. Real-time monitoring and predictive modeling assist in controlling thickness, composition, and microstructure of the coatings better for uniform performance and quality. Moreover, AI enables the design of adaptive process controls that adjust dynamically in deposition parameters based on real-time feedback, minimizing defects and maximizing yield. Such advancements are of the most value in high-precision coating applications such as semiconductor processing, automotive, and medical devices [38].

To address growing environmental demands, researchers are also working towards developing eco-friendly coating solutions with fewer toxic precursors and lower harmful byproducts. Traditional CVD and PVD processes usually rely on chemicals and gases that are detrimental to the environment and human health. Current innovations have been aimed at using greener precursors, plasma-assisted deposition techniques, and aqueous solutions to achieve high-performance coatings with minimal environmental impact. The reuse and recycling of byproducts of deposition are also enhanced through new improvements, which make the processes further sustainable. Deposition using low-temperature technologies further reduces the use of energy, making overall manufacturing more sustainable. These moves align with global sustainability goals and ensure green manufacture processes, while the high performance and durability of the coated surfaces remain intact [39].

In general, the ongoing development of CVD and PVD technologies via nanostructured coatings, hybrid deposition techniques, AI optimization, and environmentally friendly innovations is transforming surface engineering. Not only do these developments boost the functional performance of coatings, but they also confront important industrial and environmental issues, marking the way to more sustainable and high-performance material solutions in the future [40].

19.6 CONCLUSION

Chemical vapor deposition and physical vapor deposition are key processes in the production of surfaces with enhanced corrosion and wear resistance. Both processes enable high control over coating composition, microstructure, and thickness, leading to improved performance under extreme conditions. CVD, founded on chemical reactions to deposit thin films, is widely used for the production of high-purity coatings with good adhesion. However, PVD, which involves physical transfer of material in the vacuum, is characterized by hard dense coatings with improved hardness and resistance to oxidation. Continuous advancement in deposition parameters, novel material systems, and hybrid processes has significantly enhanced CVD and PVD coatings' applications in aeronautical, automotive, biomedical, and energy sectors. Incorporating nanostructured and functionally graded coatings introduces further mechanical strength and chemical stability. Future research will have to emphasize sustainable deposition techniques that minimize energy consumption and waste generation. Self-healing, antibacterial, or hydrophobic properties are found in multifunctional coatings that can overcome future industrial needs. The integration of AI and machine learning into process optimization can also improve the uniformity and performance of coatings. Although industries are in need of higher efficiency and lifespan, CVD and PVD technologies will continue to progress with long-term reliability and sustainability in advanced surface engineering applications.

REFERENCES

1. Schmitz T. *Functional coatings by physical vapor deposition (PVD) for biomedical applications* (Doctoral dissertation, Universität Würzburg).
2. Deng Y, Chen W, Li B, Wang C, Kuang T, Li Y. Physical vapor deposition technology for coated cutting tools: A review. Ceramics International. 2020 Aug 1;46(11):18373–90.
3. Farooq SA, Raina A, Mohan S, Arvind Singh R, Jayalakshmi S, Irfan UL, Haq M. Nanostructured coatings: Review on processing techniques, corrosion behaviour and tribological performance. Nanomaterials. 2022 Apr 12;12(8):1323.
4. Antunes RA, De Oliveira MC. Corrosion processes of physical vapor deposition-coated metallic implants. Critical Reviews™ in Biomedical Engineering. 2009;37(6).
5. O'Keefe MJ, Rigsbee JM. The science, technology and materials applications of physical vapor deposition processes. In Materials and Processes for Surface and Interface Engineering 1995 (pp. 151–83). Dordrecht: Springer Netherlands.
6. Adesina AY, Ogunlakin N. Physical and chemical vapour deposition coatings. In Advances in Corrosion Control of Magnesium and its Alloys 2023 (pp. 305–24). CRC Press.
7. Baptista A, Silva FJ, Porteiro J, Míguez JL, Pinto G, Fernandes L. On the physical vapour deposition (PVD): Evolution of magnetron sputtering processes for industrial applications. Procedia Manufacturing. 2018 Jan 1;17:746–57.
8. Hussein MA, Adesina AY, Kumar AM, Sorour AA, Ankah N, Al-Aqeeli N. Mechanical, in-vitro corrosion, and tribological characteristics of TiN coating produced by cathodic arc physical vapor deposition on Ti20Nb13Zr alloy for biomedical applications. Thin Solid Films. 2020 Sep 1;709:138183.
9. Palani S, Michael EG, Desta M, Atnaw SM, Banoth R, Kolanji S. Physical vapor deposition coating process in biomedical applications: An overview. Sustainable Advanced Manufacturing and Materials Processing. 2022 Nov 15:67–93.
10. Sabzi M, Mousavi Anijdan SH, Shamsodin M, Farzam M, Hojjati-Najafabadi A, Feng P, Park N, Lee U. A review on sustainable manufacturing of ceramic-based thin films by chemical vapor deposition (CVD): Reactions kinetics and the deposition mechanisms. Coatings. 2023 Jan 14;13(1):188.
11. Ichou H, Arrousse N, Berdimurodov E, Aliev N. Exploring the advancements in physical vapor deposition coating: A review. Journal of Bio-and Tribo-Corrosion. 2024 Mar;10(1):3.

12. Sugiura H, Kondo H, Tsutsumi T, Ishikawa K, Hori M. Effects of ion bombardment energy flux on chemical compositions and structures of hydrogenated amorphous carbon films grown by a radical-injection plasma-enhanced chemical vapor deposition. C. 2019 Jan 24;5(1):8.
13. Kapade TT, Shah SK, Shah DS, Kumar NS, Bhor JK, Panwar VS, Singh G. Thermal analysis of v-cut twisted tape fitted heat exchanger: A numerical approach. International Journal on Interactive Design and Manufacturing (IJIDeM). 2024;1–17.
14. Keawhan C, Wongpanya P, Witit-Anun N, Songsiriritthigul P. Corrosion behavior of AISI 4140 steel surface coated by physical vapor deposition. Journal of Metals, Materials and Minerals. 2012 Jun 30;22(1).
15. Dan A, Bijalwan PK, Pathak AS, Bhagat AN. A review on physical vapor deposition-based metallic coatings on steel as an alternative to conventional galvanized coatings. Journal of Coatings Technology and Research. 2022 Mar;19(2):403–38.
16. Fotovvati B, Namdari N, Dehghanghadikolaei A. On coating techniques for surface protection: A review. Journal of Manufacturing and Materials Processing. 2019 Mar 25;3(1):28.
17. Muhammed M, Javidani M, Ebrahimi Sadrabadi T, Heidari M, Levasseur T, Jahazi M. A comprehensive review of cathodic arc evaporation physical vapour deposition (CAE-PVD) coatings for enhanced tribological performance. Coatings. 2024 Feb 20;14(3):246.
18. Aizawa T, Yoshino T, Ito KI, Fukuda T. Thick β-SiC CVD-coated SiC die system for dry cold forging of metals. Crystals. 2020 Jun 24;10(6):539.
19. Rubshtein AP, Gao K, Vladimirov AB, Plotnikov SA, Zhang B, Zhang J. Structure, wear and corrosion behaviours of Cr–Al–C and multilayer [Cr–Al–C/aC] n coatings fabricated by physical vapour deposition and plasma-assisted chemical vapour deposition techniques. Surface and Coatings Technology. 2019 Nov 15;377:124912.
20. Geyao L, Yang D, Wanglin C, Chengyong W. Development and application of physical vapor deposited coatings for medical devices: A review. Procedia CIRP. 2020 Jan 1;89:250–62.
21. Singh G, Singh R, Gul J. Machinability behavior of human implant materials. Journal of Electrochemical Science and Engineering. 2023;13(1):99–114.
22. Baptista A, Silva F, Porteiro J, Míguez J, Pinto G. Sputtering physical vapour deposition (PVD) coatings: A critical review on process improvement and market trend demands. Coatings. 2018 Nov 14;8(11):402.
23. Singh G, Vasudev H, Arora H. A short note on the processing of materials through microwave route. In Advances in Materials Processing: Select Proceedings of ICFMMP 2020 (pp. 101–111). Springer Singapore.
24. Dini JW. Properties of coatings: Comparisons of electroplated, physical vapor deposited, chemical vapor deposited, and plasma sprayed coatings. Material and Manufacturing Process. 1997 May 1;12(3):437–72.
25. Pahade VS, Chavan PS, Baisane VP. A review paper on vapour deposition coating. International Journal of Applied Science and Engineering. 2016;3(6):257640.
26. Singh G, Vasudev H, Arora H. A short note on the processing of materials through microwave route. In Advances in Materials Processing: Select Proceedings of ICFMMP 2020 (pp. 101–11). Springer Singapore.
27. Abebe Y, Sivaprakasam P, Desta M, Udayaprakash J, Saravanan MP. Tribological behavior of physical vapor deposition coating for punch and dies: An overview. International Journal of Vehicle Structures & Systems (IJVSS). 2023 Apr 1;15(3).
28. D'Avico L, Beltrami R, Lecis N, Trasatti SP. Corrosion behavior and surface properties of PVD coatings for mold technology applications. Coatings. 2018 Dec 22;9(1):7.
29. Choy KL, ed. *Chemical Vapour Deposition (CVD): Advances, Technology and Applications*. CRC Press, 2019 Jun 7.

30. Verma PK, Vasudev H, Kumar V. Biocompatibility and corrosion performance of HA-based LVOF sprayed and post-annealed coatings on SS 254 surgical-grade bio-implant material. Ceramics International. 2025 Jan 13. https://www.sciencedirect.com/science/article/abs/pii/S0272884225001774
31. Yushkov YG, Oks EM, Tyunkov AV, Zolotukhin DB. Dielectric coating deposition regimes during electron-beam evaporation of ceramics in the fore-vacuum pressure range. Coatings. 2022 Jan 24;12(2):130.
32. Martinho RP, Silva FJ, Martins C, Lopes H. Comparative study of PVD and CVD cutting tools performance in milling of duplex stainless steel. The International Journal of Advanced Manufacturing Technology. 2019 Jun 19;102:2423–39.
33. Mallick R, Kumar R, Panda A, Sahoo AK. Hard turning performance evaluation using CVD and PVD coated carbide tools: A comparative study. Surface Review and Letters. 2022 Feb 2;29(2):2250020.
34. Singh G, Singh H, Sharma Y, Vasudev H, Prakash C. Analysis and optimization of various process parameters and effect on the hardness of SS-304 stainless steel welded joints. International Journal on Interactive Design and Manufacturing (IJIDeM). 2024;18(5):2889–96.
35. Lukaszkowicz K. Review of nanocomposite thin films and coatings deposited by PVD and CVD technology. Nanomaterials. 2011 Dec 22:145–63.
36. Raja Abdullah RI, Redzuwan BI, Abdul Aziz MS, Kasim MS. Comparative study of tool wear in milling titanium alloy (Ti-6Al-4V) using PVD and CVD coated cutting tool. Industrial Lubrication and Tribology. 2017 May 8;69(3):363–70.
37. Kalita K, Ghadai RK, Chakraborty S. Parametric optimization of CVD process for DLC Thin film coatings: A comparative analysis. Sādhanā. 2022 Jun;47(2):57.
38. Dobrzańska-Danikiewicz AD, Gołombek K, Pakuła D, Mikuła J, Staszuk M, Żukowska LW. Long-term development directions of PVD/CVD coatings deposited onto sintered tool materials. Archives of Materials Science and Engineering. 2011 Jun;49(2):69–96.
39. Dobrzański LA, Staszuk M, Honysz R. Application of artificial intelligence methods in PVD and CVD coatings properties modelling. Archives of Materials Science and Engineering. 2012 Dec;58(2):152–57.
40. Pancielejko M. Adhesion of PVD and CVD hard coatings as an essential parameter that determines the durability of coated tools. Archives of Materials Science and Engineering. 2019;96(1).

Index